Parkinson's Disease: Genetics and Pathogenesis

Parkinson's Disease: Genetics and Pathogenesis

Editors

Suzanne Lesage
Joanne Trinh

Basel • Beijing • Wuhan • Barcelona • Belgrade • Novi Sad • Cluj • Manchester

Editors
Suzanne Lesage
Sorbonne Université
Paris, France

Joanne Trinh
University of Lübeck
Lübeck, Germany

Editorial Office
MDPI
St. Alban-Anlage 66
4052 Basel, Switzerland

This is a reprint of articles from the Special Issue published online in the open access journal *Genes* (ISSN 2073-4425) (available at: https://www.mdpi.com/journal/genes/special_issues/Parkinson_Genetic).

For citation purposes, cite each article independently as indicated on the article page online and as indicated below:

Lastname, A.A.; Lastname, B.B. Article Title. *Journal Name* **Year**, *Volume Number*, Page Range.

ISBN 978-3-0365-8880-3 (Hbk)
ISBN 978-3-0365-8881-0 (PDF)
doi.org/10.3390/books978-3-0365-8881-0

© 2023 by the authors. Articles in this book are Open Access and distributed under the Creative Commons Attribution (CC BY) license. The book as a whole is distributed by MDPI under the terms and conditions of the Creative Commons Attribution-NonCommercial-NoDerivs (CC BY-NC-ND) license.

Contents

About the Editors vii

Suzanne Lesage and Joanne Trinh
Special Issue "Parkinson's Disease: Genetics and Pathogenesis"
Reprinted from: *Genes* 2023, 14, 737, doi:10.3390/genes14030737 1

Jeffrey Kim, Etienne W. Daadi, Thomas Oh, Elyas S. Daadi and Marcel M. Daadi
Human Induced Pluripotent Stem Cell Phenotyping and Preclinical Modeling of Familial
Parkinson's Disease
Reprinted from: *Genes* 2022, 13, 1937, doi:10.3390/genes13111937 5

**Graziella Mangone, Marion Houot, Rahul Gaurav, Susana Boluda, Nadya Pyatigorskaya,
Alizé Chalancon, et al.**
Relationship between Substantia Nigra Neuromelanin Imaging and Dual Alpha-Synuclein
Labeling of Labial Minor in Salivary Glands in Isolated Rapid Eye Movement Sleep Behavior
Disorder and Parkinson's Disease
Reprinted from: *Genes* 2022, 13, 1715, doi:10.3390/genes13101715 39

**Ester Pantaleo, Alfonso Monaco, Nicola Amoroso, Angela Lombardi, Daniele Urso,
Benedetta Tafuri, et al.**
A Machine Learning Approach to Parkinson's Disease Blood Transcriptomics
Reprinted from: *Genes* 2022, 13, 727, doi:10.3390/genes13050727 51

**Aymeric Lanore, Suzanne Lesage, Louise-Laure Mariani, Poornima Jayadev Menon,
Philippe Ravassard, Helene Cheval, et al.**
Does the Expression and Epigenetics of Genes Involved in Monogenic Forms of Parkinson's
Disease Influence Sporadic Forms?
Reprinted from: *Genes* 2022, 13, 479, doi:10.3390/genes13030479 73

Fangzhi Jia, Avi Fellner and Kishore Raj Kumar
Monogenic Parkinson's Disease: Genotype, Phenotype, Pathophysiology, and Genetic Testing
Reprinted from: *Genes* 2022, 13, 471, doi:10.3390/genes13030471 93

**Theresa Lüth, Joshua Laß, Susen Schaake, Inken Wohlers, Jelena Pozojevic,
Roland Dominic G. Jamora, et al.**
Elucidating Hexanucleotide Repeat Number and Methylation within the X-Linked
Dystonia-Parkinsonism (XDP)-Related SVA Retrotransposon in *TAF1* with Nanopore
Sequencing
Reprinted from: *Genes* 2022, 13, 126, doi:10.3390/genes13010126 119

**Sebastian Koch, Björn-Hergen Laabs, Meike Kasten, Eva-Juliane Vollstedt, Jos Becktepe,
Norbert Brüggemann, et al.**
Validity and Prognostic Value of a Polygenic Risk Score for Parkinson's Disease
Reprinted from: *Genes* 2021, 12, 1859, doi:10.3390/genes12121859 131

Jannik Prasuhn and Norbert Brüggemann
Gene Therapeutic Approaches for the Treatment of Mitochondrial Dysfunction in Parkinson's
Disease
Reprinted from: *Genes* 2021, 12, 1840, doi:10.3390/genes12111840 153

**Sungyang Jo, Kye Won Park, Yun Su Hwang, Seung Hyun Lee, Ho-Sung Ryu and
Sun Ju Chung**
Microarray Genotyping Identifies New Loci Associated with Dementia in Parkinson's Disease
Reprinted from: *Genes* 2021, 12, 1975, doi:10.3390/genes12121975 173

Inas Elsayed, Alejandro Martinez-Carrasco, Mario Cornejo-Olivas and Sara Bandres-Ciga
Mapping the Diverse and Inclusive Future of Parkinson's Disease Genetics and Its Widespread Impact
Reprinted from: *Genes* **2021**, *12*, 1681, doi:10.3390/genes12111681 . . . . **185**

Tatiana Usenko, Anastasia Bezrukova, Katerina Basharova, Alexandra Panteleeva, Mikhail Nikolaev, Alena Kopytova, et al.
Comparative Transcriptome Analysis in Monocyte-Derived Macrophages of Asymptomatic *GBA* Mutation Carriers and Patients with GBA-Associated Parkinson's Disease
Reprinted from: *Genes* **2021**, *12*, 1545, doi:10.3390/genes12101545 . . . . **197**

Hila Kobo, Orly Goldstein, Mali Gana-Weisz, Anat Bar-Shira, Tanya Gurevich, Avner Thaler, et al.
C9orf72-G_4C_2 Intermediate Repeats and Parkinson's Disease; A Data-Driven Hypothesis
Reprinted from: *Genes* **2021**, *12*, 1210, doi:10.3390/genes12081210 . . . . **215**

About the Editors

Suzanne Lesage

Dr. Suzanne Lesage has a long-term expertise in the fields of Genetics, Molecular Biology, and Bioinformatics. She already has a long experience with the clinical and genetic aspects of Parkinson's disease, combining experience in phenotypic evaluation, linkage analyses, omics data processing, and functional studies. As PI or co-investigator, she works in a multidisciplinary group, consisting of neurologists, geneticists, neurobiologists, biostatisticians, and bioinformatics specialists, working on the molecular bases and pathogenic mechanisms of hereditary movement disorders. She also coordinates the scientific work of her group through international consortia. The work of her research team has made possible and improved clinical diagnosis of Parkinson's disease, and more precise analyses of symptoms for phenotype/genotype correlations. Altogether, the integrated genetic and clinical studies of her team and international collaborators, particularly from North Africa, have enabled her group to successfully publish their work in > 200 original papers and reviews in both major generalist and journals, one-third of which were with impact factor > 10.

Joanne Trinh

PD Dr. Joanne Trinh began her studies in the field of movement disorders while training for her BSc and PhD in Medical Genetics at the University of British Columbia. Following a scientific fellowship at the University of Antwerp, Belgium, Dr. Trinh returned to Canada to complete her work on genetic modifiers in LRRK2 parkinsonism. She subsequently joined the Institute of Neurogenetics in Lübeck, where she received habilitation and a Faculty position. She is now head of the "Integrative Omics in Parkinson's disease" research group, which investigates genetic modifiers in nuclear and mitochondrial DNA and lifestyle and environmental factors in parkinsonism. Her research group in Lübeck will continue to use big-data approaches, with an emphasis of long-read sequencing, to elucidate the causes of neurological disease.

Editorial
Special Issue "Parkinson's Disease: Genetics and Pathogenesis"

Suzanne Lesage [1,*] and Joanne Trinh [2]

[1] Sorbonne Université, Institut du Cerveau-Paris Brain Institute, ICM, INSERM U 1127, CNRS UMR 7225, Assistance Publique-Hôpitaux de Paris (AP-HP), 75013 Paris, France
[2] Institute of Neurogenetics, University of Lübeck, 23562 Lübeck, Germany; joanne.trinh@neuro.uni-luebeck.de
* Correspondence: suzanne.lesage@upmc.fr; Tel.: +33-(0)-1-57-27-46-80

Citation: Lesage, S.; Trinh, J. Special Issue "Parkinson's Disease: Genetics and Pathogenesis". *Genes* 2023, 14, 737. https://doi.org/10.3390/genes14030737

Received: 8 March 2023
Accepted: 15 March 2023
Published: 17 March 2023

Copyright: © 2023 by the authors. Licensee MDPI, Basel, Switzerland. This article is an open access article distributed under the terms and conditions of the Creative Commons Attribution (CC BY) license (https://creativecommons.org/licenses/by/4.0/).

Parkinson's disease (PD) is a common and incurable neurodegenerative disease, affecting 1% of the population over the age of 65. The disease has clinical and pathological definitions through its cardinal motor manifestations and substantia nigra neuronal loss associated with intraneuronal Lewy bodies, respectively. Nevertheless, the molecular mechanisms that lead to neurodegeneration remain elusive. It is becoming increasingly apparent that genetic factors contribute to its complex pathogenesis. More than 23 loci and 13 genes, including *LRRK2*, *SNCA*, *GBA1*, *PRKN*, *PINK1*, and *PARK7/DJ-1*, clearly linked to inherited forms of Parkinsonism, have been identified to date. The knowledge acquired from their protein products revealed pathways of neurodegeneration that Mendelian and sporadic Parkinsonism may share. These pathways include synaptic, lysosomal, mitochondrial, and immune-mediated mechanisms of pathogenesis.

This Special Issue, "Parkinson's Disease: Genetics and Pathogenesis", collects 12 high-quality papers, including 7 original research articles and 5 reviews, that seek to deepen the knowledge of multiple aspects related to Parkinsonism.

Two reviews by Jia et al. [1] and Elsayed et al. [2] provide a comprehensive overview of the current knowledge of PD genetics in the genotype–phenotype relationship and associated pathophysiology with a focus on genetic testing and its current challenges and limitations. In addition, Jia et al. [1] discuss the role of heterozygous mutations in genes associated with autosomal recessive PD and the impact of digenism (i.e., dual *LRRK2* and *GBA1* mutation carriers) on the clinical outcomes. It is now recognized that critical genetic differences exist according to ethnicities and regions. Including ethnic diversity, specifically under-represented populations, in PD genetics research is essential to provide novel insights regarding the generalized genetic map of the disease. It will also improve our understanding of the disease biology, pathogenesis, and health care of PD patients. In the future, global efforts will play a key role in exploiting genomic data to identify rare genetic causes of PD or to replicate important gene discoveries. Furthermore, newer global initiatives such as the Global Parkinson's disease program (GP2) [3] will offer diverse and expansive representation of under-represented populations from different ethnic groups and geographical regions. A third review by Kim et al. [4] covers the most common mutations in PD-related genes, such as *LRRK2*, *SNCA*, *GBA1*, *PRKN*, *PINK1*, and *PARK7/DJ-1*, the function of these protein products, and the consequences of their mutations on the pathophysiological mechanisms leading to PD. They emphasize further consequences of these mutations using induced pluripotent stem cells (iPSCs) for a disease-in-a-dish approach and genetic animal models.

To identify potential early predictive biomarkers in PD, Mangone et al. [5] investigated the presence of immuno-stained misfolded α-Synuclein in minor salivary gland biopsies with *substantia nigra pars compacta* (SNc) damage measured by magnetic resonance imaging. They studied 27 idiopathic PD, 16 with isolated rapid eye movement sleep disorders, a prodromal form of α-synucleinopathies, and 18 healthy controls. The authors concluded that the α-Synuclein detection in minor salivary gland biopsies lacks sensitivity and specificity and does not correlate with SNc damage. In a second original paper, Usenko et al. [6]

compared the gene expression profile in monocyte-derived macrophages from four healthy controls, five PD patients, and four asymptomatic relatives carrying either heterozygous *GBA1* L444P or N370S mutations. They found dysregulated genes involved in neuronal functions, inflammation and zinc metabolism in *GBA1*-PD patients, independent of the nature of *GBA1* mutations compared to the other two groups. In particular, altered expression of *DUSP1* encoding the mitogen-activated protein kinase 1 (MKP-1) phosphatase implicated in regulating apoptosis, endoplasmic reticulum stress, cell cycle, and autophagy can be considered a potential biomarker for *GBA1*-related PD. Taking advantage of newer technologies, Pantaleo et al. [7] used whole blood transcriptome data and advanced machine learning approaches for the future selection and classification of 390 early (drug-naïve) PD patients against 189 age-matched healthy controls. The authors identified approximately 500 genes implicated in a certain number of significant functions and pathways. Some have already been linked to the pathogenesis of PD (e.g., oxidative stress, inflammation, and vesicular dysfunction) and associations between PD and diseases (e.g., diabetes mellitus or inflammatory bowel disease). The narrative review by Prasuhn and Brüggemann [8] highlights the importance of one of the known PD-associated pathways: mitochondrial dysfunction as a molecular cause in monogenic and idiopathic PD. They focus on gene therapeutic targets and challenges necessary to translate molecular findings into potential clinical applications, highlighting different treatment strategies.

Epigenetic modifications cause functional gene regulation during development, adult life, and aging and have been recently implicated in neurodegenerative diseases, such as PD. The regulation of genes responsible for monogenic forms of PD may be involved in sporadic PD. Lanoré et al. [9] reviewed the epigenetic mechanisms regulating gene expression, including DNA methylation, histone modification and epigenetic changes by non-coding RNAs. An example is *SNCA*, encoding α-Synuclein, with the understanding of its regulation being a longstanding central focus for the community working on PD. The accumulation of this protein in the Lewy bodies or neurites, the identification of mutations in the coding regions of the gene or multiplications (duplications or triplications) of the whole gene in familial PD, and the strong association of single nucleotide polymorphisms (SNPs) with sporadic PD indicate the importance of this protein in the pathogenesis of the disease. Interestingly, *SNCA* contains several transcriptionally activated histone modifications and associated potential transcription factor binding sites in the non-coding regions of the gene that strongly suggest alternative regulation pathways. Thus, studies report that DNA methylation of *SNCA* may modulate its expression, particularly hypomethylation in intron 1 of *SNCA*, which was observed in several brain regions or in peripheral tissues of sporadic PD patients and an increased *SNCA* expression. On the other hand, in postmortem midbrain samples, an enrichment of three histone modification marks, such as H3K4me3, H3K27ac, and H3K27me3, was reported in *SNCA* regulatory regions. Finally, micro-RNAs (miRNAs), such as miR-7, miR-153, and miR-34b/c, bind to the 3′-UTR of *SNCA* mRNA, destabilizing the mRNA and reducing its levels.

Genome-wide association study (GWAS) has widened our understanding of the genetics of PD and has identified more than 90 genetic loci associated with PD [10]. Jo et al. [11] performed a GWAS on dementia in 318 PD patients with dementia, 326 PD patients without dementia, and 648 healthy controls, all of Korean origin. The data analysis led to identifying the new loci of *MUL1* associated with dementia in PD, suggesting an essential role of mitochondrial dysfunction in the development of dementia in patients with PD. Two other loci containing *ZHX2* and *ERP29* were also found to be associated with dementia in PD patients. In the original research publication, Koch et al. [12] used the development of polygenic risk scores (PRSs) to summarize the effect of genetic background on an individual's disease risk in a single number. The authors were able to replicate the performance of the PD-PRS developed by Nalls et al. [13] in an independent dataset, suggesting that the PRS may be a meaningful research tool to investigate and adjust for the polygenic component of PD. However, this tool is not relevant for individual risk prediction.

Emerging studies revealed that expansions or intermediate repeats of simple short DNA sequences could cause or act as risk factors for different neurological diseases, including PD, depending on the number of repetitions. In the original paper by Kobo et al. [14], stratified analysis in 1106 Ashkenazi PD patients and 600 ethnically matched controls suggest that intermediate-size hexanucleotide repeats (20-60 repeats) in *c9orf72* are a risk factor for PD in individuals without common Ashkenazi Jewish founder mutations in *LRRK2*, *GBA1*, or *SMPD1* compared with those with these mutations. The authors propose a model that may drive the risk for PD by the number of repeats and the genotypes of 44 informative single nucleotide variants (SNVs) within the risk-haplotype, affecting the *c9orf72* RNA expression levels. In the second original research paper, Lüth et al. [15] established a straightforward Nanopore long-read deep sequencing workflow to quantify the hexanucleotide repeat number in the *TAF1* SINE-VNTR-Alu (SVA) insertion in patients with X-linked dystonia-Parkinsonism (XDP). In addition, the authors utilized this novel technology to investigate variations within the SVA locus other than the repeat motif and to detect CpG methylation using a Cas9-targeted approach across a large-22 kb region containing the *TAF1* SVA.

Overall, this Special Issue highlights the richness of studies bringing recent advances in our knowledge on the genetic architecture contributing to PD. This volume should be an important contribution to the field by improving our understanding of the pathophysiology and thus will help with the efforts to develop targeted therapies and personalized medicine.

Author Contributions: Conceptualization, S.L.; writing-original draft preparation, S.L.; writing-review and editing, S.L. and J.T. All authors have read and agreed to the published version of the manuscript.

Funding: S.L. acknowledges funding from the Fondation pour la Recherche Médicale (FRM, MND202004011718).

Acknowledgments: We would like to thank Poornima Menon for the manuscript reading.

Conflicts of Interest: The authors declare no conflict of interest.

References

1. Jia, F.; Fellner, A.; Kumar, K.R. Monogenic Parkinson's Disease: Genotype, Phenotype, Pathophysiology, and Genetic Testing. *Genes* **2022**, *13*, 471. [CrossRef] [PubMed]
2. Elsayed, I.; Martinez-Carrasco, A.; Cornejo-Olivas, M.; Bandres-Ciga, S. Mapping the Diverse and Inclusive Future of Parkinson's Disease Genetics and Its Widespread Impact. *Genes* **2021**, *12*, 1681. [CrossRef] [PubMed]
3. Global Parkinson's Genetics Program. GP2: The Global Parkinson's Genetics Program. *Mov. Disord.* **2021**, *36*, 842–851. [CrossRef] [PubMed]
4. Kim, J.; Daadi, E.W.; Oh, T.; Daadi, E.S.; Daadi, M.M. Human Induced Pluripotent Stem Cell Phenotyping and Preclinical Modeling of Familial Parkinson's Disease. *Genes* **2022**, *13*, 1937. [CrossRef] [PubMed]
5. Mangone, G.; Houot, M.; Gaurav, R.; Boluda, S.; Pyatigorskaya, N.; Chalancon, A.; Seilhean, D.; Prigent, A.; Lehéricy, S.; Arnulf, I.; et al. Relationship between Substantia Nigra Neuromelanin Imaging and Dual Alpha-Synuclein Labeling of Labial Minor in Salivary Glands in Isolated Rapid Eye Movement Sleep Behavior Disorder and Parkinson's Disease. *Genes* **2022**, *13*, 1715. [CrossRef] [PubMed]
6. Usenko, T.; Bezrukova, A.; Basharova, K.; Panteleeva, A.; Nikolaev, M.; Kopytova, A.; Miliukhina, I.; Emelyanov, A.; Zakharova, E.; Pchelina, S. Comparative Transcriptome Analysis in Monocyte-Derived Macrophages of Asymptomatic GBA Mutation Carriers and Patients with GBA-Associated Parkinson's Disease. *Genes* **2021**, *12*, 1545. [CrossRef] [PubMed]
7. Pantaleo, E.; Monaco, A.; Amoroso, N.; Lombardi, A.; Bellantuono, L.; Urso, D.; Lo Giudice, C.; Picardi, E.; Tafuri, B.; Nigro, S.; et al. Machine Learning Approach to Parkinson's Disease Blood Transcriptomics. *Genes* **2022**, *13*, 727. [CrossRef] [PubMed]
8. Prasuhn, J.; Brüggemann, N. Gene Therapeutic Approaches for the Treatment of Mitochondrial Dysfunction in Parkinson's Disease. *Genes* **2021**, *12*, 1840. [CrossRef] [PubMed]
9. Lanore, A.; Lesage, S.; Mariani, L.L.; Menon, P.J.; Ravassard, P.; Cheval, C.; Corti, O.; Brice, A.; Corvol, J.C. Does the Expression and Epigenetics of Genes Involved in Monogenic Forms of Parkinson's Disease Influence Sporadic Forms? *Genes* **2022**, *13*, 479. [CrossRef] [PubMed]
10. Chang, D.; Nalls, M.A.; Hallgrímsdóttir, I.B.; Hunkapiller, J.; van der Brug, M.; Cai, F.; International Parkinson's Disease Genomics Consortium; 23andMe Research Team; Kerchner, G.A.; Ayalon, G.; et al. A meta-analysis of genome-wide association studies identifies 17 new Parkinson's disease risk loci. *Nat. Genet.* **2017**, *49*, 1511–1516. [CrossRef] [PubMed]

11. Jo, S.; Park, K.W.; Hwang, Y.S.; Lee, S.H.; Ryu, H.S.; Chung, S.J. Microarray Genotyping Identifies New Loci Associated with Dementia in Parkinson's Disease. *Genes* **2021**, *12*, 1975. [CrossRef] [PubMed]
12. Koch, S.; Laabs, B.H.; Kasten, M.; Vollstedt, E.J.; Becktepe, J.; Brüggemann, N.; Franke, A.; Krämer, U.M.; Kuhlenbäumer, G.; Lieb, W.; et al. Validity and Prognostic Value of a Polygenic Risk Score for Parkinson's Disease. *Genes* **2021**, *12*, 1859. [CrossRef] [PubMed]
13. Nalls, M.A.; Blauwendraat, C.; Vallerga, C.L.; Heilbron, K.; Bandres-Ciga, S.; Chang, D.; Tan, M.; Kia, D.A.; Noyce, A.J.; Xue, A.; et al. Identification of novel risk loci, causal insights, and heritable risk for Parkinson's disease: A meta-analysis of genome-wide association studies. *Lancet Neurol.* **2019**, *18*, 1091–1102. [CrossRef] [PubMed]
14. Kobo, H.; Goldstein, O.; Gana-Weisz, M.; Bar-Shira, A.; Gurevich, T.; Thaler, A.; Mirelman, A.; Giladi, N.; Orr-Urtreger, A. C9orf72-G4C2 Intermediate Repeats and Parkinson's Disease; A Data-Driven Hypothesis. *Genes* **2021**, *12*, 1210. [CrossRef] [PubMed]
15. Lüth, T.; Laβ, J.; Schaake, S.; Wohlers, I.; Pozojevic, J.; Jamora, R.D.G.; Rosales, R.L.; Brüggemann, N.; Saranza, G.; Diesta, C.C.E.; et al. Elucidating Hexanucleotide Repeat Number and Methylation within the X-Linked Dystonia-Parkinsonism (XDP)-Related SVA Retrotransposon in TAF1 with Nanopore Sequencing. *Genes* **2022**, *13*, 126. [CrossRef] [PubMed]

Disclaimer/Publisher's Note: The statements, opinions and data contained in all publications are solely those of the individual author(s) and contributor(s) and not of MDPI and/or the editor(s). MDPI and/or the editor(s) disclaim responsibility for any injury to people or property resulting from any ideas, methods, instructions or products referred to in the content.

Review

Human Induced Pluripotent Stem Cell Phenotyping and Preclinical Modeling of Familial Parkinson's Disease

Jeffrey Kim [1,2], Etienne W. Daadi [1], Thomas Oh [1], Elyas S. Daadi [1] and Marcel M. Daadi [1,2,3,*]

[1] Southwest National Primate Research Center, Texas Biomedical Research Institute, San Antonio, TX 78227, USA
[2] Cell Systems and Anatomy, San Antonio, TX 78229, USA
[3] Department of Radiology, Long School of Medicine, University of Texas Health at San Antonio, San Antonio, TX 78229, USA
* Correspondence: mdaadi@txbiomed.org

Abstract: Parkinson's disease (PD) is primarily idiopathic and a highly heterogenous neurodegenerative disease with patients experiencing a wide array of motor and non-motor symptoms. A major challenge for understanding susceptibility to PD is to determine the genetic and environmental factors that influence the mechanisms underlying the variations in disease-associated traits. The pathological hallmark of PD is the degeneration of dopaminergic neurons in the substantia nigra pars compacta region of the brain and post-mortem Lewy pathology, which leads to the loss of projecting axons innervating the striatum and to impaired motor and cognitive functions. While the cause of PD is still largely unknown, genome-wide association studies provide evidence that numerous polymorphic variants in various genes contribute to sporadic PD, and 10 to 15% of all cases are linked to some form of hereditary mutations, either autosomal dominant or recessive. Among the most common mutations observed in PD patients are in the genes LRRK2, SNCA, GBA1, PINK1, PRKN, and PARK7/DJ-1. In this review, we cover these PD-related mutations, the use of induced pluripotent stem cells as a disease in a dish model, and genetic animal models to better understand the diversity in the pathogenesis and long-term outcomes seen in PD patients.

Keywords: Parkinson's disease; genetic basis for pathophysiology; induced pluripotent stem cells; brain organoids; in vitro models of familial Parkinson's disease; personalized medicine

1. Parkinson's Disease Pathology, Symptoms, Treatments

Parkinson's disease (PD) is a chronic and progressive neurodegenerative disorder that arises from the loss of dopaminergic neurons in the substantia nigra pars compacta (SNpc) of the midbrain and other organ systems [1,2]. It was first characterized by Dr. James Parkinson in 1817 in his "An essay on the Shaking Palsy", where he described patients with trembling extremities, hunched posture, and a shuffling gait [3]. It is the second most prevalent neurodegenerative disease that affects 1% of the population older than 60 years of age [4,5]. The number of PD patients has doubled from 1990 to 2015 and is projected to double again by 2040 [6]. At the onset of the disease, 30–70% of midbrain dopaminergic neurons are lost [7] yet the disease progresses slowly over years with a prodromal stage that may last 10–15 years before the presentation of motor deficits and proper diagnosis [8,9].

The loss of dopaminergic neurons in the SNpc severs the nigrostriatal pathway, which is one of four major dopamine pathways in the brain. When the connection between the SNpc and the dorsal striatum is lost, the ability to produce controlled movement becomes impaired. The symptoms of the disease include uncontrollable resting tremors, postural imbalance, cogwheel rigidity, bradykinesia, akinesia, and cognitive impairment [10–16]. Depression, dementia, and hallucinations are also observed [17–25]. Available treatments only provide symptomatic therapy. Levodopa or dopamine agonists are used to replace lost dopamine. However, long-term use of levodopa may lead to adverse side effects, such as dyskinesia [26]. Monoamine oxidase B (MAO-B) and catechol-O-methyl transferase (COMT) inhibitors are provided in conjunction with levodopa to inactivate dopamine metabolism and degradation [27]. When the patient no longer responds positively to dopamine replacement treatment, a neurosurgical procedure known as deep brain stimulation can be performed to alleviate symptoms [28–32]. This procedure utilizes an electrode that is surgically implanted into the ventral intermediate nucleus of the thalamus, globus pallidus, and subthalamic nucleus to reduce severe motor complications [33,34]. To date, there is no curative treatment available for PD.

For the vast majority of patients, the cause of PD remains unknown, and it is primarily an idiopathic disease. Nevertheless, there are many environmental risk factors that are known contributors to PD, including metals such as iron, copper, manganese, lead, and mercury, as well as toxins such as 1-methyl-4-phenyl-1,2,3,6-tetrahydropyridine (MPTP), rotenone, paraquat, dieldrin, hexachlorohexanes, and 2-4-dichlorophenoxyacetic acid [35]. The use of amphetamines and methamphetamines also increases risk [36]. However, age remains to be the strongest risk factor for PD [37].

At least 5% of cases are linked to specific genetic mutations [38]. Mutations in genes, such as leucine-rich repeat kinase 2 (LRRK2), α-synuclein (SNCA), glucosylceramidase β 1 (GBA1), phosphatase and tensing homolog-induced kinase 1 (PINK1), parkin (PRKN)and PARK7 (also referred to as DJ-1), are associated with increased PD risk [39]. Much of what we know about these mutations was discovered by using in vivo transgenic animal models and in vitro stem cell models [40–43]. The consequences of these mutations lead to aberrations in oxidative stress, mitochondrial dysfunction, perturbed protein quality control, protein aggregation, and altered kinase activity. Some mutations may also lead to the early onset of PD.

2. Leucine-Rich-Repeat Kinase 2 (LRRK2)

The LRRK2 protein has Roc-COR-kinase domains and exhibits profound kinase and GTPase activity [44–46]. It exists as an oligomeric structure with minimal kinase activity. Once bound to guanosine triphosphate (GTP), it dissociates to form an intermediate structure whereupon it autophosphorylates to form a homodimer kinase [47]. It has functions linked to transcription, translation, autophagy, mitochondrial function, cytoskeletal remodeling vesicular transport, dopamine homeostasis, and synaptogenesis [48–53]. It has also been shown in human brains to be constitutively expressed in neurons and glia [54]. LRRK2 interacts with 14-3-3 proteins via phosphorylated Ser910 and Ser935 [55]. The binding of 14-3-3 to LRRK2 is thought to be disrupted with PD-related mutations [56]. Knockout of wild-type LRRK2 causes impairment of protein degradation pathways, accumulation of α-synuclein (α-syn), and apoptotic cell death in aged mice, signifying LRRK2 is essential in those pathways [57]. Furthermore, increased LRRK2 kinase activity was observed in post-mortem brain tissue and immune cells of idiopathic PD patients, providing clinical significance of LRRK2 as a therapeutic target [58–60].

LRRK2 is known to be phosphorylated at these sites: Ser910, Ser935, Ser955, Ser973, Ser1292, and Thr826 [61]. Specifically, the Ser1292 residue is the site of autophosphorylation [62]. Autophosphorylation is significantly increased in disease-causing variants [63]. The kinase activity is also required for the cytotoxicity of LRRK2 mutants [64,65]. Autophosphorylation was observed to have decreased when LRRK2 was chemically inhibited [55]. Pharmacological inhibition was reported to protect against the toxic phenotype of hyperkinase activity [66]. However, LRRK2 kinase inhibition was shown to have off-target effects in peripheral tissues in animal models [67–69].

LRRK2 was reported to phosphorylate cytoskeletal proteins, such as tau, microtubule affinity regulating kinase 1 (MARK1), tubulins, rho guanine nucleotide exchange factor 7 (ARHGEF7), and ezrin-radixin-moesin (ERM) [70–76]. LRRK2 also exhibits physical interaction with F-actin and microtubules [66,77]. Pathogenic LRRK2 is believed to alter cytoskeletal scaffolding leading to deficits in neurite outgrowth and axonal transport [78,79]. In fact, phosphorylation of ezrin, radixin, and moesin (*ERM*) family proteins by pathogenic LRRK2 promoted the cytoskeletal rearrangement [76]. Rab10 is another phosphorylation target of LRRK2, specifically at Thr73 [80]. LRRK2 variants augmented Rab10 phosphorylation which affected vesicular transport [81]. Increased Rab10 phosphorylation was observed in dopaminergic neurons in the SNpc of idiopathic PD patients, which makes this an excellent and indirect measure for the LRRK2 kinase activity [82]. LRRK2 was also shown to control the vesicle trafficking [83,84]. Snapin, EndophilinA, and Rab5a were identified as possible LRRK2 substrates involved in synaptic vesicle exo- and endocytosis [85–88]. Evidence suggests that LRRK2 could inhibit autophagy, causing accumulation of autophagic organelles, that is reversed through kinase inhibition [89–91]. Hyperphosphorylation also affected 4E-BP1, leading to excessive protein translation and neurodegeneration in Drosophila model [50]. However, Drosophila only possesses LRRK1 and not LRRK2 and such findings were not strongly translated to mammalian models [92]. Although, LRRK2 phosphorylates ribosomal protein S15, regulating protein synthesis in the G2019S LRRK2 transgenic Drosophila and human dopaminergic neurons [93].

3. Common LRRK2 Mutations That Lead to Parkinson's Disease

The most prevalent genetic mutation has been identified in the LRRK2 gene [94,95]. This particular mutation, known as the G2019S, accounts for at least 1% of all PD cases and 5–15% of familial PD cases [96]. Clinically, symptoms of LRRK2 mutant carriers are indistinguishable from idiopathic PD [97]. LRRK2 variants either increase or decrease the risk of developing PD. The common variants in Asia associated with increased risk of PD are A419V, R1628P, and G2385R [98–100]. These missense variants increase the risk for disease by about two-fold [101]. Interestingly, the variants N551K, R1398H, and K1423K appear to reduce the risk of PD [102,103]. There are eight isoforms of LRRK2 that lead to autosomal dominant familial PD: R1441C/H/G/S, Y1699C, N1437H, I2020T, and G2019S [104–107]. Each of these isoforms is a single amino acid substitution. The R1441C/H/G/S, N1437H, and Y1699C variants are located in the GTPase domain and reduce the GTP hydrolysis [108–110]. The variants in the GTPase and kinase domains are linked to neurotoxicity [52,63]. The I2020T and G2019S mutations are found on the kinase domain and lead to increased LRRK2 kinase activity [63–65,107,111,112]. Although the effect on kinase activity is not as pronounced in I2020T as it is in the G2019S variant [113], it has been well established that the mutations within the kinase domain augment kinase function and lead to Parkinsonism.

4. The G2019S LRRK2 Mutation

The most prevalent of all LRRK2 mutations is the G2019S point mutation located in the kinase domain of the protein [95,114]. This mutation exhibits high prevalence in Ashkenazi Jewish, North African Berber, and Arab PD patients [115–117]. Interestingly, it is not as common among Asian PD patients [39,118–123]. It is an autosomal dominant point mutation that resides in the kinase domain. This in turn leads to the hyper-kinase activity of LRRK2 exhibited in enhanced autophosphorylation at S1292, up to four-fold increase [63–65,124]. However, the mutation does not alter the gene expression of LRRK2. The symptoms that arise in disease-manifesting carriers closely resemble those associated with idiopathic PD [125,126]. Kinase inhibition blocks neurotoxicity in vitro and in vivo [66]. Interestingly, there is no marked phenotypic difference between heterozygous and homozygous G2019S mutant carriers [127]. Dopaminergic neurons derived from induced pluripotent stem cells (iPSC) harboring the G2019S mutation demonstrate increased susceptibility to oxidative stress and early neuritic branching defects [128]. G2019S LRRK2 is also linked to an upregulation of the p53-p21 pathway, contributing to cellular senescence and accumulation of α-syn [129]. However, the full impact of G2019S mutation is still largely unknown [97].

5. G2019S LRRK2 Induced Pluripotent Stem Cell Models

In vitro disease modeling has been revolutionized with the use of iPSCs [130–134]. To generate iPSCs, somatic cells are reprogrammed with the transfection of pluripotent factors octamer-binding protein 3/4 (Oct3/4), SRY-box transcription factor 2 (Sox2), c-Myc, and Krüppel-like factor 4 (Klf4) [135]. The reprogramming of adult somatic cells allows for the creation of patient-specific iPSCs that harbor known pathological genetic mutations [136]. The reprogrammed cells express embryonic stem cell markers and are capable of differentiating into any cell type of the three germ layers. Patient iPSCs can be genetically modified to correct mutations with CRISPR/Cas9, transcription activator-like effector nucleases (TALENS), or zinc-finger nuclease [137–142]. iPSCs can then be differentiated into a cell type of choice and sequenced to identify differentially expressed genes. Patient-derived iPSCs are a powerful tool to study a disease in a dish, and indeed many groups have used neurons derived from PD patient iPSCs to advance our understanding of the cause of neuronal degeneration.

The first LRRK2 iPSCs were generated from the PD patient fibroblasts of a 63-year-old male homozygous carrier and a 42-year-old male heterozygous for G2019S LRRK2 [143]. Accumulation of α-syn and neurite retraction have been observed in LRRK2 G2019S cells [128,144,145]. Endogenous α-syn accumulation also occurred when exogenous α-syn in the form of recombinant human preformed fibrils was cultured in iPSC-derived neurons [146]. Neural stem cells derived from G2019S iPSCs exhibited increased susceptibility to proteasomal stress and passage-dependent deficiencies in nuclear-envelope organization, clonal expansion, and neuronal differentiation [143]. Specifically, neuronal differentiation was affected as dopaminergic neurons showed early neuritic branching defects [147]. Sensory neurons derived from iPSCs also showed shortened neurites as well as microtubule-rich axon aggregation and altered calcium dynamics [148]. Furthermore, G2019S iPSC-derived dopaminergic neurons also displayed lower baseline ER-Ca^{2+} levels with Ca^{2+} influx increased, and Ca^{2+} buffering capacity decreased after membrane depolarization [149].

Single-cell RNA-sequencing (scRNA-seq) of G2019S PD patient iPSC-derived neural stem cells revealed affected genes involved in mitochondrial function, DNA repair, protein degradation, oxidative stress, lysosome biogenesis, ubiquitin–proteasome system, endosome function, autophagy, and mitochondrial quality control [150]. It was also recently shown in another scRNA-seq experiment that neuroepithelial stem cells (NESC) derived from G2019S iPSCs exhibited mitochondrial defects [151]. They found that the G2019S NESCs exhibited fragmented mitochondria, impaired mitochondria function, and impaired autophagosomal-lysosomal pathway, and the cells were more prone to release reactive oxygen species (ROS). They later expanded on their scRNA-seq data to show that G2019S LRRK2 NESCs initiated early cell-cycle exit and earlier neural differentiation than wild type (WT), leading to increased cell apoptosis [152]. This was in part due to the downregulation of the core regulatory circuit transcription factor Nuclear Receptor Subfamily 2 Group F Member 1 (NR2F1) and altered distal super-enhancer activity. Importantly, using a single-cell longitudinal imaging platform, early studies demonstrated that Nrf2 expression in neurons directly mitigates toxicity induced by α-synuclein and mutations in LRRK2 and that this effect is time-dependent [153]. Furthermore, mitochondrial genome damage and mitochondrial transport-related PD pathogenesis were also present [154,155]. It is also believed that G2019S LRRK2 disrupts the interaction between the mitochondrial transport protein, Miro, to PINK1 and Parkin, arresting the movement of damaged mitochondria along the cytoskeleton and delaying mitophagy [156]. Dopaminergic neurons derived from G2019S iPSCs show this abnormal mitochondrial trafficking and distribution as well [157]. This group also revealed that despite high levels of sirtuin, there was a reduction of sirtuin deacetylase activity, nicotinamide adenine dinucleotide (NAD+), and protein lysine deacetylase activity, leading to bioenergy deficits.

In addition to mitochondrial deficits, hyper-kinase activity also disrupts the endocytosis of synaptic vesicles in iPSC-derived ventral midbrain neurons [158]. Transcriptomics and proteomics revealed that clathrin-mediated endocytosis was disrupted [159]. Specifically, endothelial cytokines I–III, dynamin-1, and various Rab proteins were significantly downregulated. Evidence suggests that the synaptic defects are due to LRRK2 phosphorylation of auxilin (DNAJC6) which causes differential clathrin binding and disrupts endocytosis [160]. This impaired endocytosis led to the accumulation of oxidized dopamine and caused reduced glucocerebrosidase activity and increased α-syn accumulation.

As mentioned before, LRRK2 G2019S mutant neurons were found to exhibit increased accumulation and release of α-syn [161]. Mutant iPSC-derived midbrain dopaminergic neurons also showed higher basal levels of LC3 II. This aberrant autophagy was reversed with the inhibition of mitochondrial fission with fission dynamin-related protein 1 (DRP1) peptide inhibitor p110 [162]. Additionally, LRRK2 phosphorylation of leucyl-tRNA synthetase (LRS) was shown to reduce leucine binding and impair autophagy, leading to protein misfolding and endoplasmic reticulum stress [163].

LRRK2 further plays a role in the innate and acquired immunity of the peripheral and central nervous system [164]. Interestingly, LRRK2 is highly expressed in macrophages and microglia [164]. Evidence suggests that LRRK2 controls the secretion of inflammatory mediators [164–166]. LRRK2 GTPase function is also implicated in the inflammatory response [167–169]. Indeed, LRRK2 mutant microglia and astrocytes exhibit increased inflammatory cytokine and chemokine production [170–174]. Knockout or pharmacological inhibition of LRRK2 alleviates this inflammatory response [175–179]. Furthermore, LRRK2 mutations that lead to α-syn accumulation also impair nuclear factor kappa B (NF-κB) signaling in iPSC-derived neurons [180]. G2019S iPSC-derived astrocytes exhibited downregulation of matrix metalloproteinase 2 (mmp2) and transforming growth factor β1 (TGFβ1) [181]. Finally, LRRK2 may also play a role in hematopoiesis as G2019S LRRK2 iPSC-derived monocytes undergo accelerated production while CD14+CD16+ monocytes are reduced [182]. These mutant monocytes also exhibit migratory deficits.

6. G2019S LRRK2 Animal Models

LRRK2 animal models have provided insights into the regulation of protein translation, vesicle trafficking, neurite outgrowth, autophagy, and cytoskeletal dynamics [93,183,184]. Different models exhibit different clinical aberrations, such as degeneration of midbrain dopaminergic neurons, accumulation of α-syn, abnormal dopamine secretion, and behavioral deficits. LRRK2 overexpression in Drosophila led to the age-dependent loss of dopaminergic neurons and reduction of locomotor activity [185–187]. G2019S LRRK2 induced loss of photoreceptors and the impaired visual system was also observed [188]. The mutation also induced the mislocalization of tau in dendrites, causing degeneration [189]. Overexpression of G2019S LRRK2 in C. elegans also resulted in dopaminergic neuron degeneration, enhanced vulnerability to mitochondrial dysfunction, and inhibition of autophagy [190,191]. Transgenic mice expressing G2019S mutations exhibited motor deficits but with minimal evidence of neurodegeneration [192]. Only two groups have demonstrated the G2019S LRRK2 model of age-dependent loss of dopaminergic neurons in the SNpc [193,194]. Transgenic rats expressing the G2019S mutation exhibited oxidative stress in the striatum and SNpc with increased inducible nitric oxide synthase expression and abnormal morphology of SNpc dopaminergic neurons [195,196]. Unfortunately, genetic animal models display inconsistent phenotypes and do not fully replicate the human condition (neurodegeneration, Lewy body formation, and significant motor deficits), thus better models are needed [197].

7. Synuclein α (SNCA)

The SNCA gene encodes the presynaptic protein α-Synuclein (α-syn), which has been found to be localized in the nuclear membrane and synaptic vesicles [198]. The function of this protein is not well understood but current evidence suggests it participates in the axonal transport of synaptic vesicles by binding and transporting fatty acids [199–201]. It is also believed to participate in the differentiation and survival of dopaminergic neuron progenitor cells of mice and humans [202,203]. In fact, α-syn is expressed in the SNpc, especially in neurons containing neuromelanin, an insoluble granular pigment [204].

The misfolding of α-syn can lead to aberrant aggregation in the form of insoluble filaments and deposits in nerve cells [205]. These α-syn aggregates are also a major component of Lewy bodies, cellular inclusions in the neuronal cytoplasm that may lead to PD pathogenesis [206]. Lewy bodies impair neuronal communication and may even spread to healthy neurons [207,208]. They are also known to increase oxidative stress, disrupt axonal transport for neurotransmitter vesicles, and contribute to transcriptional dysregulation, protein sequestration, mitochondrial and synaptic dysfunction, and inhibition of the ubiquitin–proteasome system [207–213]. Accumulation of α-syn also affects the lysosomal clearance of protein aggregates, thus resulting in a vicious cycle perpetuating the toxic effects of α-syn aggregates [214]. It has been shown that the majority of α-syn in Lewy bodies in postmortem PD brain tissue appears as phosphorylated S129 α-syn [215,216]. Evidence suggests that S129 phosphorylation promotes α-syn aggregation and neurotoxicity [217]. Aggregated forms of α-syn are in fact more prone to S129 phosphorylation and accumulation during disease progression [218–220]. Polo-like kinases, casein kinases, and G protein-coupled receptor kinases have been shown to modulate the phosphorylation of α-syn at S129 [220–225]. Interestingly, numerous reports suggest that LRRK2 may be involved in phosphorylating S129 and α-syn aggregation [226,227]. Ultimately, it is still uncertain whether S129 α-syn is indeed neurotoxic [228].

Braak et al. hypothesized that neurodegeneration occurs in a predetermined sequence caused by an unknown pathogen in the gut or nasal cavity initiating sporadic PD [229,230]. This is associated with a specific α-syn spreading pattern, which may be why PD patients exhibit gastrointestinal and olfactory problems [231–233]. Indeed, Lewy body pathology has been confirmed in neurons of the olfactory tract and enteric nervous system [234–237]. Therefore, it is possible that α-syn propagates in a prion-like fashion [238]. In fact, the cell-to-cell transmission of α-syn has been observed after transgenic α-syn overexpression or exposure to preformed fibrils of α-syn and homogenates from postmortem PD patients [239–244]. Therefore, the pathological aggregation of misfolded α-syn is hypothesized to be critical in PD pathogenesis.

An increased risk of developing PD in humans can result due to an overexpression of the SNCA gene because of locus triplication [245]. Alternatively, SNCA missense mutations have also been shown to increase PD risk. Mutations in the SNCA gene were first identified as causing autosomal dominant PD in a large Italian family known as the Contursi kindred [246]. These patients carried the A53T point mutation and exhibited hallmark characteristics of PD including Lewy body pathology and positive response to L-dopa treatment. However, early onset and rapid disease progression were also observed. Other missense mutations include A30P [247–249], E46K [250], A53E [251,252], A53V [253], G51D [254,255], H50Q [256], and A18T and A29S [257]. Copy number variations have also been reported in PD patients [245]. Carriers of the A30P mutation were also associated with early onset, but a milder disease progression was observed compared to those with the A53T mutation [247]. The A53T and A30P mutations increase the likelihood of α-syn protein oligomerization instead of fibrillation [258], which is believed to accelerate α-syn aggregation [259]. Patients with A53T and E46K developed dementia consistent with Lewy body dementia [250,260] as the E46K mutation also promotes α-syn aggregation similar to A53T [261]. Interestingly, patients with the E46K mutation have also experienced visual hallucinations [250]. The A18T and A29S missense mutations were found in patients with sporadic PD [257].

8. SNCA Induced Pluripotent Stem Cell Models

Dopaminergic neurons derived from iPSCs of a PD patient with SNCA triplication have been studied. The level of α-syn protein in these dopaminergic neurons was twice the amount compared to those derived from normal iPSCs [262]. These neurons exhibited changes in growth, viability, cellular energy metabolism, and stress resistance when challenged with starvation or toxins [263], and increased oxidative stress [264]. SNCA triplication also led to a reduced capacity for iPSCs to differentiate into neurons, decreased neurite outgrowth, and lower neuronal activity compared to normal neurons [265]. The mRNA levels of nuclear receptor-related 1 protein (NURR1), G-protein-regulated inward-rectifier potassium channel 2 (GIRK-2), and tyrosine hydroxylase (TH) were also significantly reduced. Furthermore, lysosomal dysfunction has also been also induced by the α-syn accumulation [266]. Aggregates appeared to interact with ATP synthase and lead to premature mitochondrial permeability transition pore opening, making neurons more vulnerable to cell death [267]. Interestingly, SNCA triplication also affects non-neuronal cells, such as microglia. Microglia derived from SNCA triplication iPSCs exhibited impaired phagocytosis compared to isogenic controls [268].

An isogenic gene-corrected iPSC line for A53T was generated with zinc-finger nuclease-mediated genomic editing [140]. Gene correction reversed nitrosative stress and endoplasmic reticulum stress in iPSC-derived neurons [269]. A53T mutation increased apoptotic cell death in iPSC-derived midbrain dopaminergic neurons by increasing S-nitrosylation of MEF2C, affecting the transcriptional regulation of PGC1a, a master regulator of mitochondrial biogenesis [270]. A53T iPSC-derived neurons also displayed irregular protein aggregation, compromised neuritic outgrowth, contorted or fragmented axons with varicosities containing α-syn and Tau, and disrupted synaptic connectivity [271]. Interestingly, A53T midbrain dopaminergic neurons contained higher concentrations of α-syn monomers relative to tetramers when compared to isogenic controls, ultimately decreasing solubility α-syn [272]. Oligomeric α-syn has emerged as the key mediator for α-syn accumulation [273]. In fact, A53T and SNCA triplication iPSC-derived dopaminergic neurons exhibited increased α-syn oligomerization in a proximity ligation assay [274]. Increased sensitivity to mitochondrial toxins and nitrosative stress-induced neuronal loss were also observed in A53T dopaminergic neurons [275]. Furthermore, transcriptomic analysis of A53T and SNCA triplication iPSC-derived dopaminergic neurons revealed perturbations in the expression of genes likened to mitochondrial function, which was consistent with a reduction in mitochondrial respiration, impaired mitochondrial membrane potential, aberrant mitochondrial morphology, and decreased levels of phosphorylated DPR1 Ser616 [274]. They also observed increased endoplasmic reticulum stress and impaired cholesterol and lipid homeostasis. Single-cell transcriptomic analysis of A53T iPSC-derived dopaminergic neurons compared to an isogenic gene-corrected counterpart revealed perturbations in glycolysis, cholesterol metabolism, synaptic signaling, and ubiquitin–proteasomal degradation [276]. Apart from neuronal lineage cells, O4+ oligodendrocyte linage cells derived from A53T iPSCs also exhibited impaired maturation compared to controls [277].

9. SNCA Animal Models

Transgenic models with SNCA mutations have typically failed to display clear dopaminergic neurodegeneration or parkinsonian motor deficits [278]. However, there has been apparent α-syn aggregation and altered neuronal functions in these models. For example, transgenic mice expressing truncated α-syn show a reduced number of nigro-striatal neurons due to cell loss during early development [279]. However, the clinical relevance of these transgenic models remains questionable as the mutated α-syn protein is affecting early developmental stages rather than the later onset of neurodegeneration as seen in human patients [280].

SNCA overexpression models in rodents have been shown to affect the development and maintenance of dopaminergic neurons [281]. Overexpression has led to the formation of α-syn aggregates in the brain causing motor and olfactory deficits but not dopaminergic neurodegeneration [282]. E46K SNCA rats also display α-syn aggregation, altered metabolism of striatal dopamine, and increased oxidative stress but again not dopaminergic neurodegeneration [283].

Viral vectors such as adeno-associated virus (AAV) and lentivirus (LV) have also been used for transfecting rodents with SNCA. The AAV6 serotype generated an 80% loss of dopaminergic neurons and profound motor deficits [284]. This model also showed progressive neurodegeneration over a 2-to-4-month period. Using an LV vector led to α-syn aggregation but with no apparent loss of dopaminergic neurons nor behavioral changes [285]. Alternatively, AAV2/7-α-syn transduction in mouse SNpc produced dose-dependent dopaminergic neurodegeneration and motor deficits [286]. AAV2/2 was used to deliver WT and A53T α-syn into marmosets, leading to 30–60% nigral dopaminergic neurodegeneration and subsequent striatal dopamine depletion with mild motor deficits [287]. AAV1/2-A53T α-syn in macaques displayed 30% nigrostriatal dopaminergic neurodegeneration, 50% dopamine depletion, and 40% DAT reduction [288].

10. Glucocerebrosidase 1 (GBA1)

GBA1 encodes a lysosomal protein β-glucocerebrosidase (GCase). Mutations in this gene result in the accumulation of glycolipid substrates in lysosomes which disrupts lysosomal function and can lead to an autosomal recessive lysosomal storage disorder known as Gaucher's disease (GD) [289]. Although a small minority of GBA1 mutation carriers develop PD [290,291], mutations in GBA1 increase the risk of developing PD [292,293]. In fact, GBA1 mutations are common genetic risk factors for PD, where 7–10% of patients with PD are carriers of a GBA1 mutation [291,294]. The L444P point mutation has been identified as being associated with PD and GD patients [291,295]. Furthermore, the L444P mutation appears to have a higher risk of developing PD compared to other GBA1 point mutations [296–298]. Interestingly, there is no difference in risk between patients with either homozygous or heterozygous mutations, although homozygous carriers tend to have PD onset 6 to 11 years earlier than heterozygous counterparts [290,294,298,299]. The clinical features of GBA1-associated PD are similar to those associated with idiopathic PD, including olfactory deficits and sleep disturbance but with earlier onset and accelerated autonomic, cognitive, and motor decline [298]. These patients also present with a more pronounced loss of nigrostriatal dopaminergic neurons and greater Lewy body pathology compared to those with idiopathic PD. GCase deficiency and lysosomal dysfunction as a result of GBA1 mutations are thought to be important pathogenic mechanisms for PD [300]. GBA1 mutations appear to impair α-syn degradation in the lysosome due to perturbed GCase activity [301,302]. Interestingly, GCase activity is also reduced in postmortem brain tissue of PD patients without GBA1 mutations [303,304]. It is possible that the GlcCer substrate accumulation may promote the pathogenic conversion of α-syn into its insoluble form [305,306]. Indeed, the GlcCer substrate also stabilizes a-syn oligomeric intermediates and induces rapid polymerization of fibrils [302].

11. GBA1 Induced Pluripotent Stem Cell Models

Interestingly, WT GCase activity is reduced in brain tissue and iPSC-derived neurons of idiopathic PD patients and other genetic forms of PD without GBA1 mutation [160,303,304,307,308]. It was noted that mitochondrial oxidative stress leads to the accumulation of oxidized dopamine, resulting in reduced GCase activity, lysosomal dysfunction, and α-syn accumulation [309]. The iPSCs were generated from patients with GD and PD harboring GBA1 mutations and differentiated into midbrain DA neurons. The iPSC-derived neurons exhibited reduced GCase activity and protein levels, increased glucosylceramide and α-syn levels as well as autophagic and lysosomal defects [310]. The mutant neurons also exhibited dysregulation of calcium homeostasis and increased vulnerability to stress responses involving the elevation of cystolic calcium. Gene correction of the mutation rescues the observed pathological phenotypes [310]. N370S iPSC-derived dopaminergic neurons exhibited disruption of the autophagy pathway and ER stress leading to elevated extracellular a-syn [311]. GBA-PD patient-derived dopaminergic neurons with heterozygous N370S and L444P mutations display stress responses, sphingolipid accumulation, mitochondrial dysfunction, increased mitochondrial ROS, and changes in NAD+ metabolism, ameliorated with NAD+ precursor nicotinamide riboside [312]. Heterozygous-null GBA1 iPSC-derived cortical neurons and astrocytes exhibit reduced lysosome number, increased lysosomal pH, reduced lysosomal cathepsin protease activity, and increased accumulation of soluble and insoluble α-syn without changes in α-syn mRNA levels [313,314].

GCase chaperones were able to recover GCase activity and reduce α-syn levels in iPSC-derived DA neurons and mouse models [308]. GBA1 iPSC-derived neurons exhibited prolonged mitochondria–lysosome contacts due to defective modulation of the untethering protein TBC1D15, which mediates Rab7 GTP hydrolysis for contact untethering, ultimately leading to disrupted mitochondrial distribution and function [315]. This defect was rescued with a GCase modulator, indicating deficits were due to a lack of GCase activity. Another GCase chaperone S-181 was tested on 84GG GBA1 patient-derived dopaminergic neurons partially restored lysosomal function and lowered accumulation of oxidized dopamine, glucosylceramide, and α-syn [307].

Interestingly, G2019S LRRK2 iPSC-derived dopaminergic neurons also exhibited reduced GCase activity [316]. In fact, pharmacological inhibition of LRRK2 kinase activity increased GCase activity in both iPSC-derived dopaminergic neurons carrying LRRK2 and GBA1 mutations. The increase in GCase activity was sufficient to partially rescue the accumulation of oxidized dopamine and α-syn. Heterozygous-null GBA-1 iPSC-derived cortical neurons did not exhibit any differences in WT LRRK2 kinase activity [314]. However, LRRK2 inhibition rescued lysosomal number and Cathepsin L activity, and partial lysosome re-acidification. Interestingly, they did not see any change in GCase activity with LRRK2 inhibition, contradicting previous findings [316]. The possible interplay between LRRK2 and GBA1 is fascinating and needs to be further explored.

12. GBA1 Animal Models

There are over 200 known GD-associated mutations; therefore, it is difficult to determine which mutations are particularly responsible for PD susceptibility [295]. Homozygous L444P mice generated partial gene duplication and either died soon after birth due to compromised epidermal permeability barrier caused by defective glucosylceramide metabolism or exhibited systemic inflammation [317,318]. L444P conditional knock-in mice lived longer and exhibited increased striatal α-syn levels and astrogliosis at 1 year of age, although motor performance was not assessed [319]. Heterozygous L444P mice demonstrated impaired neuronal autophagy and mitophagy as well as mitochondrial dysfunction [320,321]. Heterozygous L444P mice also demonstrated impaired α-syn degradation and increased α-syn levels but interestingly did not form α-syn aggregates [320–322]. Furthermore, heterozygous L444P mice did not exhibit nigrostriatal neurodegeneration, neuroinflammation, or impairments in olfaction, coordination, and cognition [321–323]. The D409H mutation is a rare but severe mutation found in GBA1 PD patients [324]. α-Syn aggregation was observed in the cerebellum and brainstem of homozygous D409H mice [325]. However, heterozygous and homozygous D409H mice did not exhibit nigrostriatal neurodegeneration, neuroinflammation, or motor deficits [326]. Treatment with a GCase chaperone resulted in the activation of WT GCase and reduction of GCase lipid substrates and α-syn in brain tissue [307]. Homozygous V394L mice have 27% of WT GCase activity but do not exhibit α-syn aggregation [325]. Homozygous R643C mice exhibited increased nigrostriatal α-syn and neuroinflammation [319]. Homozygous N370S mutation has been neonatal lethal despite mild phenotype in patients [327]. Furthermore, several GBA1-PD models combine GBA1 mutation with overexpression of α-syn mutations to induce α-syn aggregation and pronounced PD symptoms despite the fact that these two mutations are not reported in PD patients [323,326,328].

13. PTEN-Induced Kinase 1 (PINK1)

Phosphatase and tensin homolog (PTEN)-induced putative kinase 1 (PINK1) is a mitochondria-targeted Ser/Thr protein kinase. In normal physiological conditions, PINK1 is imported into the mitochondria via the translocase of the outer membrane and translocase of the inner membrane [329]. PINK1 is cleaved in the transmembrane segment by the mitochondrial intramembrane protease PARL, where it is then retranslocated to the cytosol for proteasomal degradation [330]. Once the mitochondrial inner membrane becomes depolarized, PINK1 mitochondrial transport is arrested and phosphorylates Parkin on the ubiquitin-like domain and activates the E3 ligase activity of Parkin [331,332]. Parkin ubiquitinates mitochondrial outer membrane proteins and induces autophagic clearance of depolarized mitochondria [333–335]. PINK1 is localized on the mitochondria and may exhibit a protective effect, but as a result of a mutation that protection is lost, resulting in increased susceptibility to cellular stress [336]. It is also known to phosphorylate TNF receptor-associated protein 1 (TRAP1), 5-hydroxytryptamine receptor 2a (5-HT2A), and Parkin [337–339]. PINK1 mutations are the second most common autosomal-recessive form of early onset PD [329,340,341]. Furthermore, Pink1 mutation carriers may develop psychiatric comorbidity alongside gait disturbances [342,343].

14. PINK1 Induced Pluripotent Stem Cell Models

PINK1 kinase activity was significantly reduced in G411S PINK1 mutant iPSC-derived neurons [344]. In iPSC-derived neurons, PINK1 mutations reduced complex I activity, which lead to a reduction in mitochondrial membrane potential [345]. PINK1 PD patient iPSC-derived neurons treated with valinomycin, which triggers rapid loss of mitochondrial membrane potential, impaired recruitment of overexpressed Parkin to mitochondria [346,347]. Interestingly, loss of PINK1 abolished the degradation of mitochondrial protein only in fibroblasts, but not in isogenic iPSC-derived neurons, which do not exhibit significant mitophagy even with parkin overexpression and valinomycin treatment [346]. Phosphorylation of Ser250 in NdufA10 regulates the activity of ubiquinone reductase in mitochondrial complex I. A phosphomimetic mutant of NdufA10 reversed the deficiencies in complex I activity and ATP synthesis in PINK1 iPSC-derived neurons [345]. Furthermore, direct supplementation of cardiolipin, a mitochondrial inner membrane-specific lipid, to isolated mitochondria rescues PINK1-induced complex I defects [348]. Additionally, antioxidant treatment with coenzyme Q10 can rescue the cellular vulnerability associated with mitochondrial dysfunction in iPSC-derived neurons from PINK1 PD patients [349]. Apart from abnormal mitochondrial morphology, PINK1 PD patient iPSC-derived midbrain dopaminergic neurons exhibited α-syn accumulation and increased cytosolic dopamine levels [350]. Interestingly, increased expression of LRRK2 mRNA and protein was observed in PINK1 iPSC-derived neurons [351]. Moreover, transient overexpression of WT PINK1 can downregulate LRRK2 expression. The implication of this study suggests a convergent pathway between these two genes in PD pathogenesis.

15. PINK1 Animal Models

PINK1 knockout mice appear to have an age-dependent and moderate reduction in striatal dopamine levels accompanied by low locomotor activity due to deficient mitochondrial respiration and increased sensitivity to oxidative stress [352,353]. This model does not exhibit significant neurodegeneration, Lewy body formation, or loss of dopamine. Another PINK1 knockout mouse model has been reported to show olfactory and gait disturbances, similar to prodromal symptoms of human PD patients [354]. Pink1 deficient mice also have impaired dopamine release [355] and exhibited impaired complex I function, mitochondrial depolarization reduced ATP synthesis, and increased sensitivity to apoptotic stress [356]. As such, PINK1 knockout mice may be useful as a model for prodromal PD [357,358]. On the other hand, PINK1 null mice with an exon 4–5 deletion showed progressive loss of striatal dopamine, but nigrostriatal neurodegeneration was not observed [359].

The phenotype for rats differs from mice, with a closer resemblance to PD pathology. PINK1 knockout rats exhibit age-dependent loss of nigral dopaminergic neurons beginning at age 6–8 months [360,361]. Motor deficits including reduced rearing frequency and distance traveled in an open field, reduced hind limb grip strength, and increased foot slips and traversal time on a tapered balance beam were apparent [360,362]. These rats also exhibited mitochondrial respiration deficits and α-syn aggregation, although different from Lewy body pathology [361,362].

16. Parkin (PRKN)

Parkin is a RING domain-containing E3 ubiquitin ligase important for mitochondria quality control through mitophagy [363]. Autosomal recessive juvenile parkinsonism was first identified in Japanese patients with early onset PD [364]. Currently, there are more than 100 known mutations identified in Parkin [365–368]. Parkin mutations account for 50% of familial and 15% of sporadic cases of European PD patients with onset before 45 years of age [369,370]. Parkin mutations are also the most common form of juvenile PD or early onset PD [365,366,369,371–373]. Carriers of Parkin mutations exhibit earlier and more symmetrical onset, slower disease progression, and greater response to L-dopa, but seem to develop dyskinesia earlier as well [369,374,375]. Furthermore, cognitive impairment is rare, and dementia or depression are not present in Parkin mutation patients [369,376,377]. Most PD patients with Parkin mutations do not develop Lewy body pathology [378]. Furthermore, it appears that missense mutations are equally as detrimental to truncation and deletion mutations [379].

17. PRKN-Induced Pluripotent Stem Cell Models

Parkin patient iPSC-derived neurons present abnormal or enlarged mitochondria accompanied by increased oxidative stress and enhanced activity of nuclear factor erythroid 2-related factor 2 (Nrf2) pathway [380]. Proteomics analysis of Parkin knockout iPSC-derived neurons revealed disturbances in oxidative stress defense, mitochondrial respiration and morphology, cell cycle control, and cell viability [381]. Structural and functional analysis verified an increase in mitochondrial area and the presence of elongated mitochondria as well as impaired glycolysis and lactate-supported respiration. This abnormal mitochondria phenotype, such as elongated shape and larger volume, has been observed by other groups as well [350,380,382]. However, two different studies using qPCR showed that there was no significant difference in mitochondria DNA copy number [380,383]. Additionally, it seems Parkin mutations in iPSC-derived neurons may alter mitochondrial morphology in a portion of cells, particularly TH+ dopaminergic neurons [382]. In fact, Parkin patient iPSC-derived dopaminergic neurons exhibited smaller and less functional mitochondria than those in non-dopaminergic neurons [384] and exhibited increased cytotoxicity with known PD environmental risk factors, such as exposure to heavy metals such as copper and cadmium [385]. They also showed a significant increase in mitochondrial fragmentation, initial ROS generation, and loss of mitochondrial membrane potential following copper exposure. Transfected Parkin cell lines have shown that Parkin is recruited to the mitochondria to ubiquitinate a variety of substrates for the induction of mitophagy [363,386]. However, the recruitment of endogenous Parkin to mitochondria has not been robustly seen in cell lines [387,388], mice [389,390], or iPSC-derived human neurons [346,383].

The precision of dopaminergic transmission is significantly disrupted by increased spontaneous dopamine release and decreased reuptake [383]. Dopamine induces oscillatory neuronal activities in Parkin iPSC-derived neurons but not in normal iPSC-derived neurons [391]. Interestingly, this phenotype mirrors the widespread rhythmic bursting of neuronal activities in the basal ganglia of PD patients [392]. Parkin iPSC-derived midbrain dopaminergic neurons exhibited reduced length and complexity of neuronal processes understood to be caused by a marked decrease in microtubule stability, as the phenotype was rescued by taxol, a microtubule-stabilizing drug, or the overexpression of Parkin [393]. This decreased neurite length and complexity was also mimicked in normal neurons treated with colchicine, a microtubule-depolymerizing agent. Reduced microtubule stability in iPSC-derived neurons was observed by other groups as well [394].

In iPSC-derived neurons from two Parkin patients, increased accumulation of α-syn was observed in the patient with Lewy body pathology, but not in the other patient who did not have Lewy body pathology [380]. There was no significant difference in α-syn protein levels in iPSC-derived midbrain dopaminergic neurons from two Parkin PD patients and two normal subjects [383]. Although, an increased level of α-syn protein was observed in two other studies using iPSC-derived neurons from PD patients with Parkin mutations [350,382]. The inconsistency of α-syn accumulation in Parkin PD patient-derived cells may suggest that α-syn accumulation is independent of Parkin as Parkin PD patients do not necessarily develop Lewy body pathology.

18. PRKN Animal Models

Transgenic rodent models with Parkin overexpression exhibited protection from 6-OHDA or α-syn overexpression in the SNpc [395–398]. Parkin knockout mice have nigrostriatal mitochondrial respiration deficits and increased markers for oxidative stress but did not exhibit any significant nigrostriatal dopaminergic neurodegeneration nor PD-like locomotor deficits [360,399–405]. Although there were instances of Parkin knockout mice exhibiting slight impairment of dopamine release [406,407], they did not show significant behavioral deficits or age-dependent nigral dopaminergic neurodegeneration [360]. They also lacked Lewy body pathology similarly to human parkin mutant carriers that rarely showed LB pathology [408]. Successful transgenic rodent models that recapitulated dopaminergic neurodegeneration were eventually developed. Indeed, Parkin-Q311X-DAT-BAC mice expressing a C-terminal truncated human mutant Parkin in dopaminergic neurons exhibited late onset and progressive hypokinetic motor deficits, age-dependent nigrostriatal dopaminergic neurodegeneration, α-syn accumulation, and a significant reduction in striatal dopamine neuron terminals and dopamine levels [409]. Overexpression of both T240R Parkin and human WT Parkin in rats with AAV2/8 induced progressive and dose-dependent dopaminergic neurodegeneration [410].

19. Parkinsonism Associated Deglycase (DJ-1 Also Known as PARK7)

The DJ-1 protein is 189 amino acid residues with three cysteines and normally forms a homodimer that exhibits antioxidant activities [411]. DJ-1 is highly expressed in astrocytes in the frontal cortex and SNpc of idiopathic PD brains and is not an essential component of Lewy bodies [412,413]. It is also involved in the regulation of apoptosis and pro-survival signaling, autophagy, inflammatory responses, and protection against oxidative stress [414,415]. In fact, DJ-1 overexpression protects against oxidative stress while DJ-1 knockout increases oxidative stress-induced cell death [416–421]. It seems that the cysteine residue at position 106 is required for DJ-1 mediated protection from oxidative stress [411,422–424]. DJ-1 levels increase in response to oxidative stress caused by dopamine to suppress ROS accumulation [425]. DJ-1 also functions as a redox-sensitive chaperone that is activated in an oxidative cytoplasmic environment and can inhibit the generation of α-syn aggregates [426,427]. It can directly interact with α-syn monomers and oligomers, where mutant DJ-1 exhibits α-syn dimerization [428]. Furthermore, DJ-1 deficiency decreases LAMP2A expression, a receptor required for CMA-mediated α-syn degradation, thus leading to α-syn accumulation [429]. It can also act as a glyoxalase III to detoxify reactive dicarbonyl species, such as glyoxal and methylglyoxal into glycolic or lactic acid in the absence of glutathione [430].

A deletion mutation in a Dutch family and a homozygous point mutation L166P in an Italian family were identified to cause parkinsonism [431]. DJ-1 mutations only account for less than 1% of all early onset PD cases [432]. The median age of onset for DJ-1 PD is 27 years [433]. PD patients carrying DJ-1 mutations exhibit early onset of dyskinesia, rigidity, and tremors and respond well to L-DOPA treatment [431,434,435]. Cognitive deficits and psychotic disturbances typically arise later in disease progression [436]. The L166P mutation abolishes DJ-1 dimerization, abrogating its neuroprotective activity [437]. Post-mortem brain tissue of a patient with L172Q DJ-1 mutation exhibited Lewy body pathology [438]. The clinical manifestations of DJ-1 patients are like those with Parkin and PINK1 mutations. However, compared to Parkin and PINK1, DJ-1 mutation carriers exhibit a higher percentage of non-motor symptoms such as anxiety, cognitive decline, depression, and psychopathic symptoms [433,439,440]. It has been reported that PD brains exhibit decreased levels of DJ-1 mRNA and protein, but also show the presence of extra-oxidized

DJ-1 isoforms [441] and that acidic isoforms of DJ-1 monomer were accumulated in sporadic PD brains [442]. It remains unclear how DJ-1 contributes to PD pathogenesis.

20. DJ-1 Induced Pluripotent Stem Cell Models

Homozygous DJ-1 mutant iPSC-derived midbrain dopaminergic neurons exhibited increased dopamine oxidation and neuromelanin-like pigmented aggregates compared to an isogenic gene-corrected control [309]. Oxidative stress from dopamine metabolism triggered mitochondrial oxidative stress, which was significantly attenuated by blocking dopamine synthesis [309]. Dopamine-induced oxidative stress inactivated GCase inhibited lysosomal function and led to increased expression of α-syn [309]. DJ-1 knockout iPSC-derived dopaminergic neurons also exhibited enhanced α-syn fibril-induced aggregation and neuronal death [443]. Other DJ-1 patient-specific iPSCs have been generated very recently, but phenotypic analysis has yet to be performed [444–446].

21. DJ-1 Animal Models

DJ-1 knockout mice show mild motor deficits and altered nigrostriatal synaptic physiology, but without dopaminergic neuron loss or significant change in dopamine levels [417,447,448]. However, one report showed that the DJ-1 knockout mouse line exhibited a loss of dopaminergic neurons in the ventral tegmental area and slight behavioral changes, such as diminished rearing behavior and impaired object recognition [449]. It was also observed that DJ-1 deficient mice exhibit alteration in dopamine metabolism, specifically an increase in dopamine reuptake causing an accumulation of striatal dopamine [450]. The higher overall level of oxidized dopamine did not result in neurodegeneration as seen in PD patients and this may be due to the fact that human SNpc dopaminergic neurons have higher overall dopamine levels than those in mice. A DJ-1 nullizygous mouse was fully backcrossed with a C57BL/6 background and displayed dramatic early onset unilateral loss of nigrostriatal dopaminergic neurons that progressed bilaterally with aging [451]. These mice also exhibited age-dependent bilateral degeneration in the locus ceruleus nucleus and displayed mild motor deficits. DJ-1 knockout mice also exhibited more microglial activation, especially in response to lipopolysaccharide insult [452]. On the other hand, DJ-1 knockout rats showed significant age-dependent nigrostriatal dopaminergic neuron loss (~50%) between 6 and 8 months of age, accompanied by motor deficits [360]. Furthermore, mitochondria from DJ-1 knockout rats showed altered respiration compared to that of WT rats [453].

22. Overall Discussion

As PD is primarily an idiopathic disease, the major challenge is in understanding the interplay between genetic and environmental factors that influence susceptibility to PD. In the last two decades, research into PD genetics has deepened our understanding of PD risk, onset, progression, and therapeutic approaches. These studies have revealed pathogenic pathways of neurodegeneration shared between inherited and sporadic PD. No animal PD model is perfect as it is challenging to recapitulate the age of onset, the timing of disease progression, and the spectrum of pathologies present in PD patients. Single model modalities, such as neurotoxin models recapitulate neurodegeneration and PD-like symptoms and are appropriate for addressing specific questions tailored to this model, as is the case for the genetic models. Current research has been utilizing iPSCs for a disease-in-a-dish approach. Isogenic gene-corrected cell lines offer unequivocal perturbations caused by genetic mutations. Since iPSC-based in vitro models are patient-derived they present many advantages including the ability to generate a variety of neural lineages in 3D complex systems that serve to model the in vivo brain tissue cytoarchitecture for studying pathogenesis. PD is not a single disease entity, but rather made up of subtypes based on differences in the spectrum of symptoms and the nature and distribution of Lewy body pathologies.Thus, it is necessary to combine iPSC-based in vitro models, ex vivo post-mortem brain specimens, and in vivo models to further our understanding of the different cellular and molecular mechanisms underlying PD pathogenesis.

Author Contributions: J.K., E.W.D., T.O., E.S.D. and M.M.D. provided summaries of literature searches and presentations, performed research related to the topic; provided comments on the manuscript, J.K. and M.M.D. wrote the manuscript. All authors have read and agreed to the published version of the manuscript.

Funding: National Institute on Aging R56 AG059284 and the NIH Primate Center Base grant (Office of Research Infrastructure Programs/OD P51 OD011133).

Acknowledgments: This work was also supported by the Worth Family Fund, The Perry and Ruby Stevens Charitable Foundation, The Robert J. Jr. and Helen C. Kleberg Foundation, The Marmion Family Fund, The William and Ella Owens Medical Research Foundation, the National Institute on Aging R56 AG059284 and the NIH Primate Center Base grant (Office of Research Infrastructure Programs/OD P51 OD011133).

Conflicts of Interest: The authors declare no conflict of interest.

References

1. Braak, H.; Braak, E.; Yilmazer, D.; Schultz, C.; de Vos, R.A.; Jansen, E.N. Nigral and extranigral pathology in Parkinson's disease. *J. Neural Transm. Suppl.* **1995**, *46*, 15–31. [PubMed]
2. Jellinger, K.A. Pathology of Parkinson's disease. Changes other than the nigrostriatal pathway. *Mol. Chem. Neuropathol.* **1991**, *14*, 153–197. [PubMed]
3. Parkinson, T. Outlines of Zoonosological Tables. *Lond. Med. Phys. J.* 1817, *38*, 449–453. [PubMed]
4. Hirsch, L.; Jette, N.; Frolkis, A.; Steeves, T.; Pringsheim, T. The Incidence of Parkinson's Disease: A Systematic Review and Meta-Analysis. *Neuroepidemiology* **2016**, *46*, 292–300. [CrossRef] [PubMed]
5. Shalash, A.S.; Hamid, E.; Elrassas, H.H.; Bedair, A.S.; Abushouk, A.I.; Khamis, M.; Hashim, M.; Ahmed, N.S.; Ashour, S.; Elbalkimy, M. Non-Motor Symptoms as Predictors of Quality of Life in Egyptian Patients With Parkinson's Disease: A Cross-Sectional Study Using a Culturally Adapted 39-Item Parkinson's Disease Questionnaire. *Front. Neurol.* **2018**, *9*, 357. [CrossRef]
6. Dorsey, E.R.; Sherer, T.; Okun, M.S.; Bloem, B.R. The Emerging Evidence of the Parkinson Pandemic. *J. Park. Dis.* **2018**, *8*, S3–S8. [CrossRef]
7. Riederer, P.; Wuketich, S. Time course of nigrostriatal degeneration in parkinson's disease. A detailed study of influential factors in human brain amine analysis. *J. Neural Transm.* **1976**, *38*, 277–301. [CrossRef]
8. Schaeffer, E.; Postuma, R.B.; Berg, D. Prodromal PD: A new nosological entity. *Prog. Brain Res.* **2020**, *252*, 331–356. [CrossRef]
9. Poewe, W.; Seppi, K.; Tanner, C.M.; Halliday, G.M.; Brundin, P.; Volkmann, J.; Schrag, A.E.; Lang, A.E. Parkinson disease. *Nat. Rev. Dis. Prim.* **2017**, *3*, 17013. [CrossRef]
10. Hughes, A.J.; Daniel, S.E.; Kilford, L.; Lees, A.J. Accuracy of clinical diagnosis of idiopathic Parkinson's disease: A clinicopathological study of 100 cases. *J. Neurol. Neurosurg. Psychiatry* **1992**, *55*, 181–184. [CrossRef]
11. Jankovic, J. Parkinson's disease: Clinical features and diagnosis. *J. Neurol. Neurosurg. Psychiatry* **2008**, *79*, 368–376. [CrossRef]
12. Gelb, D.J.; Oliver, E.; Gilman, S. Diagnostic criteria for Parkinson Disease. *Arch. Neurol.* **1999**, *56*, 7. [CrossRef]
13. Reichmann, H. Clinical criteria for the diagnosis of Parkinson's disease. *Neurodegener Dis.* **2010**, *7*, 284–290. [CrossRef]
14. Virmani, T.; Moskowitz, C.B.; Vonsattel, J.P.; Fahn, S. Clinicopathological characteristics of freezing of gait in autopsy-confirmed Parkinson's disease. *Mov. Disord.* **2015**, *30*, 1874–1884. [CrossRef]
15. Tolosa, E.; Compta, Y. Dystonia in Parkinson's disease. *J. Neurol.* **2006**, *253* (Suppl. S7), VII7–VII13. [CrossRef]
16. Doherty, K.M.; van de Warrenburg, B.P.; Peralta, M.C.; Silveira-Moriyama, L.; Azulay, J.P.; Gershanik, O.S.; Bloem, B.R. Postural deformities in Parkinson's disease. *Lancet Neurol.* **2011**, *10*, 538–549. [CrossRef]
17. Emre, M.; Aarsland, D.; Brown, R.; Burn, D.J.; Duyckaerts, C.; Mizuno, Y.; Broe, G.A.; Cummings, J.; Dickson, D.W.; Gauthier, S.; et al. Clinical diagnostic criteria for dementia associated with Parkinson's disease. *Mov. Disord.* **2007**, *22*, 1689–1707. [CrossRef]
18. Remy, P.; Doder, M.; Lees, A.; Turjanski, N.; Brooks, D. Depression in Parkinson's disease: Loss of dopamine and noradrenaline innervation in the limbic system. *Brain* **2005**, *128*, 1314–1322. [CrossRef]
19. Onofrj, M.; Thomas, A.; Bonanni, L. New approaches to understanding hallucinations in Parkinson's disease: Phenomenology and possible origins. *Expert. Rev. Neurother.* **2007**, *7*, 1731–1750. [CrossRef]
20. Williams-Gray, C.H.; Foltynie, T.; Lewis, S.J.; Barker, R.A. Cognitive deficits and psychosis in Parkinson's disease: A review of pathophysiology and therapeutic options. *CNS Drugs* **2006**, *20*, 477–505. [CrossRef]
21. Williams-Gray, C.H.; Foltynie, T.; Brayne, C.E.; Robbins, T.W.; Barker, R.A. Evolution of cognitive dysfunction in an incident Parkinson's disease cohort. *Brain* **2007**, *130*, 1787–1798. [CrossRef] [PubMed]
22. McAuley, J.H.; Gregory, S. Prevalence and clinical course of olfactory hallucinations in idiopathic Parkinson's disease. *J. Park. Dis.* **2012**, *2*, 199–205. [CrossRef] [PubMed]

23. Inzelberg, R.; Kipervasser, S.; Korczyn, A.D. Auditory hallucinations in Parkinson's disease. *J. Neurol. Neurosurg. Psychiatry* **1998**, *64*, 533–535. [CrossRef] [PubMed]
24. Fenelon, G.; Thobois, S.; Bonnet, A.M.; Broussolle, E.; Tison, F. Tactile hallucinations in Parkinson's disease. *J. Neurol.* **2002**, *249*, 1699–1703. [CrossRef] [PubMed]
25. Reijnders, J.S.; Ehrt, U.; Weber, W.E.; Aarsland, D.; Leentjens, A.F. A systematic review of prevalence studies of depression in Parkinson's disease. *Mov. Disord.* **2008**, *23*, 183–189. [CrossRef]
26. Marsden, C.D.; Parkes, J.D. Success and Problems of Long-Term Levodopa Therapy in Parkinson'S Disease. *Lancet* **1977**, *309*, 345–349. [CrossRef]
27. Thanvi, B.R.; Lo, T.C. Long term motor complications of levodopa: Clinical features, mechanisms, and management strategies. *Postgrad. Med. J.* **2004**, *80*, 452–458. [CrossRef]
28. Deuschl, G.; Schade-Brittinger, C.; Krack, P.; Volkmann, J.; Schäfer, H.; Bötzel, K.; Daniels, C.; Deutschländer, A.; Gruber, D.; Hamel, W.; et al. A Randomized Trial of Deep-Brain Stimulation for Parkinson's Disease. *N. Engl. J. Med.* **2006**, *355*, 896–908. [CrossRef]
29. Krack, P.; Pollak, P.; Limousin, P.; Hoffmann, D.; Benazzouz, A.; Benabid, A.L. Inhibition of levodopa effects by internal pallidal stimulation. *Mov. Disord.* **1998**, *13*, 648–652. [CrossRef]
30. Kumar, R.; Lozano, A.M.; Kim, Y.J.; Hutchison, W.D.; Sime, E.; Halket, E.; Lang, A.E. Double-blind evaluation of subthalamic nucleus deep brain stimulation in advanced Parkinson's disease. *Neurology* **1998**, *51*, 850–855. [CrossRef]
31. Limousin, P.; Krack, P.; Pollak, P.; Benazzouz, A.; Ardouin, C.; Hoffmann, D.; Benabid, A.L. Electrical stimulation of the subthalamic nucleus in advanced Parkinson's disease. *N. Engl. J. Med.* **1998**, *339*, 1105–1111. [CrossRef]
32. Wichmann, T.; DeLong, M.R. Deep Brain Stimulation for Movement Disorders of Basal Ganglia Origin: Restoring Function or Functionality? *Neurotherapeutics* **2016**, *13*, 264–283. [CrossRef]
33. Linazasoro, G.; Van Blercom, N.; Lasa, A. Unilateral subthalamic deep brain stimulation in advanced Parkinson's disease. *Mov. Disord.* **2003**, *18*, 713–716. [CrossRef]
34. Diamond, A.; Jankovic, J. The effect of deep brain stimulation on quality of life in movement disorders. *J. Neurol. Neurosurg. Psychiatry* **2005**, *76*, 1188–1193. [CrossRef]
35. Goldman, S.M. Environmental toxins and Parkinson's disease. *Annu. Rev. Pharmacol. Toxicol.* **2014**, *54*, 141–164. [CrossRef]
36. Callaghan, R.C.; Cunningham, J.K.; Sykes, J.; Kish, S.J. Increased risk of Parkinson's disease in individuals hospitalized with conditions related to the use of methamphetamine or other amphetamine-type drugs. *Drug Alcohol Depend.* **2012**, *120*, 35–40. [CrossRef]
37. Antony, P.M.; Diederich, N.J.; Kruger, R.; Balling, R. The hallmarks of Parkinson's disease. *FEBS J.* **2013**, *280*, 5981–5993. [CrossRef]
38. Verstraeten, A.; Theuns, J.; Van Broeckhoven, C. Progress in unraveling the genetic etiology of Parkinson disease in a genomic era. *Trends Genet.* **2015**, *31*, 140–149. [CrossRef]
39. Kalinderi, K.; Bostantjopoulou, S.; Fidani, L. The genetic background of Parkinson's disease: Current progress and future prospects. *Acta Neurol. Scand.* **2016**, *134*, 314–326. [CrossRef]
40. Cenci, M.A.; Bjorklund, A. Animal models for preclinical Parkinson's research: An update and critical appraisal. *Prog. Brain Res.* **2020**, *252*, 27–59. [CrossRef]
41. Bose, A.; Petsko, G.A.; Studer, L. Induced pluripotent stem cells: A tool for modeling Parkinson's disease. *Trends Neurosci.* **2022**, *45*, 608–620. [CrossRef]
42. Antonov, S.A.; Novosadova, E.V. Current State-of-the-Art and Unresolved Problems in Using Human Induced Pluripotent Stem Cell-Derived Dopamine Neurons for Parkinson's Disease Drug Development. *Int. J. Mol. Sci.* **2021**, *22*, 3381. [CrossRef] [PubMed]
43. Creed, R.B.; Goldberg, M.S. New Developments in Genetic rat models of Parkinson's Disease. *Mov. Disord.* **2018**, *33*, 717–729. [CrossRef] [PubMed]
44. Mills, R.D.; Mulhern, T.D.; Liu, F.; Culvenor, J.G.; Cheng, H.C. Prediction of the repeat domain structures and impact of parkinsonism-associated variations on structure and function of all functional domains of leucine-rich repeat kinase 2 (LRRK2). *Hum. Mutat.* **2014**, *35*, 395–412. [CrossRef] [PubMed]
45. Rudenko, I.N.; Kaganovich, A.; Langston, R.G.; Beilina, A.; Ndukwe, K.; Kumaran, R.; Dillman, A.A.; Chia, R.; Cookson, M.R. The G2385R risk factor for Parkinson's disease enhances CHIP-dependent intracellular degradation of LRRK2. *Biochem. J.* **2017**, *474*, 1547–1558. [CrossRef] [PubMed]
46. Wallings, R.L.; Tansey, M.G. LRRK2 regulation of immune-pathways and inflammatory disease. *Biochem. Soc. Trans.* **2019**, *47*, 1581–1595. [CrossRef]
47. Webber, P.J.; Smith, A.D.; Sen, S.; Renfrow, M.B.; Mobley, J.A.; West, A.B. Autophosphorylation in the Leucine-Rich Repeat Kinase 2 (LRRK2) GTPase Domain Modifies Kinase and GTP-Binding Activities. *J. Mol. Biol.* **2011**, *412*, 94–110. [CrossRef]
48. Smith, W.W.; Pei, Z.; Jiang, H.; Moore, D.J.; Liang, Y.; West, A.B.; Dawson, V.L.; Dawson, T.M.; Ross, C.A. Leucine-rich repeat kinase 2 (LRRK2) interacts with parkin, and mutant LRRK2 induces neuronal degeneration. *Proc. Natl. Acad. Sci. USA* **2005**, *102*, 18676–18681. [CrossRef]
49. Ho, C.C.-Y.; Rideout, H.J.; Ribe, E.; Troy, C.M.; Dauer, W.T. The Parkinson Disease Protein LRRK2 Transduces Death Signals via FADD and Caspase-8 in a Cellular Model of Neurodegeneration. *J. Neurosci.* **2009**, *29*, 1011–1016. [CrossRef]
50. Imai, Y.; Gehrke, S.; Wang, H.Q.; Takahashi, R.; Hasegawa, K.; Oota, E.; Lu, B. Phosphorylation of 4E-BP by LRRK2 affects the maintenance of dopaminergic neurons in Drosophila. *EMBO J.* **2008**, *27*, 2432–2443. [CrossRef]

51. Kanao, T.; Venderova, K.; Park, D.S.; Unterman, T.; Lu, B.; Imai, Y. Activation of FoxO by LRRK2 induces expression of proapoptotic proteins and alters survival of postmitotic dopaminergic neuron in Drosophila. *Hum. Mol. Genet.* **2010**, *19*, 3747–3758. [CrossRef]
52. Cookson, M.R. The role of leucine-rich repeat kinase 2 (LRRK2) in Parkinson's disease. *Nat. Rev. Neurosci.* **2010**, *11*, 791–797. [CrossRef]
53. Migheli, R.; Del Giudice, M.G.; Spissu, Y.; Sanna, G.; Xiong, Y.; Dawson, T.M.; Dawson, V.L.; Galioto, M.; Rocchitta, G.; Biosa, A.; et al. LRRK2 Affects Vesicle Trafficking, Neurotransmitter Extracellular Level and Membrane Receptor Localization. *PLoS ONE* **2013**, *8*, e77198. [CrossRef]
54. Miklossy, J.; Arai, T.; Guo, J.-P.; Klegeris, A.; Yu, S.; McGeer, E.G.; McGeer, P.L. LRRK2 Expression in Normal and Pathologic Human Brain and in Human Cell Lines. *J. Neuropathol. Exp. Neurol.* **2006**, *65*, 953–963. [CrossRef]
55. Dzamko, N.; Deak, M.; Hentati, F.; Reith, A.D.; Prescott, A.R.; Alessi, D.R.; Nichols, R.J. Inhibition of LRRK2 kinase activity leads to dephosphorylation of Ser(910)/Ser(935), disruption of 14-3-3 binding and altered cytoplasmic localization. *Biochem. J.* **2010**, *430*, 405–413. [CrossRef]
56. Nichols, R.J.; Dzamko, N.; Morrice, N.A.; Campbell, D.G.; Deak, M.; Ordureau, A.; Macartney, T.; Tong, Y.; Shen, J.; Prescott, A.R.; et al. 14-3-3 binding to LRRK2 is disrupted by multiple Parkinson's disease-associated mutations and regulates cytoplasmic localization. *Biochem. J.* **2010**, *430*, 393–404. [CrossRef]
57. Tong, Y.; Yamaguchi, H.; Giaime, E.; Boyle, S.; Kopan, R.; Kelleher, R.J., 3rd; Shen, J. Loss of leucine-rich repeat kinase 2 causes impairment of protein degradation pathways, accumulation of α-synuclein, and apoptotic cell death in aged mice. *Proc. Natl. Acad. Sci. USA* **2010**, *107*, 9879–9884. [CrossRef]
58. Cook, D.A.; Kannarkat, G.T.; Cintron, A.F.; Butkovich, L.M.; Fraser, K.B.; Chang, J.; Grigoryan, N.; Factor, S.A.; West, A.B.; Boss, J.M.; et al. LRRK2 levels in immune cells are increased in Parkinson's disease. *NPJ Park. Dis.* **2017**, *3*, 11. [CrossRef]
59. Dzamko, N.; Gysbers, A.M.; Bandopadhyay, R.; Bolliger, M.F.; Uchino, A.; Zhao, Y.; Takao, M.; Wauters, S.; van de Berg, W.D.; Takahashi-Fujigasaki, J.; et al. LRRK2 levels and phosphorylation in Parkinson's disease brain and cases with restricted Lewy bodies. *Mov. Disord.* **2017**, *32*, 423–432. [CrossRef]
60. Atashrazm, F.; Hammond, D.; Perera, G.; Bolliger, M.F.; Matar, E.; Halliday, G.M.; Schule, B.; Lewis, S.J.G.; Nichols, R.J.; Dzamko, N. LRRK2-mediated Rab10 phosphorylation in immune cells from Parkinson's disease patients. *Mov. Disord.* **2019**, *34*, 406–415. [CrossRef]
61. Nichols, R.J. LRRK2 Phosphorylation. *Adv. Neurobiol.* **2017**, *14*, 51–70. [CrossRef] [PubMed]
62. Sheng, Z.; Zhang, S.; Bustos, D.; Kleinheinz, T.; Le Pichon, C.E.; Dominguez, S.L.; Solanoy, H.O.; Drummond, J.; Zhang, X.; Ding, X.; et al. Ser1292 autophosphorylation is an indicator of LRRK2 kinase activity and contributes to the cellular effects of PD mutations. *Sci. Transl. Med.* **2012**, *4*, 164ra161. [CrossRef] [PubMed]
63. West, A.B.; Moore, D.J.; Biskup, S.; Bugayenko, A.; Smith, W.W.; Ross, C.A.; Dawson, V.L.; Dawson, T.M. Parkinson's disease-associated mutations in leucine-rich repeat kinase 2 augment kinase activity. *Proc. Natl. Acad. Sci. USA* **2005**, *102*, 16842–16847. [CrossRef] [PubMed]
64. Greggio, E.; Jain, S.; Kingsbury, A.; Bandopadhyay, R.; Lewis, P.; Kaganovich, A.; van der Brug, M.P.; Beilina, A.; Blackinton, J.; Thomas, K.J.; et al. Kinase activity is required for the toxic effects of mutant LRRK2/dardarin. *Neurobiol. Dis.* **2006**, *23*, 329–341. [CrossRef] [PubMed]
65. Smith, W.W.; Pei, Z.; Jiang, H.; Dawson, V.L.; Dawson, T.M.; Ross, C.A. Kinase activity of mutant LRRK2 mediates neuronal toxicity. *Nat. Neurosci.* **2006**, *9*, 1231–1233. [CrossRef]
66. Lee, S.; Liu, H.P.; Lin, W.Y.; Guo, H.; Lu, B. LRRK2 kinase regulates synaptic morphology through distinct substrates at the presynaptic and postsynaptic compartments of the Drosophila neuromuscular junction. *J. Neurosci.* **2010**, *30*, 16959–16969. [CrossRef]
67. Fuji, R.N.; Flagella, M.; Baca, M.; Baptista, M.A.; Brodbeck, J.; Chan, B.K.; Fiske, B.K.; Honigberg, L.; Jubb, A.M.; Katavolos, P.; et al. Effect of selective LRRK2 kinase inhibition on nonhuman primate lung. *Sci. Transl. Med.* **2015**, *7*, 273ra215. [CrossRef]
68. Baptista, M.A.; Dave, K.D.; Frasier, M.A.; Sherer, T.B.; Greeley, M.; Beck, M.J.; Varsho, J.S.; Parker, G.A.; Moore, C.; Churchill, M.J.; et al. Loss of leucine-rich repeat kinase 2 (LRRK2) in rats leads to progressive abnormal phenotypes in peripheral organs. *PLoS ONE* **2013**, *8*, e80705. [CrossRef]
69. Herzig, M.C.; Kolly, C.; Persohn, E.; Theil, D.; Schweizer, T.; Hafner, T.; Stemmelen, C.; Troxler, T.J.; Schmid, P.; Danner, S.; et al. LRRK2 protein levels are determined by kinase function and are crucial for kidney and lung homeostasis in mice. *Hum. Mol. Genet.* **2011**, *20*, 4209–4223. [CrossRef]
70. Kawakami, F.; Yabata, T.; Ohta, E.; Maekawa, T.; Shimada, N.; Suzuki, M.; Maruyama, H.; Ichikawa, T.; Obata, F. LRRK2 phosphorylates tubulin-associated tau but not the free molecule: LRRK2-mediated regulation of the tau-tubulin association and neurite outgrowth. *PLoS ONE* **2012**, *7*, e30834. [CrossRef]
71. Bailey, R.M.; Covy, J.P.; Melrose, H.L.; Rousseau, L.; Watkinson, R.; Knight, J.; Miles, S.; Farrer, M.J.; Dickson, D.W.; Giasson, B.I.; et al. LRRK2 phosphorylates novel tau epitopes and promotes tauopathy. *Acta Neuropathol.* **2013**, *126*, 809–827. [CrossRef]
72. Krumova, P.; Reyniers, L.; Meyer, M.; Lobbestael, E.; Stauffer, D.; Gerrits, B.; Muller, L.; Hoving, S.; Kaupmann, K.; Voshol, J.; et al. Chemical genetic approach identifies microtubule affinity-regulating kinase 1 as a leucine-rich repeat kinase 2 substrate. *FASEB J.* **2015**, *29*, 2980–2992. [CrossRef]

73. Gillardon, F. Leucine-rich repeat kinase 2 phosphorylates brain tubulin-beta isoforms and modulates microtubule stability—A point of convergence in Parkinsonian neurodegeneration? *J. Neurochem.* **2009**, *110*, 1514–1522. [CrossRef]
74. Haebig, K.; Gloeckner, C.J.; Miralles, M.G.; Gillardon, F.; Schulte, C.; Riess, O.; Ueffing, M.; Biskup, S.; Bonin, M. ARHGEF7 (Beta-PIX) acts as guanine nucleotide exchange factor for leucine-rich repeat kinase 2. *PLoS ONE* **2010**, *5*, e13762. [CrossRef]
75. Jaleel, M.; Nichols, R.J.; Deak, M.; Campbell, D.G.; Gillardon, F.; Knebel, A.; Alessi, D.R. LRRK2 phosphorylates moesin at threonine-558: Characterization of how Parkinson's disease mutants affect kinase activity. *Biochem. J.* **2007**, *405*, 307–317. [CrossRef]
76. Parisiadou, L.; Xie, C.; Cho, H.J.; Lin, X.; Gu, X.L.; Long, C.X.; Lobbestael, E.; Baekelandt, V.; Taymans, J.M.; Sun, L.; et al. Phosphorylation of ezrin/radixin/moesin proteins by LRRK2 promotes the rearrangement of actin cytoskeleton in neuronal morphogenesis. *J. Neurosci.* **2009**, *29*, 13971–13980. [CrossRef]
77. Gandhi, P.N.; Wang, X.; Zhu, X.; Chen, S.G.; Wilson-Delfosse, A.L. The Roc domain of leucine-rich repeat kinase 2 is sufficient for interaction with microtubules. *J. Neurosci. Res.* **2008**, *86*, 1711–1720. [CrossRef]
78. MacLeod, D.; Dowman, J.; Hammond, R.; Leete, T.; Inoue, K.; Abeliovich, A. The familial Parkinsonism gene LRRK2 regulates neurite process morphology. *Neuron* **2006**, *52*, 587–593. [CrossRef]
79. Godena, V.K.; Brookes-Hocking, N.; Moller, A.; Shaw, G.; Oswald, M.; Sancho, R.M.; Miller, C.C.; Whitworth, A.J.; de Vos, K.J. Increasing microtubule acetylation rescues axonal transport and locomotor deficits caused by LRRK2 Roc-COR domain mutations. *Nat. Commun.* **2014**, *5*, 5245. [CrossRef]
80. Steger, M.; Tonelli, F.; Ito, G.; Davies, P.; Trost, M.; Vetter, M.; Wachter, S.; Lorentzen, E.; Duddy, G.; Wilson, S.; et al. Phosphoproteomics reveals that Parkinson's disease kinase LRRK2 regulates a subset of Rab GTPases. *Elife* **2016**, *5*, e12813. [CrossRef]
81. Rivero-Rios, P.; Romo-Lozano, M.; Fernandez, B.; Fdez, E.; Hilfiker, S. Distinct Roles for RAB10 and RAB29 in Pathogenic LRRK2-Mediated Endolysosomal Trafficking Alterations. *Cells* **2020**, *9*, 1719. [CrossRef] [PubMed]
82. Di Maio, R.; Hoffman, E.K.; Rocha, E.M.; Keeney, M.T.; Sanders, L.H.; De Miranda, B.R.; Zharikov, A.; Van Laar, A.; Stepan, A.F.; Lanz, T.A.; et al. LRRK2 activation in idiopathic Parkinson's disease. *Sci. Transl. Med.* **2018**, *10*, eaar5429. [CrossRef] [PubMed]
83. Piccoli, G.; Condliffe, S.B.; Bauer, M.; Giesert, F.; Boldt, K.; de Astis, S.; Meixner, A.; Sarioglu, H.; Vogt-Weisenhorn, D.M.; Wurst, W.; et al. LRRK2 controls synaptic vesicle storage and mobilization within the recycling pool. *J. Neurosci.* **2011**, *31*, 2225–2237. [CrossRef] [PubMed]
84. Cirnaru, M.D.; Marte, A.; Belluzzi, E.; Russo, I.; Gabrielli, M.; Longo, F.; Arcuri, L.; Murru, L.; Bubacco, L.; Matteoli, M.; et al. LRRK2 kinase activity regulates synaptic vesicle trafficking and neurotransmitter release through modulation of LRRK2 macro-molecular complex. *Front. Mol. Neurosci.* **2014**, *7*, 49. [CrossRef]
85. Yun, H.J.; Park, J.; Ho, D.H.; Kim, H.; Kim, C.H.; Oh, H.; Ga, I.; Seo, H.; Chang, S.; Son, I.; et al. LRRK2 phosphorylates Snapin and inhibits interaction of Snapin with SNAP-25. *Exp. Mol. Med.* **2013**, *45*, e36. [CrossRef]
86. Matta, S.; van Kolen, K.; da Cunha, R.; van den Bogaart, G.; Mandemakers, W.; Miskiewicz, K.; de Bock, P.J.; Morais, V.A.; Vilain, S.; Haddad, D.; et al. LRRK2 controls an EndoA phosphorylation cycle in synaptic endocytosis. *Neuron* **2012**, *75*, 1008–1021. [CrossRef]
87. Arranz, A.M.; Delbroek, L.; van Kolen, K.; Guimaraes, M.R.; Mandemakers, W.; Daneels, G.; Matta, S.; Calafate, S.; Shaban, H.; Baatsen, P.; et al. LRRK2 functions in synaptic vesicle endocytosis through a kinase-dependent mechanism. *J. Cell. Sci.* **2015**, *128*, 541–552. [CrossRef]
88. Yun, H.J.; Kim, H.; Ga, I.; Oh, H.; Ho, D.H.; Kim, J.; Seo, H.; Son, I.; Seol, W. An early endosome regulator, Rab5b, is an LRRK2 kinase substrate. *J. Biochem.* **2015**, *157*, 485–495. [CrossRef]
89. Schapansky, J.; Nardozzi, J.D.; Felizia, F.; LaVoie, M.J. Membrane recruitment of endogenous LRRK2 precedes its potent regulation of autophagy. *Hum. Mol. Genet.* **2014**, *23*, 4201–4214. [CrossRef]
90. Gomez-Suaga, P.; Fdez, E.; Blanca Ramirez, M.; Hilfiker, S. A Link between Autophagy and the Pathophysiology of LRRK2 in Parkinson's Disease. *Park. Dis.* **2012**, *2012*, 324521. [CrossRef]
91. Manzoni, C.; Mamais, A.; Dihanich, S.; McGoldrick, P.; Devine, M.J.; Zerle, J.; Kara, E.; Taanman, J.W.; Healy, D.G.; Marti-Masso, J.F.; et al. Pathogenic Parkinson's disease mutations across the functional domains of LRRK2 alter the autophagic/lysosomal response to starvation. *Biochem. Biophys. Res. Commun.* **2013**, *441*, 862–866. [CrossRef]
92. Trancikova, A.; Mamais, A.; Webber, P.J.; Stafa, K.; Tsika, E.; Glauser, L.; West, A.B.; Bandopadhyay, R.; Moore, D.J. Phosphorylation of 4E-BP1 in the mammalian brain is not altered by LRRK2 expression or pathogenic mutations. *PLoS ONE* **2012**, *7*, e47784. [CrossRef]
93. Martin, I.; Abalde-Atristain, L.; Kim, J.W.; Dawson, T.M.; Dawson, V.L. Aberrant protein synthesis in G2019S LRRK2 Drosophila Parkinson disease-related phenotypes. *Fly* **2014**, *8*, 165–169. [CrossRef]
94. Paisan-Ruiz, C.; Jain, S.; Evans, E.W.; Gilks, W.P.; Simon, J.; van der Brug, M.; Lopez de Munain, A.; Aparicio, S.; Gil, A.M.; Khan, N.; et al. Cloning of the gene containing mutations that cause PARK8-linked Parkinson's disease. *Neuron* **2004**, *44*, 595–600. [CrossRef]
95. Zimprich, A.; Biskup, S.; Leitner, P.; Lichtner, P.; Farrer, M.; Lincoln, S.; Kachergus, J.; Hulihan, M.; Uitti, R.J.; Calne, D.B.; et al. Mutations in LRRK2 cause autosomal-dominant parkinsonism with pleomorphic pathology. *Neuron* **2004**, *44*, 601–607. [CrossRef]
96. Pankratz, N.; Foroud, T. Genetics of Parkinson disease. *Genet. Med.* **2007**, *9*, 801–811. [CrossRef]
97. Klein, C.; Westenberger, A. Genetics of Parkinson's disease. *Cold Spring Harb. Perspect Med.* **2012**, *2*, a008888. [CrossRef]

98. Li, X.X.; Liao, Q.; Xia, H.; Yang, X.L. Association between Parkinson's disease and G2019S and R1441C mutations of the LRRK2 gene. *Exp. Ther. Med.* **2015**, *10*, 1450–1454. [CrossRef]
99. Fu, X.; Zheng, Y.; Hong, H.; He, Y.; Zhou, S.; Guo, C.; Liu, Y.; Xian, W.; Zeng, J.; Li, J.; et al. LRRK2 G2385R and LRRK2 R1628P increase risk of Parkinson's disease in a Han Chinese population from Southern Mainland China. *Park. Relat. Disord.* **2013**, *19*, 397–398. [CrossRef]
100. Funayama, M.; Li, Y.; Tomiyama, H.; Yoshino, H.; Imamichi, Y.; Yamamoto, M.; Murata, M.; Toda, T.; Mizuno, Y.; Hattori, N. Leucine-rich repeat kinase 2 G2385R variant is a risk factor for Parkinson disease in Asian population. *Neuroreport* **2007**, *18*, 273–275. [CrossRef]
101. Di Fonzo, A.; Wu-Chou, Y.H.; Lu, C.S.; van Doeselaar, M.; Simons, E.J.; Rohe, C.F.; Chang, H.C.; Chen, R.S.; Weng, Y.H.; Vanacore, N.; et al. A common missense variant in the LRRK2 gene, Gly2385Arg, associated with Parkinson's disease risk in Taiwan. *Neurogenetics* **2006**, *7*, 133–138. [CrossRef] [PubMed]
102. Tan, E.K.; Peng, R.; Teo, Y.Y.; Tan, L.C.; Angeles, D.; Ho, P.; Chen, M.L.; Lin, C.H.; Mao, X.Y.; Chang, X.L.; et al. Multiple LRRK2 variants modulate risk of Parkinson disease: A Chinese multicenter study. *Hum. Mutat.* **2010**, *31*, 561–568. [CrossRef] [PubMed]
103. Shu, L.; Zhang, Y.; Sun, Q.; Pan, H.; Tang, B. A Comprehensive Analysis of Population Differences in LRRK2 Variant Distribution in Parkinson's Disease. *Front. Aging Neurosci.* **2019**, *11*, 13. [CrossRef] [PubMed]
104. Mata, I.F.; Kachergus, J.M.; Taylor, J.P.; Lincoln, S.; Aasly, J.; Lynch, T.; Hulihan, M.M.; Cobb, S.A.; Wu, R.M.; Lu, C.S.; et al. Lrrk2 pathogenic substitutions in Parkinson's disease. *Neurogenetics* **2005**, *6*, 171–177. [CrossRef] [PubMed]
105. Khan, N.L.; Jain, S.; Lynch, J.M.; Pavese, N.; Abou-Sleiman, P.; Holton, J.L.; Healy, D.G.; Gilks, W.P.; Sweeney, M.G.; Ganguly, M.; et al. Mutations in the gene LRRK2 encoding dardarin (PARK8) cause familial Parkinson's disease: Clinical, pathological, olfactory and functional imaging and genetic data. *Brain* **2005**, *128*, 2786–2796. [CrossRef]
106. Kachergus, J.; Mata, I.F.; Hulihan, M.; Taylor, J.P.; Lincoln, S.; Aasly, J.; Gibson, J.M.; Ross, O.A.; Lynch, T.; Wiley, J.; et al. Identification of a novel LRRK2 mutation linked to autosomal dominant parkinsonism: Evidence of a common founder across European populations. *Am. J. Hum. Genet.* **2005**, *76*, 672–680. [CrossRef]
107. Lu, C.S.; Simons, E.J.; Wu-Chou, Y.H.; Fonzo, A.D.; Chang, H.C.; Chen, R.S.; Weng, Y.H.; Rohe, C.F.; Breedveld, G.J.; Hattori, N.; et al. The LRRK2 I2012T, G2019S, and I2020T mutations are rare in Taiwanese patients with sporadic Parkinson's disease. *Park. Relat. Disord.* **2005**, *11*, 521–522. [CrossRef]
108. Lewis, P.A.; Greggio, E.; Beilina, A.; Jain, S.; Baker, A.; Cookson, M.R. The R1441C mutation of LRRK2 disrupts GTP hydrolysis. *Biochem. Biophys. Res. Commun.* **2007**, *357*, 668–671. [CrossRef]
109. Liao, J.; Wu, C.X.; Burlak, C.; Zhang, S.; Sahm, H.; Wang, M.; Zhang, Z.Y.; Vogel, K.W.; Federici, M.; Riddle, S.M.; et al. Parkinson disease-associated mutation R1441H in LRRK2 prolongs the "active state" of its GTPase domain. *Proc. Natl. Acad. Sci. USA* **2014**, *111*, 4055–4060. [CrossRef]
110. Puschmann, A.; Englund, E.; Ross, O.A.; Vilarino-Guell, C.; Lincoln, S.J.; Kachergus, J.M.; Cobb, S.A.; Tornqvist, A.L.; Rehncrona, S.; Widner, H.; et al. First neuropathological description of a patient with Parkinson's disease and LRRK2 p.N1437H mutation. *Park. Relat. Disord.* **2012**, *18*, 332–338. [CrossRef]
111. Funayama, M.; Hasegawa, K.; Kowa, H.; Saito, M.; Tsuji, S.; Obata, F. A new locus for Parkinson's Disease (PARK8) maps to chromosome 12p11.2-q13.1. *Ann. Neurol.* **2002**, *51*, 296–301. [CrossRef]
112. Gloeckner, C.J.; Kinkl, N.; Schumacher, A.; Braun, R.J.; O'Neill, E.; Meitinger, T.; Kolch, W.; Prokisch, H.; Ueffing, M. The Parkinson disease causing LRRK2 mutation I2020T is associated with increased kinase activity. *Hum. Mol. Genet.* **2006**, *15*, 223–232. [CrossRef]
113. Greggio, E.; Cookson, M.R. Leucine-rich repeat kinase 2 mutations and Parkinson's disease: Three questions. *ASN Neuro.* **2009**, *1*, e00002. [CrossRef]
114. Gilks, W.P.; Abou-Sleiman, P.M.; Gandhi, S.; Jain, S.; Singleton, A.; Lees, A.J.; Shaw, K.; Bhatia, K.P.; Bonifati, V.; Quinn, N.P.; et al. A common LRRK2 mutation in idiopathic Parkinson's disease. *Lancet* **2005**, *365*, 415–416. [CrossRef]
115. Lesage, S.; Durr, A.; Tazir, M.; Lohmann, E.; Leutenegger, A.L.; Janin, S.; Pollak, P.; Brice, A.; French Parkinson's Disease Genetics Study Group. LRRK2 G2019S as a cause of Parkinson's disease in North African Arabs. *N. Engl. J. Med.* **2006**, *354*, 422–423. [CrossRef]
116. Ozelius, L.J.; Senthil, G.; Saunders-Pullman, R.; Ohmann, E.; Deligtisch, A.; Tagliati, M.; Hunt, A.L.; Klein, C.; Henick, B.; Hailpern, S.M.; et al. LRRK2 G2019S as a cause of Parkinson's disease in Ashkenazi Jews. *N. Engl. J. Med.* **2006**, *354*, 424–425. [CrossRef]
117. Dachsel, J.C.; Farrer, M.J. LRRK2 and Parkinson disease. *Arch. Neurol.* **2010**, *67*, 542–547. [CrossRef]
118. Tan, E.K.; Shen, H.; Tan, L.C.; Farrer, M.; Yew, K.; Chua, E.; Jamora, R.D.; Puvan, K.; Puong, K.Y.; Zhao, Y.; et al. The G2019S LRRK2 mutation is uncommon in an Asian cohort of Parkinson's disease patients. *Neurosci. Lett.* **2005**, *384*, 327–329. [CrossRef]
119. Fung, H.C.; Chen, C.M.; Hardy, J.; Hernandez, D.; Singleton, A.; Wu, Y.R. Lack of G2019S LRRK2 mutation in a cohort of Taiwanese with sporadic Parkinson's disease. *Mov. Disord.* **2006**, *21*, 880–881. [CrossRef]
120. Tomiyama, H.; Li, Y.; Funayama, M.; Hasegawa, K.; Yoshino, H.; Kubo, S.; Sato, K.; Hattori, T.; Lu, C.S.; Inzelberg, R.; et al. Clinicogenetic study of mutations in LRRK2 exon 41 in Parkinson's disease patients from 18 countries. *Mov. Disord.* **2006**, *21*, 1102–1108. [CrossRef]
121. Kalinderi, K.; Fidani, L.; Bostantjopoulou, S.; Katsarou, Z.; Kotsis, A. The G2019S LRRK2 mutation is uncommon amongst Greek patients with sporadic Parkinson's disease. *Eur. J. Neurol.* **2007**, *14*, 1088–1090. [CrossRef] [PubMed]

122. Tan, E.K.; Zhao, Y.; Skipper, L.; Tan, M.G.; Di Fonzo, A.; Sun, L.; Fook-Chong, S.; Tang, S.; Chua, E.; Yuen, Y.; et al. The LRRK2 Gly2385Arg variant is associated with Parkinson's disease: Genetic and functional evidence. *Hum. Genet.* **2007**, *120*, 857–863. [CrossRef] [PubMed]
123. Gandhi, P.N.; Chen, S.G.; Wilson-Delfosse, A.L. Leucine-rich repeat kinase 2 (LRRK2): A key player in the pathogenesis of Parkinson's disease. *J. Neurosci. Res.* **2009**, *87*, 1283–1295. [CrossRef]
124. Luzón-Toro, B.; de la Torre, E.R.; Delgado, A.; Pérez-Tur, J.; Hilfiker, S. Mechanistic insight into the dominant mode of the Parkinson's disease-associated G2019S LRRK2 mutation. *Hum. Mol. Genet.* **2007**, *16*, 2031–2039. [CrossRef] [PubMed]
125. Marras, C.; Alcalay, R.N.; Caspell-Garcia, C.; Coffey, C.; Chan, P.; Duda, J.E.; Facheris, M.F.; Fernandez-Santiago, R.; Ruiz-Martinez, J.; Mestre, T.; et al. Motor and nonmotor heterogeneity of LRRK2-related and idiopathic Parkinson's disease. *Mov. Disord.* **2016**, *31*, 1192–1202. [CrossRef]
126. Beilina, A.; Rudenko, I.N.; Kaganovich, A.; Civiero, L.; Chau, H.; Kalia, S.K.; Kalia, L.V.; Lobbestael, E.; Chia, R.; Ndukwe, K.; et al. Unbiased screen for interactors of leucine-rich repeat kinase 2 supports a common pathway for sporadic and familial Parkinson disease. *Proc. Natl. Acad. Sci. USA* **2014**, *111*, 2626–2631. [CrossRef]
127. Ishihara, L.; Warren, L.; Gibson, R.; Amouri, R.; Lesage, S.; Durr, A.; Tazir, M.; Wszolek, Z.K.; Uitti, R.J.; Nichols, W.C.; et al. Clinical features of Parkinson disease patients with homozygous leucine-rich repeat kinase 2 G2019S mutations. *Arch. Neurol.* **2006**, *63*, 1250–1254. [CrossRef]
128. Reinhardt, P.; Schmid, B.; Burbulla, L.F.; Schondorf, D.C.; Wagner, L.; Glatza, M.; Hoing, S.; Hargus, G.; Heck, S.A.; Dhingra, A.; et al. Genetic correction of a LRRK2 mutation in human iPSCs links parkinsonian neurodegeneration to ERK-dependent changes in gene expression. *Cell. Stem. Cell.* **2013**, *12*, 354–367. [CrossRef]
129. Ho, D.H.; Seol, W.; Son, I. Upregulation of the p53–p21 pathway by G2019S LRRK2 contributes to the cellular senescence and accumulation of α-synuclein. *Cell. Cycle* **2019**, *18*, 467–475. [CrossRef]
130. Ross, C.A.; Akimov, S.S. Human-induced pluripotent stem cells: Potential for neurodegenerative diseases. *Hum. Mol. Genet.* **2014**, *23*, 17–26. [CrossRef]
131. Pu, J.; Jiang, H.; Zhang, B.; Feng, J. Redefining Parkinson's disease research using induced pluripotent stem cells. *Curr. Neurol. Neurosci. Rep.* **2012**, *12*, 392–398. [CrossRef]
132. Soldner, F.; Jaenisch, R. Medicine iPSC disease modeling. *Science* **2012**, *338*, 1155–1156. [CrossRef]
133. Shi, Y.; Inoue, H.; Wu, J.C.; Yamanaka, S. Induced pluripotent stem cell technology: A decade of progress. *Nat. Rev. Drug Discov.* **2017**, *16*, 115–130. [CrossRef]
134. Gunaseeli, I.; Doss, M.X.; Antzelevitch, C.; Hescheler, J.; Sachinidis, A. Induced pluripotent stem cells as a model for accelerated patient- and disease-specific drug discovery. *Curr. Med. Chem.* **2010**, *17*, 759–766. [CrossRef]
135. Takahashi, K.; Yamanaka, S. Induction of pluripotent stem cells from mouse embryonic and adult fibroblast cultures by defined factors. *Cell* **2006**, *126*, 663–676. [CrossRef]
136. Tiscornia, G.; Vivas, E.L.; Belmonte, J.C.I. Diseases in a dish: Modeling human genetic disorders using induced pluripotent cells. *Nat. Med.* **2011**, *17*, 1570–1576. [CrossRef]
137. Grobarczyk, B.; Franco, B.; Hanon, K.; Malgrange, B. Generation of Isogenic Human iPS Cell Line Precisely Corrected by Genome Editing Using the CRISPR/Cas9 System. *Stem. Cell. Rev. Rep.* **2015**, *11*, 774–787. [CrossRef]
138. Bibikova, M.; Carroll, D.; Segal, D.J.; Trautman, J.K.; Smith, J.; Kim, Y.G.; Chandrasegaran, S. Stimulation of homologous recombination through targeted cleavage by chimeric nucleases. *Mol. Cell. Biol.* **2001**, *21*, 289–297. [CrossRef]
139. Komor, A.C.; Kim, Y.B.; Packer, M.S.; Zuris, J.A.; Liu, D.R. Programmable editing of a target base in genomic DNA without double-stranded DNA cleavage. *Nature* **2016**, *533*, 420–424. [CrossRef]
140. Soldner, F.; Laganiere, J.; Cheng, A.W.; Hockemeyer, D.; Gao, Q.; Alagappan, R.; Khurana, V.; Golbe, L.I.; Myers, R.H.; Lindquist, S.; et al. Generation of isogenic pluripotent stem cells differing exclusively at two early onset Parkinson point mutations. *Cell* **2011**, *146*, 318–331. [CrossRef]
141. Gasiunas, G.; Barrangou, R.; Horvath, P.; Siksnys, V. Cas9-crRNA ribonucleoprotein complex mediates specific DNA cleavage for adaptive immunity in bacteria. *Proc. Natl. Acad. Sci. USA* **2012**, *109*, E2579–E2586. [CrossRef] [PubMed]
142. Jinek, M.; Chylinski, K.; Fonfara, I.; Hauer, M.; Doudna, J.A.; Charpentier, E. A programmable dual-RNA-guided DNA endonuclease in adaptive bacterial immunity. *Science* **2012**, *337*, 816–821. [CrossRef] [PubMed]
143. Liu, G.H.; Qu, J.; Suzuki, K.; Nivet, E.; Li, M.; Montserrat, N.; Yi, F.; Xu, X.; Ruiz, S.; Zhang, W.; et al. Progressive degeneration of human neural stem cells caused by pathogenic LRRK2. *Nature* **2012**, *491*, 603–607. [CrossRef] [PubMed]
144. Sanchez-Danes, A.; Richaud-Patin, Y.; Carballo-Carbajal, I.; Jimenez-Delgado, S.; Caig, C.; Mora, S.; Di Guglielmo, C.; Ezquerra, M.; Patel, B.; Giralt, A.; et al. Disease-specific phenotypes in dopamine neurons from human iPS-based models of genetic and sporadic Parkinson's disease. *EMBO Mol. Med.* **2012**, *4*, 380–395. [CrossRef] [PubMed]
145. Nguyen, H.N.; Byers, B.; Cord, B.; Shcheglovitov, A.; Byrne, J.; Gujar, P.; Kee, K.; Schüle, B.; Dolmetsch, R.E.; Langston, W.; et al. LRRK2 Mutant iPSC-Derived DA Neurons Demonstrate Increased Susceptibility to Oxidative Stress. *Cell. Stem. Cell.* **2011**, *8*, 267–280. [CrossRef]
146. Bieri, G.; Brahic, M.; Bousset, L.; Couthouis, J.; Kramer, N.J.; Ma, R.; Nakayama, L.; Monbureau, M.; Defensor, E.; Schule, B.; et al. LRRK2 modifies α-syn pathology and spread in mouse models and human neurons. *Acta Neuropathol.* **2019**, *137*, 961–980. [CrossRef]

147. Borgs, L.; Peyre, E.; Alix, P.; Hanon, K.; Grobarczyk, B.; Godin, J.D.; Purnelle, A.; Krusy, N.; Maquet, P.; Lefebvre, P.; et al. Dopaminergic neurons differentiating from LRRK2 G2019S induced pluripotent stem cells show early neuritic branching defects. *Sci. Rep.* **2016**, *6*, 33377. [CrossRef]
148. Schwab, A.J.; Ebert, A.D. Neurite Aggregation and Calcium Dysfunction in iPSC-Derived Sensory Neurons with Parkinson's Disease-Related LRRK2 G2019S Mutation. *Stem. Cell. Rep.* **2015**, *5*, 1039–1052. [CrossRef]
149. Korecka, J.A.; Talbot, S.; Osborn, T.M.; de Leeuw, S.M.; Levy, S.A.; Ferrari, E.J.; Moskites, A.; Atkinson, E.; Jodelka, F.M.; Hinrich, A.J.; et al. Neurite Collapse and Altered ER Ca^{2+} Control in Human Parkinson Disease Patient iPSC-Derived Neurons with LRRK2 G2019S Mutation. *Stem. Cell. Rep.* **2019**, *12*, 29–41. [CrossRef]
150. Kim, J.; Daadi, M.M. Non-cell autonomous mechanism of Parkinson's disease pathology caused by G2019S LRRK2 mutation in Ashkenazi Jewish patient: Single cell analysis. *Brain Res.* **2019**, *1722*, 146342. [CrossRef]
151. Walter, J.; Bolognin, S.; Antony, P.M.A.; Nickels, S.L.; Poovathingal, S.K.; Salamanca, L.; Magni, S.; Perfeito, R.; Hoel, F.; Qing, X.; et al. Neural Stem Cells of Parkinson's Disease Patients Exhibit Aberrant Mitochondrial Morphology and Functionality. *Stem. Cell. Rep.* **2019**, *12*, 878–889. [CrossRef]
152. Walter, J.; Bolognin, S.; Poovathingal, S.K.; Magni, S.; Gerard, D.; Antony, P.M.A.; Nickels, S.L.; Salamanca, L.; Berger, E.; Smits, L.M.; et al. The Parkinson's-disease-associated mutation LRRK2-G2019S alters dopaminergic differentiation dynamics via NR2F1. *Cell. Rep.* **2021**, *37*, 109864. [CrossRef]
153. Skibinski, G.; Hwang, V.; Ando, D.M.; Daub, A.; Lee, A.K.; Ravisankar, A.; Modan, S.; Finucane, M.M.; Shaby, B.A.; Finkbeiner, S. Nrf2 mitigates LRRK2- and α-synuclein-induced neurodegeneration by modulating proteostasis. *Proc. Natl. Acad. Sci. USA* **2017**, *114*, 1165–1170. [CrossRef]
154. Sanders, L.H.; Laganiere, J.; Cooper, O.; Mak, S.K.; Vu, B.J.; Huang, Y.A.; Paschon, D.E.; Vangipuram, M.; Sundararajan, R.; Urnov, F.D.; et al. LRRK2 mutations cause mitochondrial DNA damage in iPSC-derived neural cells from Parkinson's disease patients: Reversal by gene correction. *Neurobiol. Dis.* **2014**, *62*, 381–386. [CrossRef]
155. Howlett, E.H.; Jensen, N.; Belmonte, F.; Zafar, F.; Hu, X.; Kluss, J.; Schule, B.; Kaufman, B.A.; Greenamyre, J.T.; Sanders, L.H. LRRK2 G2019S-induced mitochondrial DNA damage is LRRK2 kinase dependent and inhibition restores mtDNA integrity in Parkinson's disease. *Hum. Mol. Genet.* **2017**, *26*, 4340–4351. [CrossRef]
156. Hsieh, C.H.; Shaltouki, A.; Gonzalez, A.E.; Bettencourt da Cruz, A.; Burbulla, L.F.; St Lawrence, E.; Schule, B.; Krainc, D.; Palmer, T.D.; Wang, X. Functional Impairment in Miro Degradation and Mitophagy Is a Shared Feature in Familial and Sporadic Parkinson's Disease. *Cell. Stem. Cell.* **2016**, *19*, 709–724. [CrossRef]
157. Schwab, A.J.; Sison, S.L.; Meade, M.R.; Broniowska, K.A.; Corbett, J.A.; Ebert, A.D. Decreased Sirtuin Deacetylase Activity in LRRK2 G2019S iPSC-Derived Dopaminergic Neurons. *Stem. Cell. Rep.* **2017**, *9*, 1839–1852. [CrossRef]
158. Pan, P.Y.; Li, X.; Wang, J.; Powell, J.; Wang, Q.; Zhang, Y.; Chen, Z.; Wicinski, B.; Hof, P.; Ryan, T.A.; et al. Parkinson's Disease-Associated LRRK2 Hyperactive Kinase Mutant Disrupts Synaptic Vesicle Trafficking in Ventral Midbrain Neurons. *J. Neurosci.* **2017**, *37*, 11366–11376. [CrossRef]
159. Connor-Robson, N.; Booth, H.; Martin, J.G.; Gao, B.; Li, K.; Doig, N.; Vowles, J.; Browne, C.; Klinger, L.; Juhasz, P.; et al. An integrated transcriptomics and proteomics analysis reveals functional endocytic dysregulation caused by mutations in LRRK2. *Neurobiol. Dis.* **2019**, *127*, 512–526. [CrossRef]
160. Nguyen, M.; Krainc, D. LRRK2 phosphorylation of auxilin mediates synaptic defects in dopaminergic neurons from patients with Parkinson's disease. *Proc. Natl. Acad. Sci. USA* **2018**, *115*, 5576–5581. [CrossRef]
161. Schapansky, J.; Khasnavis, S.; DeAndrade, M.P.; Nardozzi, J.D.; Falkson, S.R.; Boyd, J.D.; Sanderson, J.B.; Bartels, T.; Melrose, H.L.; LaVoie, M.J. Familial knockin mutation of LRRK2 causes lysosomal dysfunction and accumulation of endogenous insoluble α-synuclein in neurons. *Neurobiol. Dis.* **2018**, *111*, 26–35. [CrossRef] [PubMed]
162. Su, Y.C.; Qi, X. Inhibition of excessive mitochondrialfissionreduced aberrant autophagy and neuronal damage caused by LRRK2 G2019S mutation. *Hum. Mol. Genet.* **2013**, *22*, 4545–4561. [CrossRef] [PubMed]
163. Ho, D.H.; Kim, H.; Nam, D.; Sim, H.; Kim, J.; Kim, H.G.; Son, I.; Seol, W. LRRK2 impairs autophagy by mediating phosphorylation of leucyl-tRNA synthetase. *Cell. Biochem. Funct.* **2018**, *36*, 431–442. [CrossRef] [PubMed]
164. Lee, H.; James, W.S.; Cowley, S.A. LRRK2 in peripheral and central nervous system innate immunity: Its link to Parkinson's disease. *Biochem. Soc. Trans.* **2017**, *45*, 131–139. [CrossRef] [PubMed]
165. Filippini, A.; Gennarelli, M.; Russo, I. Leucine-rich repeat kinase 2-related functions in GLIA: An update of the last years. *Biochem. Soc. Trans.* **2021**, *49*, 1375–1384. [CrossRef]
166. Russo, I.; Bubacco, L.; Greggio, E. LRRK2 and neuroinflammation: Partners in crime in Parkinson's disease? *J. Neuroinflamm.* **2014**, *11*, 52. [CrossRef]
167. Li, T.; Yang, D.; Zhong, S.; Thomas, J.M.; Xue, F.; Liu, J.; Kong, L.; Voulalas, P.; Hassan, H.E.; Park, J.S.; et al. Novel LRRK2 GTP-binding inhibitors reduced degeneration in Parkinson's disease cell and mouse models. *Hum. Mol. Genet.* **2014**, *23*, 6212–6222. [CrossRef]
168. Li, T.; He, X.; Thomas, J.M.; Yang, D.; Zhong, S.; Xue, F.; Smith, W.W. A novel GTP-binding inhibitor, FX2149, attenuates LRRK2 toxicity in Parkinson's disease models. *PLoS ONE* **2015**, *10*, e0122461. [CrossRef]
169. Li, T.; Ning, B.; Kong, L.; Dai, B.; He, X.; Thomas, J.M.; Sawa, A.; Ross, C.A.; Smith, W.W. A LRRK2 GTP Binding Inhibitor, 68, Reduces LPS-Induced Signaling Events and TNF-alpha Release in Human Lymphoblasts. *Cells* **2021**, *10*, 480. [CrossRef]

170. Caesar, M.; Felk, S.; Zach, S.; Bronstad, G.; Aasly, J.O.; Gasser, T.; Gillardon, F. Changes in matrix metalloprotease activity and progranulin levels may contribute to the pathophysiological function of mutant leucine-rich repeat kinase 2. *Glia* **2014**, *62*, 1075–1092. [CrossRef]
171. Gillardon, F.; Schmid, R.; Draheim, H. Parkinson's disease-linked leucine-rich repeat kinase 2(R1441G) mutation increases proinflammatory cytokine release from activated primary microglial cells and resultant neurotoxicity. *Neuroscience* **2012**, *208*, 41–48. [CrossRef]
172. Ho, D.H.; Seol, W.; Eun, J.H.; Son, I.H. Phosphorylation of p53 by LRRK2 induces microglial tumor necrosis factor α-mediated neurotoxicity. *Biochem. Biophys. Res. Commun.* **2017**, *482*, 1088–1094. [CrossRef]
173. Sonninen, T.M.; Hamalainen, R.H.; Koskuvi, M.; Oksanen, M.; Shakirzyanova, A.; Wojciechowski, S.; Puttonen, K.; Naumenko, N.; Goldsteins, G.; Laham-Karam, N.; et al. Metabolic alterations in Parkinson's disease astrocytes. *Sci. Rep.* **2020**, *10*, 14474. [CrossRef]
174. Russo, I.; Di Benedetto, G.; Kaganovich, A.; Ding, J.; Mercatelli, D.; Morari, M.; Cookson, M.R.; Bubacco, L.; Greggio, E. Leucine-rich repeat kinase 2 controls protein kinase A activation state through phosphodiesterase 4. *J. Neuroinflamm.* **2018**, *15*, 297. [CrossRef]
175. Kim, B.; Yang, M.S.; Choi, D.; Kim, J.H.; Kim, H.S.; Seol, W.; Choi, S.; Jou, I.; Kim, E.Y.; Joe, E.H. Impaired inflammatory responses in murine Lrrk2-knockdown brain microglia. *PLoS ONE* **2012**, *7*, e34693. [CrossRef]
176. Kim, J.; Pajarillo, E.; Rizor, A.; Son, D.S.; Lee, J.; Aschner, M.; Lee, E. LRRK2 kinase plays a critical role in manganese-induced inflammation and apoptosis in microglia. *PLoS ONE* **2019**, *14*, e0210248. [CrossRef]
177. Russo, I.; Kaganovich, A.; Ding, J.; Landeck, N.; Mamais, A.; Varanita, T.; Biosa, A.; Tessari, I.; Bubacco, L.; Greggio, E.; et al. Transcriptome analysis of LRRK2 knock-out microglia cells reveals alterations of inflammatory- and oxidative stress-related pathways upon treatment with α-synuclein fibrils. *Neurobiol. Dis.* **2019**, *129*, 67–78. [CrossRef]
178. Moehle, M.S.; Webber, P.J.; Tse, T.; Sukar, N.; Standaert, D.G.; DeSilva, T.M.; Cowell, R.M.; West, A.B. LRRK2 Inhibition Attenuates Microglial Inflammatory Responses. *J. Neurosci.* **2012**, *32*, 1602–1611. [CrossRef]
179. Munoz, L.; Kavanagh, M.E.; Phoa, A.F.; Heng, B.; Dzamko, N.; Chen, E.J.; Doddareddy, M.R.; Guillemin, G.J.; Kassiou, M. Optimisation of LRRK2 inhibitors and assessment of functional efficacy in cell-based models of neuroinflammation. *Eur. J. Med. Chem.* **2015**, *95*, 29–34. [CrossRef] [PubMed]
180. De Maturana, R.L.; Lang, V.; Zubiarrain, A.; Sousa, A.; Vazquez, N.; Gorostidi, A.; Aguila, J.; Lopez de Munain, A.; Rodriguez, M.; Sanchez-Pernaute, R. Mutations in LRRK2 impair NF-kappaB pathway in iPSC-derived neurons. *J. Neuroinflamm.* **2016**, *13*, 295. [CrossRef]
181. Booth, H.D.E.; Wessely, F.; Connor-Robson, N.; Rinaldi, F.; Vowles, J.; Browne, C.; Evetts, S.G.; Hu, M.T.; Cowley, S.A.; Webber, C.; et al. RNA sequencing reveals MMP2 and TGFB1 downregulation in LRRK2 G2019S Parkinson's iPSC-derived astrocytes. *Neurobiol. Dis.* **2019**, *129*, 56–66. [CrossRef] [PubMed]
182. Speidel, A.; Felk, S.; Reinhardt, P.; Sterneckert, J.; Gillardon, F. Leucine-Rich Repeat Kinase 2 Influences Fate Decision of Human Monocytes Differentiated from Induced Pluripotent Stem Cells. *PLoS ONE* **2016**, *11*, e0165949. [CrossRef] [PubMed]
183. Cookson, M.R. LRRK2 Pathways Leading to Neurodegeneration. *Curr. Neurol. Neurosci. Rep.* **2015**, *15*, 42. [CrossRef] [PubMed]
184. Nikonova, E.V.; Xiong, Y.; Tanis, K.Q.; Dawson, V.L.; Vogel, R.L.; Finney, E.M.; Stone, D.J.; Reynolds, I.J.; Kern, J.T.; Dawson, T.M. Transcriptional responses to loss or gain of function of the leucine-rich repeat kinase 2 (LRRK2) gene uncover biological processes modulated by LRRK2 activity. *Hum. Mol. Genet.* **2012**, *21*, 163–174. [CrossRef] [PubMed]
185. Liu, Z.; Wang, X.; Yu, Y.; Li, X.; Wang, T.; Jiang, H.; Ren, Q.; Jiao, Y.; Sawa, A.; Moran, T.; et al. A Drosophila model for LRRK2-linked parkinsonism. *Proc. Natl. Acad. Sci. USA* **2008**, *105*, 2693–2698. [CrossRef]
186. Venderova, K.; Kabbach, G.; Abdel-Messih, E.; Zhang, Y.; Parks, R.J.; Imai, Y.; Gehrke, S.; Ngsee, J.; Lavoie, M.J.; Slack, R.S.; et al. Leucine-Rich Repeat Kinase 2 interacts with Parkin, DJ-1 and PINK-1 in a Drosophila melanogaster model of Parkinson's disease. *Hum. Mol. Genet.* **2009**, *18*, 4390–4404. [CrossRef]
187. Ng, C.H.; Mok, S.Z.; Koh, C.; Ouyang, X.; Fivaz, M.L.; Tan, E.K.; Dawson, V.L.; Dawson, T.M.; Yu, F.; Lim, K.L. Parkin protects against LRRK2 G2019S mutant-induced dopaminergic neurodegeneration in Drosophila. *J. Neurosci.* **2009**, *29*, 11257–11262. [CrossRef]
188. Hindle, S.; Afsari, F.; Stark, M.; Middleton, C.A.; Evans, G.J.; Sweeney, S.T.; Elliott, C.J. Dopaminergic expression of the Parkinsonian gene LRRK2-G2019S leads to non-autonomous visual neurodegeneration, accelerated by increased neural demands for energy. *Hum. Mol. Genet.* **2013**, *22*, 2129–2140. [CrossRef]
189. Lin, C.H.; Tsai, P.I.; Wu, R.M.; Chien, C.T. LRRK2 G2019S Mutation Induces Dendrite Degeneration through Mislocalization and Phosphorylation of Tau by Recruiting Autoactivated GSK3. *J. Neurosci.* **2010**, *30*, 13138–13149. [CrossRef]
190. Saha, S.; Guillily, M.D.; Ferree, A.; Lanceta, J.; Chan, D.; Ghosh, J.; Hsu, C.H.; Segal, L.; Raghavan, K.; Matsumoto, K.; et al. LRRK2 modulates vulnerability to mitochondrial dysfunction in Caenorhabditis elegans. *J. Neurosci.* **2009**, *29*, 9210–9218. [CrossRef]
191. Saha, S.; Liu-Yesucevitz, L.; Wolozin, B. Regulation of autophagy by LRRK2 in Caenorhabditis elegans. *Neurodegener. Dis.* **2014**, *13*, 110–113. [CrossRef]
192. Li, X.; Patel, J.C.; Wang, J.; Avshalumov, M.V.; Nicholson, C.; Buxbaum, J.D.; Elder, G.A.; Rice, M.E.; Yue, Z. Enhanced striatal dopamine transmission and motor performance with LRRK2 overexpression in mice is eliminated by familial Parkinson's disease mutation G2019S. *J. Neurosci. Off. J. Soc. Neurosci.* **2010**, *30*, 1788–1797. [CrossRef]

193. Chen, H.; Chan, B.K.; Drummond, J.; Estrada, A.A.; Gunzner-Toste, J.; Liu, X.; Liu, Y.; Moffat, J.; Shore, D.; Sweeney, Z.K.; et al. Discovery of selective LRRK2 inhibitors guided by computational analysis and molecular modeling. *J. Med. Chem.* **2012**, *55*, 5536–5545. [CrossRef]
194. Ramonet, D.; Daher, J.P.; Lin, B.M.; Stafa, K.; Kim, J.; Banerjee, R.; Westerlund, M.; Pletnikova, O.; Glauser, L.; Yang, L.; et al. Dopaminergic neuronal loss, reduced neurite complexity and autophagic abnormalities in transgenic mice expressing G2019S mutant LRRK2. *PLoS ONE* **2011**, *6*, e18568. [CrossRef]
195. Lee, J.W.; Cannon, J.R. LRRK2 mutations and neurotoxicant susceptibility. *Exp. Biol. Med.* **2015**, *240*, 752–759. [CrossRef]
196. Walker, M.D.; Volta, M.; Cataldi, S.; Dinelle, K.; Beccano-Kelly, D.; Munsie, L.; Kornelsen, R.; Mah, C.; Chou, P.; Co, K.; et al. Behavioral deficits and striatal DA signaling in LRRK2 p.G2019S transgenic rats: A multimodal investigation including PET neuroimaging. *J. Park. Dis.* **2014**, *4*, 483–498. [CrossRef]
197. Pingale, T.; Gupta, G.L. Classic and evolving animal models in Parkinson's disease. *Pharmacol. Biochem. Behav.* **2020**, *199*, 173060. [CrossRef]
198. Totterdell, S.; Meredith, G.E. Localization of alpha-synuclein to identified fibers and synapses in the normal mouse brain. *Neuroscience* **2005**, *135*, 907–913. [CrossRef]
199. Alim, M.A.; Hossain, M.S.; Arima, K.; Takeda, K.; Izumiyama, Y.; Nakamura, M.; Kaji, H.; Shinoda, T.; Hisanaga, S.; Ueda, K. Tubulin seeds α-synuclein fibril formation. *J. Biol. Chem.* **2002**, *277*, 2112–2117. [CrossRef]
200. Cole, N.B.; Murphy, D.D.; Grider, T.; Rueter, S.; Brasaemle, D.; Nussbaum, R.L. Lipid droplet binding and oligomerization properties of the Parkinson's disease protein α-synuclein. *J. Biol. Chem.* **2002**, *277*, 6344–6352. [CrossRef]
201. Mak, S.K.; McCormack, A.L.; Langston, J.W.; Kordower, J.H.; Di Monte, D.A. Decreased α-synuclein expression in the aging mouse substantia nigra. *Exp. Neurol.* **2009**, *220*, 359–365. [CrossRef] [PubMed]
202. Michell, A.W.; Tofaris, G.K.; Gossage, H.; Tyers, P.; Spillantini, M.G.; Barker, R.A. The effect of truncated human α-synuclein (1–120) on dopaminergic cells in a transgenic mouse model of Parkinson's disease. *Cell. Transplant.* **2007**, *16*, 461–474. [CrossRef] [PubMed]
203. Schneider, B.L.; Seehus, C.R.; Capowski, E.E.; Aebischer, P.; Zhang, S.C.; Svendsen, C.N. Over-expression of alpha-synuclein in human neural progenitors leads to specific changes in fate and differentiation. *Hum. Mol. Genet.* **2007**, *16*, 651–666. [CrossRef] [PubMed]
204. Purisai, M.G.; McCormack, A.L.; Langston, W.J.; Johnston, L.C.; di Monte, D.A. Alpha-synuclein expression in the substantia nigra of MPTP-lesioned non-human primates. *Neurobiol. Dis.* **2005**, *20*, 898–906. [CrossRef] [PubMed]
205. Bodles, A.M.; Guthrie, D.J.; Greer, B.; Irvine, G.B. Identification of the region of non-Abeta component (NAC) of Alzheimer's disease amyloid responsible for its aggregation and toxicity. *J. Neurochem.* **2001**, *78*, 384–395. [CrossRef]
206. Halliday, G.M.; del Tredici, K.; Braak, H. Critical appraisal of brain pathology staging related to presymptomatic and symptomatic cases of sporadic Parkinson's disease. *J. Neural Transm. Suppl.* **2006**, *70*, 99–103. [CrossRef]
207. Burke, W.J.; Kumar, V.B.; Pandey, N.; Panneton, W.M.; Gan, Q.; Franko, M.W.; O'Dell, M.; Li, S.W.; Pan, Y.; Chung, H.D.; et al. Aggregation of α-synuclein by DOPAL, the monoamine oxidase metabolite of dopamine. *Acta Neuropathol.* **2008**, *115*, 193–203. [CrossRef]
208. Sian-Hulsmann, J.; Monoranu, C.; Strobel, S.; Riederer, P. Lewy Bodies: A Spectator or Salient Killer? *CNS Neurol. Disord. Drug Targets* **2015**, *14*, 947–955. [CrossRef]
209. Foley, P.; Riederer, P. Pathogenesis and preclinical course of Parkinson's disease. *J. Neural Transm. Suppl.* **1999**, *56*, 31–74. [CrossRef]
210. Conway, K.A.; Lee, S.J.; Rochet, J.C.; Ding, T.T.; Williamson, R.E.; Lansbury, P.T., Jr. Acceleration of oligomerization, not fibrillization, is a shared property of both α-synuclein mutations linked to early-onset Parkinson's disease: Implications for pathogenesis and therapy. *Proc. Natl. Acad. Sci. USA* **2000**, *97*, 571–576. [CrossRef]
211. Ren, P.H.; Lauckner, J.E.; Kachirskaia, I.; Heuser, J.E.; Melki, R.; Kopito, R.R. Cytoplasmic penetration and persistent infection of mammalian cells by polyglutamine aggregates. *Nat. Cell. Biol.* **2009**, *11*, 219–225. [CrossRef]
212. Sherer, T.B.; Betarbet, R.; Stout, A.K.; Lund, S.; Baptista, M.; Panov, A.V.; Cookson, M.R.; Greenamyre, J.T. An in vitro model of Parkinson's disease: Linking mitochondrial impairment to altered α-synuclein metabolism and oxidative damage. *J. Neurosci.* **2002**, *22*, 7006–7015. [CrossRef]
213. Lardenoije, R.; Iatrou, A.; Kenis, G.; Kompotis, K.; Steinbusch, H.W.; Mastroeni, D.; Coleman, P.; Lemere, C.A.; Hof, P.R.; van den Hove, D.L.; et al. The epigenetics of aging and neurodegeneration. *Prog. Neurobiol.* **2015**, *131*, 21–64. [CrossRef]
214. Wong, Y.C.; Krainc, D. α-synuclein toxicity in neurodegeneration: Mechanism and therapeutic strategies. *Nat. Med.* **2017**, *23*, 1–13. [CrossRef]
215. Surguchева, I.; Newell, K.L.; Burns, J.; Surguchov, A. New α- and γ-synuclein immunopathological lesions in human brain. *Acta Neuropathol. Commun.* **2014**, *2*, 132. [CrossRef]
216. Anderson, J.P.; Walker, D.E.; Goldstein, J.M.; de Laat, R.; Banducci, K.; Caccavello, R.J.; Barbour, R.; Huang, J.; Kling, K.; Lee, M.; et al. Phosphorylation of Ser-129 is the dominant pathological modification of α-synuclein in familial and sporadic Lewy body disease. *J. Biol. Chem.* **2006**, *281*, 29739–29752. [CrossRef]
217. Sato, H.; Kato, T.; Arawaka, S. The role of Ser129 phosphorylation of α-synuclein in neurodegeneration of Parkinson's disease: A review of in vivo models. *Rev. Neurosci.* **2013**, *24*, 115–123. [CrossRef]

218. Waxman, E.A.; Giasson, B.I. Specificity and regulation of casein kinase-mediated phosphorylation of α-synuclein. *J. Neuropathol. Exp. Neurol.* **2008**, *67*, 402–416. [CrossRef]
219. Waxman, E.A.; Giasson, B.I. Characterization of kinases involved in the phosphorylation of aggregated α-synuclein. *J. Neurosci. Res.* **2011**, *89*, 231–247. [CrossRef]
220. Mbefo, M.K.; Paleologou, K.E.; Boucharaba, A.; Oueslati, A.; Schell, H.; Fournier, M.; Olschewski, D.; Yin, G.; Zweckstetter, M.; Masliah, E.; et al. Phosphorylation of synucleins by members of the Polo-like kinase family. *J. Biol. Chem.* **2010**, *285*, 2807–2822. [CrossRef]
221. Inglis, K.J.; Chereau, D.; Brigham, E.F.; Chiou, S.S.; Schobel, S.; Frigon, N.L.; Yu, M.; Caccavello, R.J.; Nelson, S.; Motter, R.; et al. Polo-like kinase 2 (PLK2) phosphorylates α-synuclein at serine 129 in central nervous system. *J. Biol. Chem.* **2009**, *284*, 2598–2602. [CrossRef] [PubMed]
222. Pronin, A.N.; Morris, A.J.; Surguchov, A.; Benovic, J.L. Synucleins are a novel class of substrates for G protein-coupled receptor kinases. *J. Biol. Chem.* **2000**, *275*, 26515–26522. [CrossRef]
223. Paleologou, K.E.; Oueslati, A.; Shakked, G.; Rospigliosi, C.C.; Kim, H.Y.; Lamberto, G.R.; Fernandez, C.O.; Schmid, A.; Chegini, F.; Gai, W.P.; et al. Phosphorylation at S87 is enhanced in synucleinopathies, inhibits α-synuclein oligomerization, and influences synuclein-membrane interactions. *J. Neurosci.* **2010**, *30*, 3184–3198. [CrossRef] [PubMed]
224. Sakamoto, M.; Arawaka, S.; Hara, S.; Sato, H.; Cui, C.; Machiya, Y.; Koyama, S.; Wada, M.; Kawanami, T.; Kurita, K.; et al. Contribution of endogenous G-protein-coupled receptor kinases to Ser129 phosphorylation of α-synuclein in HEK293 cells. *Biochem. Biophys. Res. Commun.* **2009**, *384*, 378–382. [CrossRef] [PubMed]
225. Dzamko, N.; Zhou, J.; Huang, Y.; Halliday, G.M. Parkinson's disease-implicated kinases in the brain; insights into disease pathogenesis. *Front. Mol. Neurosci.* **2014**, *7*, 57. [CrossRef]
226. Herzig, M.C.; Bidinosti, M.; Schweizer, T.; Hafner, T.; Stemmelen, C.; Weiss, A.; Danner, S.; Vidotto, N.; Stauffer, D.; Barske, C.; et al. High LRRK2 levels fail to induce or exacerbate neuronal alpha-synucleinopathy in mouse brain. *PLoS ONE* **2012**, *7*, e36581. [CrossRef]
227. Lin, X.; Parisiadou, L.; Gu, X.L.; Wang, L.; Shim, H.; Sun, L.; Xie, C.; Long, C.X.; Yang, W.J.; Ding, J.; et al. Leucine-rich repeat kinase 2 regulates the progression of neuropathology induced by Parkinson's-disease-related mutant α-synuclein. *Neuron* **2009**, *64*, 807–827. [CrossRef]
228. Buck, K.; Landeck, N.; Ulusoy, A.; Majbour, N.K.; El-Agnaf, O.M.; Kirik, D. Ser129 phosphorylation of endogenous α-synuclein induced by overexpression of polo-like kinases 2 and 3 in nigral dopamine neurons is not detrimental to their survival and function. *Neurobiol. Dis.* **2015**, *78*, 100–114. [CrossRef]
229. Braak, H.; Del Tredici, K.; Rub, U.; de Vos, R.A.; Jansen Steur, E.N.; Braak, E. Staging of brain pathology related to sporadic Parkinson's disease. *Neurobiol. Aging* **2003**, *24*, 197–211. [CrossRef]
230. Braak, H.; Rub, U.; Gai, W.P.; del Tredici, K. Idiopathic Parkinson's disease: Possible routes by which vulnerable neuronal types may be subject to neuroinvasion by an unknown pathogen. *J. Neural Transm.* **2003**, *110*, 517–536. [CrossRef]
231. Pfeiffer, R.F. Gastrointestinal dysfunction in Parkinson's disease. *Park. Relat. Disord.* **2011**, *17*, 10–15. [CrossRef]
232. Cersosimo, M.G.; Benarroch, E.E. Pathological correlates of gastrointestinal dysfunction in Parkinson's disease. *Neurobiol. Dis.* **2012**, *46*, 559–564. [CrossRef]
233. Doty, R.L. Olfactory dysfunction in Parkinson disease. *Nat. Rev. Neurol.* **2012**, *8*, 329–339. [CrossRef]
234. Beach, T.G.; White, C.L., 3rd; Hladik, C.L.; Sabbagh, M.N.; Connor, D.J.; Shill, H.A.; Sue, L.I.; Sasse, J.; Bachalakuri, J.; Henry-Watson, J.; et al. Olfactory bulb α-synucleinopathy has high specificity and sensitivity for Lewy body disorders. *Acta Neuropathol.* **2009**, *117*, 169–174. [CrossRef]
235. Hubbard, P.S.; Esiri, M.M.; Reading, M.; McShane, R.; Nagy, Z. Alpha-synuclein pathology in the olfactory pathways of dementia patients. *J. Anat.* **2007**, *211*, 117–124. [CrossRef]
236. Shannon, K.M.; Keshavarzian, A.; Mutlu, E.; Dodiya, H.B.; Daian, D.; Jaglin, J.A.; Kordower, J.H. Alpha-synuclein in colonic submucosa in early untreated Parkinson's disease. *Mov. Disord.* **2012**, *27*, 709–715. [CrossRef]
237. Braak, H.; de Vos, R.A.; Bohl, J.; Del Tredici, K. Gastric alpha-synuclein immunoreactive inclusions in Meissner's and Auerbach's plexuses in cases staged for Parkinson's disease-related brain pathology. *Neurosci. Lett.* **2006**, *396*, 67–72. [CrossRef]
238. Oueslati, A.; Ximerakis, M.; Vekrellis, K. Protein Transmission, Seeding and Degradation: Key Steps for α-Synuclein Prion-Like Propagation. *Exp. Neurobiol.* **2014**, *23*, 324–336. [CrossRef]
239. Abdelmotilib, H.; Maltbie, T.; Delic, V.; Liu, Z.; Hu, X.; Fraser, K.B.; Moehle, M.S.; Stoyka, L.; Anabtawi, N.; Krendelchtchikova, V.; et al. α-Synuclein fibril-induced inclusion spread in rats and mice correlates with dopaminergic Neurodegeneration. *Neurobiol. Dis.* **2017**, *105*, 84–98. [CrossRef]
240. Desplats, P.; Lee, H.J.; Bae, E.J.; Patrick, C.; Rockenstein, E.; Crews, L.; Spencer, B.; Masliah, E.; Lee, S.J. Inclusion formation and neuronal cell death through neuron-to-neuron transmission of α-synuclein. *Proc. Natl. Acad. Sci. USA* **2009**, *106*, 13010–13015. [CrossRef]
241. Luk, K.C.; Kehm, V.; Carroll, J.; Zhang, B.; O'Brien, P.; Trojanowski, J.Q.; Lee, V.M. Pathological α-synuclein transmission initiates Parkinson-like neurodegeneration in nontransgenic mice. *Science* **2012**, *338*, 949–953. [CrossRef] [PubMed]
242. Mougenot, A.L.; Nicot, S.; Bencsik, A.; Morignat, E.; Verchère, J.; Lakhdar, L.; Legastelois, S.; Baron, T. Prion-like acceleration of a synucleinopathy in a transgenic mouse model. *Neurobiol. Aging* **2012**, *33*, 2225–2228. [CrossRef] [PubMed]

243. Paumier, K.L.; Luk, K.C.; Manfredsson, F.P.; Kanaan, N.M.; Lipton, J.W.; Collier, T.J.; Steece-Collier, K.; Kemp, C.J.; Celano, S.; Schulz, E.; et al. Intrastriatal injection of pre-formed mouse α-synuclein fibrils into rats triggers α-synuclein pathology and bilateral nigrostriatal degeneration. *Neurobiol. Dis.* **2015**, *82*, 185–199. [CrossRef] [PubMed]
244. Peelaerts, W.; Bousset, L.; van der Perren, A.; Moskalyuk, A.; Pulizzi, R.; Giugliano, M.; van den Haute, C.; Melki, R.; Baekelandt, V. α-Synuclein strains cause distinct synucleinopathies after local and systemic administration. *Nature* **2015**, *522*, 340–344. [CrossRef] [PubMed]
245. Singleton, A.B.; Farrer, M.; Johnson, J.; Singleton, A.; Hague, S.; Kachergus, J.; Hulihan, M.; Peuralinna, T.; Dutra, A.; Nussbaum, R.; et al. α-Synuclein locus triplication causes Parkinson's disease. *Science* **2003**, *302*, 841. [CrossRef]
246. Golbe, L.I.; di Iorio, G.; Bonavita, V.; Miller, D.C.; Duvoisin, R.C. A large kindred with autosomal dominant Parkinson's disease. *Ann. Neurol.* **1990**, *27*, 276–282. [CrossRef]
247. Kruger, R.; Kuhn, W.; Muller, T.; Woitalla, D.; Graeber, M.; Kosel, S.; Przuntek, H.; Epplen, J.T.; Schols, L.; Riess, O. Ala30Pro mutation in the gene encoding α-synuclein in Parkinson's disease. *Nat. Genet.* **1998**, *18*, 106–108. [CrossRef]
248. Kruger, R.; Kuhn, W.; Leenders, K.L.; Sprengelmeyer, R.; Muller, T.; Woitalla, D.; Portman, A.T.; Maguire, R.P.; Veenma, L.; Schroder, U.; et al. Familial parkinsonism with synuclein pathology: Clinical and PET studies of A30P mutation carriers. *Neurology* **2001**, *56*, 1355–1362. [CrossRef]
249. Seidel, K.; Schols, L.; Nuber, S.; Petrasch-Parwez, E.; Gierga, K.; Wszolek, Z.; Dickson, D.; Gai, W.P.; Bornemann, A.; Riess, O.; et al. First appraisal of brain pathology owing to A30P mutant alpha-synuclein. *Ann. Neurol.* **2010**, *67*, 684–689. [CrossRef]
250. Zarranz, J.J.; Alegre, J.; Gomez-Esteban, J.C.; Lezcano, E.; Ros, R.; Ampuero, I.; Vidal, L.; Hoenicka, J.; Rodriguez, O.; Atares, B.; et al. The new mutation, E46K, of α-synuclein causes Parkinson and Lewy body dementia. *Ann. Neurol.* **2004**, *55*, 164–173. [CrossRef]
251. Martikainen, M.H.; Paivarinta, M.; Hietala, M.; Kaasinen, V. Clinical and imaging findings in Parkinson disease associated with the A53E SNCA mutation. *Neurol. Genet.* **2015**, *1*, e27. [CrossRef]
252. Pasanen, P.; Myllykangas, L.; Siitonen, M.; Raunio, A.; Kaakkola, S.; Lyytinen, J.; Tienari, P.J.; Poyhonen, M.; Paetau, A. Novel α-synuclein mutation A53E associated with atypical multiple system atrophy and Parkinson's disease-type pathology. *Neurobiol. Aging* **2014**, *35*, 2180 e2181–e2185. [CrossRef]
253. Yoshino, H.; Hirano, M.; Stoessl, A.J.; Imamichi, Y.; Ikeda, A.; Li, Y.; Funayama, M.; Yamada, I.; Nakamura, Y.; Sossi, V.; et al. Homozygous alpha-synuclein p.A53V in familial Parkinson's disease. *Neurobiol. Aging* **2017**, *57*, 248 248.e7–248.e12. [CrossRef]
254. Kiely, A.P.; Asi, Y.T.; Kara, E.; Limousin, P.; Ling, H.; Lewis, P.; Proukakis, C.; Quinn, N.; Lees, A.J.; Hardy, J.; et al. α-Synucleinopathy associated with G51D SNCA mutation: A link between Parkinson's disease and multiple system atrophy? *Acta Neuropathol.* **2013**, *125*, 753–769. [CrossRef]
255. Kiely, A.P.; Ling, H.; Asi, Y.T.; Kara, E.; Proukakis, C.; Schapira, A.H.; Morris, H.R.; Roberts, H.C.; Lubbe, S.; Limousin, P.; et al. Distinct clinical and neuropathological features of G51D SNCA mutation cases compared with SNCA duplication and H50Q mutation. *Mol. Neurodegener.* **2015**, *10*, 41. [CrossRef]
256. Appel-Cresswell, S.; Vilarino-Guell, C.; Encarnacion, M.; Sherman, H.; Yu, I.; Shah, B.; Weir, D.; Thompson, C.; Szu-Tu, C.; Trinh, J.; et al. Alpha-synuclein p.H50Q, a novel pathogenic mutation for Parkinson's disease. *Mov. Disord.* **2013**, *28*, 811–813. [CrossRef]
257. Hoffman-Zacharska, D.; Koziorowski, D.; Ross, O.A.; Milewski, M.; Poznanski, J.A.; Jurek, M.; Wszolek, Z.K.; Soto-Ortolaza, A.; Awek, J.A.S.; Janik, P.; et al. Novel A18T and pA29S substitutions in α-synuclein may be associated with sporadic Parkinson's disease. *Park. Relat. Disord.* **2013**, *19*, 1057–1060. [CrossRef]
258. Lucking, C.B.; Brice, A. Alpha-synuclein and Parkinson's disease. *Cell. Mol. Life Sci.* **2000**, *57*, 1894–1908. [CrossRef]
259. Narhi, L.; Wood, S.J.; Steavenson, S.; Jiang, Y.; Wu, G.M.; Anafi, D.; Kaufman, S.A.; Martin, F.; Sitney, K.; Denis, P.; et al. Both familial Parkinson's disease mutations accelerate α-synuclein aggregation. *J. Biol. Chem.* **1999**, *274*, 9843–9846. [CrossRef]
260. Spira, P.J.; Sharpe, D.M.; Halliday, G.; Cavanagh, J.; Nicholson, G.A. Clinical and pathological features of a Parkinsonian syndrome in a family with an Ala53Thr α-synuclein mutation. *Ann. Neurol.* **2001**, *49*, 313–319. [CrossRef]
261. Pandey, N.; Schmidt, R.E.; Galvin, J.E. The alpha-synuclein mutation E46K promotes aggregation in cultured cells. *Exp. Neurol.* **2006**, *197*, 515–520. [CrossRef] [PubMed]
262. Devine, M.J.; Ryten, M.; Vodicka, P.; Thomson, A.J.; Burdon, T.; Houlden, H.; Cavaleri, F.; Nagano, M.; Drummond, N.J.; Taanman, J.W.; et al. Parkinson's disease induced pluripotent stem cells with triplication of the α-synuclein locus. *Nat. Commun.* **2011**, *2*, 440. [CrossRef] [PubMed]
263. Flierl, A.; Oliveira, L.M.; Falomir-Lockhart, L.J.; Mak, S.K.; Hesley, J.; Soldner, F.; Arndt-Jovin, D.J.; Jaenisch, R.; Langston, J.W.; Jovin, T.M.; et al. Higher vulnerability and stress sensitivity of neuronal precursor cells carrying an alpha-synuclein gene triplication. *PLoS ONE* **2014**, *9*, e112413. [CrossRef] [PubMed]
264. Byers, B.; Cord, B.; Nguyen, H.N.; Schule, B.; Fenno, L.; Lee, P.C.; Deisseroth, K.; Langston, J.W.; Pera, R.R.; Palmer, T.D. SNCA triplication Parkinson's patient's iPSC-derived DA neurons accumulate α-synuclein and are susceptible to oxidative stress. *PLoS ONE* **2011**, *6*, e26159. [CrossRef] [PubMed]
265. Oliveira, L.M.; Falomir-Lockhart, L.J.; Botelho, M.G.; Lin, K.H.; Wales, P.; Koch, J.C.; Gerhardt, E.; Taschenberger, H.; Outeiro, T.F.; Lingor, P.; et al. Elevated α-synuclein caused by SNCA gene triplication impairs neuronal differentiation and maturation in Parkinson's patient-derived induced pluripotent stem cells. *Cell Death Dis.* **2015**, *6*, e1994. [CrossRef]

266. Mazzulli, J.R.; Zunke, F.; Isacson, O.; Studer, L.; Krainc, D. α-Synuclein-induced lysosomal dysfunction occurs through disruptions in protein trafficking in human midbrain synucleinopathy models. *Proc. Natl. Acad. Sci. USA* **2016**, *113*, 1931–1936. [CrossRef]
267. Ludtmann, M.H.R.; Angelova, P.R.; Horrocks, M.H.; Choi, M.L.; Rodrigues, M.; Baev, A.Y.; Berezhnov, A.V.; Yao, Z.; Little, D.; Banushi, B.; et al. α-synuclein oligomers interact with ATP synthase and open the permeability transition pore in Parkinson's disease. *Nat. Commun.* **2018**, *9*, 2293. [CrossRef]
268. Haenseler, W.; Zambon, F.; Lee, H.; Vowles, J.; Rinaldi, F.; Duggal, G.; Houlden, H.; Gwinn, K.; Wray, S.; Luk, K.C.; et al. Excess α-synuclein compromises phagocytosis in iPSC-derived macrophages. *Sci. Rep.* **2017**, *7*, 9003. [CrossRef]
269. Chung, C.Y.; Khurana, V.; Auluck, P.K.; Tardiff, D.F.; Mazzulli, J.R.; Soldner, F.; Baru, V.; Lou, Y.; Freyzon, Y.; Cho, S.; et al. Identification and rescue of α-synuclein toxicity in Parkinson patient-derived neurons. *Science* **2013**, *342*, 983–987. [CrossRef]
270. Ryan, S.D.; Dolatabadi, N.; Chan, S.F.; Zhang, X.; Akhtar, M.W.; Parker, J.; Soldner, F.; Sunico, C.R.; Nagar, S.; Talantova, M.; et al. Isogenic human iPSC Parkinson's model shows nitrosative stress-induced dysfunction in MEF2-PGC1alpha transcription. *Cell* **2013**, *155*, 1351–1364. [CrossRef]
271. Kouroupi, G.; Taoufik, E.; Vlachos, I.S.; Tsioras, K.; Antoniou, N.; Papastefanaki, F.; Chroni-Tzartou, D.; Wrasidlo, W.; Bohl, D.; Stellas, D.; et al. Defective synaptic connectivity and axonal neuropathology in a human iPSC-based model of familial Parkinson's disease. *Proc. Natl. Acad. Sci. USA* **2017**, *114*, E3679–E3688. [CrossRef]
272. Dettmer, U.; Newman, A.J.; Soldner, F.; Luth, E.S.; Kim, N.C.; von Saucken, V.E.; Sanderson, J.B.; Jaenisch, R.; Bartels, T.; Selkoe, D. Parkinson-causing α-synuclein missense mutations shift native tetramers to monomers as a mechanism for disease initiation. *Nat. Commun.* **2015**, *6*, 7314. [CrossRef]
273. Ono, K. The Oligomer Hypothesis in α-Synucleinopathy. *Neurochem. Res.* **2017**, *42*, 3362–3371. [CrossRef]
274. Zambon, F.; Cherubini, M.; Fernandes, H.J.R.; Lang, C.; Ryan, B.J.; Volpato, V.; Bengoa-Vergniory, N.; Vingill, S.; Attar, M.; Booth, H.D.E.; et al. Cellular α-synuclein pathology is associated with bioenergetic dysfunction in Parkinson's iPSC-derived dopamine neurons. *Hum. Mol. Genet.* **2019**, *28*, 2001–2013. [CrossRef]
275. Stykel, M.G.; Humphries, K.; Kirby, M.P.; Czaniecki, C.; Wang, T.; Ryan, T.; Bamm, V.; Ryan, S.D. Nitration of microtubules blocks axonal mitochondrial transport in a human pluripotent stem cell model of Parkinson's disease. *FASEB J.* **2018**, *32*, 5350–5364. [CrossRef]
276. Snowden, S.G.; Fernandes, H.J.R.; Kent, J.; Foskolou, S.; Tate, P.; Field, S.F.; Metzakopian, E.; Koulman, A. Development and Application of High-Throughput Single Cell Lipid Profiling: A Study of SNCA-A53T Human Dopamine Neurons. *iScience* **2020**, *23*, 101703. [CrossRef]
277. Azevedo, C.; Teku, G.; Pomeshchik, Y.; Reyes, J.F.; Chumarina, M.; Russ, K.; Savchenko, E.; Hammarberg, A.; Lamas, N.J.; Collin, A.; et al. Parkinson's disease and multiple system atrophy patient iPSC-derived oligodendrocytes exhibit alpha-synuclein-induced changes in maturation and immune reactive properties. *Proc. Natl. Acad. Sci. USA* **2022**, *119*, e2111405119. [CrossRef]
278. Fleming, S.M.; Salcedo, J.; Hutson, C.B.; Rockenstein, E.; Masliah, E.; Levine, M.S.; Chesselet, M.F. Behavioral effects of dopaminergic agonists in transgenic mice overexpressing human wildtype α-synuclein. *Neuroscience* **2006**, *142*, 1245–1253. [CrossRef]
279. Wakamatsu, M.; Ishii, A.; Iwata, S.; Sakagami, J.; Ukai, Y.; Ono, M.; Kanbe, D.; Muramatsu, S.; Kobayashi, K.; Iwatsubo, T.; et al. Selective loss of nigral dopamine neurons induced by overexpression of truncated human α-synuclein in mice. *Neurobiol. Aging* **2008**, *29*, 574–585. [CrossRef]
280. Konnova, E.A.; Swanberg, M. Animal Models of Parkinson's Disease. In *Parkinson's Disease: Pathogenesis and Clinical Aspects*; Stoker, T.B., Greenland, J.C., Eds.; Codon Publications: Brisbane, Australia, 2018.
281. Chesselet, M.F.; Fleming, S.; Mortazavi, F.; Meurers, B. Strengths and limitations of genetic mouse models of Parkinson's disease. *Park. Relat. Disord.* **2008**, *14* (Suppl. S2), S84–S87. [CrossRef]
282. Chesselet, M.F.; Richter, F.; Zhu, C.; Magen, I.; Watson, M.B.; Subramaniam, S.R. A progressive mouse model of Parkinson's disease: The Thy1-aSyn ("Line 61") mice. *Neurotherapeutics* **2012**, *9*, 297–314. [CrossRef] [PubMed]
283. Cannon, J.R.; Geghman, K.D.; Tapias, V.; Sew, T.; Dail, M.K.; Li, C.; Greenamyre, J.T. Expression of human E46K-mutated α-synuclein in BAC-transgenic rats replicates early-stage Parkinson's disease features and enhances vulnerability to mitochondrial impairment. *Exp. Neurol.* **2013**, *240*, 44–56. [CrossRef]
284. Decressac, M.; Mattsson, B.; Lundblad, M.; Weikop, P.; Bjorklund, A. Progressive neurodegenerative and behavioural changes induced by AAV-mediated overexpression of α-synuclein in midbrain dopamine neurons. *Neurobiol. Dis.* **2012**, *45*, 939–953. [CrossRef] [PubMed]
285. Van der Perren, A.; Toelen, J.; Casteels, C.; Macchi, F.; Van Rompuy, A.S.; Sarre, S.; Casadei, N.; Nuber, S.; Himmelreich, U.; Osorio Garcia, M.I.; et al. Longitudinal follow-up and characterization of a robust rat model for Parkinson's disease based on overexpression of alpha-synuclein with adeno-associated viral vectors. *Neurobiol. Aging* **2015**, *36*, 1543–1558. [CrossRef]
286. Oliveras-Salva, M.; van der Perren, A.; Casadei, N.; Stroobants, S.; Nuber, S.; D'Hooge, R.; van den Haute, C.; Baekelandt, V. rAAV2/7 vector-mediated overexpression of alpha-synuclein in mouse substantia nigra induces protein aggregation and progressive dose-dependent neurodegeneration. *Mol. Neurodegener.* **2013**, *8*, 44. [CrossRef] [PubMed]
287. Kirik, D.; Annett, L.E.; Burger, C.; Muzyczka, N.; Mandel, R.J.; Bjorklund, A. Nigrostriatal α-synucleinopathy induced by viral vector-mediated overexpression of human α-synuclein: A new primate model of Parkinson's disease. *Proc. Natl. Acad. Sci. USA* **2003**, *100*, 2884–2889. [CrossRef]

288. Koprich, J.B.; Johnston, T.H.; Reyes, G.; Omana, V.; Brotchie, J.M. Towards a Non-Human Primate Model of Alpha-Synucleinopathy for Development of Therapeutics for Parkinson's Disease: Optimization of AAV1/2 Delivery Parameters to Drive Sustained Expression of Alpha Synuclein and Dopaminergic Degeneration in Macaque. *PLoS ONE* **2016**, *11*, e0167235. [CrossRef]
289. Aharon-Peretz, J.; Rosenbaum, H.; Gershoni-Baruch, R. Mutations in the glucocerebrosidase gene and Parkinson's disease in Ashkenazi Jews. *N. Engl. J. Med.* **2004**, *351*, 1972–1977. [CrossRef]
290. Alcalay, R.N.; Dinur, T.; Quinn, T.; Sakanaka, K.; Levy, O.; Waters, C.; Fahn, S.; Dorovski, T.; Chung, W.K.; Pauciulo, M.; et al. Comparison of Parkinson risk in Ashkenazi Jewish patients with Gaucher disease and GBA heterozygotes. *JAMA Neurol.* **2014**, *71*, 752–757. [CrossRef]
291. Sidransky, E.; Samaddar, T.; Tayebi, N. Mutations in GBA are associated with familial Parkinson disease susceptibility and age at onset. *Neurology* **2009**, *73*, 1424–1425. [CrossRef]
292. Gan-Or, Z.; Giladi, N.; Rozovski, U.; Shifrin, C.; Rosner, S.; Gurevich, T.; Bar-Shira, A.; Orr-Urtreger, A. Genotype-phenotype correlations between GBA mutations and Parkinson disease risk and onset. *Neurology* **2008**, *70*, 2277–2283. [CrossRef]
293. Lesage, S.; Condroyer, C.; Hecham, N.; Anheim, M.; Belarbi, S.; Lohman, E.; Viallet, F.; Pollak, P.; Abada, M.; Durr, A.; et al. Mutations in the glucocerebrosidase gene confer a risk for Parkinson disease in North Africa. *Neurology* **2011**, *76*, 301–303. [CrossRef]
294. Rosenbloom, B.; Balwani, M.; Bronstein, J.M.; Kolodny, E.; Sathe, S.; Gwosdow, A.R.; Taylor, J.S.; Cole, J.A.; Zimran, A.; Weinreb, N.J. The incidence of Parkinsonism in patients with type 1 Gaucher disease: Data from the ICGG Gaucher Registry. *Blood Cells Mol. Dis.* **2011**, *46*, 95–102. [CrossRef]
295. Hruska, K.S.; LaMarca, M.E.; Scott, C.R.; Sidransky, E. Gaucher disease: Mutation and polymorphism spectrum in the glucocerebrosidase gene (GBA). *Hum. Mutat.* **2008**, *29*, 567–583. [CrossRef]
296. Gan-Or, Z.; Amshalom, I.; Kilarski, L.L.; Bar-Shira, A.; Gana-Weisz, M.; Mirelman, A.; Marder, K.; Bressman, S.; Giladi, N.; Orr-Urtreger, A. Differential effects of severe vs. mild GBA mutations on Parkinson disease. *Neurology* **2015**, *84*, 880–887. [CrossRef]
297. Cilia, R.; Tunesi, S.; Marotta, G.; Cereda, E.; Siri, C.; Tesei, S.; Zecchinelli, A.L.; Canesi, M.; Mariani, C.B.; Meucci, N.; et al. Survival and dementia in GBA-associated Parkinson's disease: The mutation matters. *Ann. Neurol.* **2016**, *80*, 662–673. [CrossRef]
298. Blandini, F.; Cilia, R.; Cerri, S.; Pezzoli, G.; Schapira, A.H.V.; Mullin, S.; Lanciego, J.L. Glucocerebrosidase mutations and synucleinopathies: Toward a model of precision medicine. *Mov. Disord.* **2019**, *34*, 9–21. [CrossRef]
299. Thaler, A.; Gurevich, T.; Bar Shira, A.; Gana Weisz, M.; Ash, E.; Shiner, T.; Orr-Urtreger, A.; Giladi, N.; Mirelman, A. A "dose" effect of mutations in the GBA gene on Parkinson's disease phenotype. *Park. Relat. Disord.* **2017**, *36*, 47–51. [CrossRef]
300. Corti, O.; Lesage, S.; Brice, A. What genetics tells us about the causes and mechanisms of Parkinson's disease. *Physiol. Rev.* **2011**, *91*, 1161–1218. [CrossRef]
301. Yap, T.L.; Gruschus, J.M.; Velayati, A.; Westbroek, W.; Goldin, E.; Moaven, N.; Sidransky, E.; Lee, J.C. α-synuclein interacts with Glucocerebrosidase providing a molecular link between Parkinson and Gaucher diseases. *J. Biol. Chem.* **2011**, *286*, 28080–28088. [CrossRef]
302. Mazzulli, J.R.; Xu, Y.H.; Sun, Y.; Knight, A.L.; McLean, P.J.; Caldwell, G.A.; Sidransky, E.; Grabowski, G.A.; Krainc, D. Gaucher disease glucocerebrosidase and α-synuclein form a bidirectional pathogenic loop in synucleinopathies. *Cell* **2011**, *146*, 37–52. [CrossRef] [PubMed]
303. Murphy, K.E.; Gysbers, A.M.; Abbott, S.K.; Tayebi, N.; Kim, W.S.; Sidransky, E.; Cooper, A.; Garner, B.; Halliday, G.M. Reduced glucocerebrosidase is associated with increased α-synuclein in sporadic Parkinson's disease. *Brain* **2014**, *137*, 834–848. [CrossRef] [PubMed]
304. Chiasserini, D.; Paciotti, S.; Eusebi, P.; Persichetti, E.; Tasegian, A.; Kurzawa-Akanbi, M.; Chinnery, P.F.; Morris, C.M.; Calabresi, P.; Parnetti, L.; et al. Selective loss of glucocerebrosidase activity in sporadic Parkinson's disease and dementia with Lewy bodies. *Mol. Neurodegener.* **2015**, *10*, 15. [CrossRef] [PubMed]
305. Zunke, F.; Moise, A.C.; Belur, N.R.; Gelyana, E.; Stojkovska, I.; Dzaferbegovic, H.; Toker, N.J.; Jeon, S.; Fredriksen, K.; Mazzulli, J.R. Reversible Conformational Conversion of α-Synuclein into Toxic Assemblies by Glucosylceramide. *Neuron* **2018**, *97*, 92–107.e110. [CrossRef] [PubMed]
306. Suzuki, M.; Sango, K.; Wada, K.; Nagai, Y. Pathological role of lipid interaction with α-synuclein in Parkinson's disease. *Neurochem. Int.* **2018**, *119*, 97–106. [CrossRef]
307. Burbulla, L.F.; Jeon, S.; Zheng, J.; Song, P.; Silverman, R.B.; Krainc, D. A modulator of wild-type glucocerebrosidase improves pathogenic phenotypes in dopaminergic neuronal models of Parkinson's disease. *Sci. Transl. Med.* **2019**, *11*, eaau6870. [CrossRef]
308. Mazzulli, J.R.; Zunke, F.; Tsunemi, T.; Toker, N.J.; Jeon, S.; Burbulla, L.F.; Patnaik, S.; Sidransky, E.; Marugan, J.J.; Sue, C.M.; et al. Activation of β-Glucocerebrosidase Reduces Pathological α-Synuclein and Restores Lysosomal Function in Parkinson's Patient Midbrain Neurons. *J. Neurosci.* **2016**, *36*, 7693–7706. [CrossRef]
309. Burbulla, L.F.; Song, P.; Mazzulli, J.R.; Zampese, E.; Wong, Y.C.; Jeon, S.; Santos, D.P.; Blanz, J.; Obermaier, C.D.; Strojny, C.; et al. Dopamine oxidation mediates mitochondrial and lysosomal dysfunction in Parkinson's disease. *Science* **2017**, *357*, 1255–1261. [CrossRef]

310. Schondorf, D.C.; Aureli, M.; McAllister, F.E.; Hindley, C.J.; Mayer, F.; Schmid, B.; Sardi, S.P.; Valsecchi, M.; Hoffmann, S.; Schwarz, L.K.; et al. iPSC-derived neurons from GBA1-associated Parkinson's disease patients show autophagic defects and impaired calcium homeostasis. *Nat. Commun.* **2014**, *5*, 4028. [CrossRef]
311. Fernandes, H.J.; Hartfield, E.M.; Christian, H.C.; Emmanoulidou, E.; Zheng, Y.; Booth, H.; Bogetofte, H.; Lang, C.; Ryan, B.J.; Sardi, S.P.; et al. ER Stress and Autophagic Perturbations Lead to Elevated Extracellular α-Synuclein in GBA-N370S Parkinson's iPSC-Derived Dopamine Neurons. *Stem. Cell. Rep.* **2016**, *6*, 342–356. [CrossRef]
312. Schondorf, D.C.; Ivanyuk, D.; Baden, P.; Sanchez-Martinez, A.; de Cicco, S.; Yu, C.; Giunta, I.; Schwarz, L.K.; Di Napoli, G.; Panagiotakopoulou, V.; et al. The NAD+ Precursor Nicotinamide Riboside Rescues Mitochondrial Defects and Neuronal Loss in iPSC and Fly Models of Parkinson's Disease. *Cell. Rep.* **2018**, *23*, 2976–2988. [CrossRef]
313. Sanyal, A.; DeAndrade, M.P.; Novis, H.S.; Lin, S.; Chang, J.; Lengacher, N.; Tomlinson, J.J.; Tansey, M.G.; LaVoie, M.J. Lysosome and Inflammatory Defects in GBA1-Mutant Astrocytes Are Normalized by LRRK2 Inhibition. *Mov. Disord.* **2020**, *35*, 760–773. [CrossRef]
314. Sanyal, A.; Novis, H.S.; Gasser, E.; Lin, S.; LaVoie, M.J. LRRK2 Kinase Inhibition Rescues Deficits in Lysosome Function Due to Heterozygous GBA1 Expression in Human iPSC-Derived Neurons. *Front. Neurosci.* **2020**, *14*, 442. [CrossRef]
315. Kim, S.; Wong, Y.C.; Gao, F.; Krainc, D. Dysregulation of mitochondria-lysosome contacts by GBA1 dysfunction in dopaminergic neuronal models of Parkinson's disease. *Nat. Commun.* **2021**, *12*, 1807. [CrossRef]
316. Ysselstein, D.; Nguyen, M.; Young, T.J.; Severino, A.; Schwake, M.; Merchant, K.; Krainc, D. LRRK2 kinase activity regulates lysosomal glucocerebrosidase in neurons derived from Parkinson's disease patients. *Nat. Commun.* **2019**, *10*, 5570. [CrossRef]
317. Liu, Y.; Suzuki, K.; Reed, J.D.; Grinberg, A.; Westphal, H.; Hoffmann, A.; Doring, T.; Sandhoff, K.; Proia, R.L. Mice with type 2 and 3 Gaucher disease point mutations generated by a single insertion mutagenesis procedure. *Proc. Natl. Acad. Sci. USA* **1998**, *95*, 2503–2508. [CrossRef]
318. Mizukami, H.; Mi, Y.; Wada, R.; Kono, M.; Yamashita, T.; Liu, Y.; Werth, N.; Sandhoff, R.; Sandhoff, K.; Proia, R.L. Systemic inflammation in glucocerebrosidase-deficient mice with minimal glucosylceramide storage. *J. Clin. Invest.* **2002**, *109*, 1215–1221. [CrossRef]
319. Ginns, E.I.; Mak, S.K.; Ko, N.; Karlgren, J.; Akbarian, S.; Chou, V.P.; Guo, Y.; Lim, A.; Samuelsson, S.; LaMarca, M.L.; et al. Neuroinflammation and α-synuclein accumulation in response to glucocerebrosidase deficiency are accompanied by synaptic dysfunction. *Mol. Genet. Metab.* **2014**, *111*, 152–162. [CrossRef]
320. Li, H.; Ham, A.; Ma, T.C.; Kuo, S.H.; Kanter, E.; Kim, D.; Ko, H.S.; Quan, Y.; Sardi, S.P.; Li, A.; et al. Mitochondrial dysfunction and mitophagy defect triggered by heterozygous GBA mutations. *Autophagy* **2019**, *15*, 113–130. [CrossRef]
321. Yun, S.P.; Kim, D.; Kim, S.; Kim, S.; Karuppagounder, S.S.; Kwon, S.H.; Lee, S.; Kam, T.I.; Lee, S.; Ham, S.; et al. α-Synuclein accumulation and GBA deficiency due to L444P GBA mutation contributes to MPTP-induced parkinsonism. *Mol. Neurodegener.* **2018**, *13*, 1. [CrossRef]
322. Migdalska-Richards, A.; Wegrzynowicz, M.; Rusconi, R.; Deangeli, G.; Di Monte, D.A.; Spillantini, M.G.; Schapira, A.H.V. The L444P Gba1 mutation enhances alpha-synuclein induced loss of nigral dopaminergic neurons in mice. *Brain* **2017**, *140*, 2706–2721. [CrossRef] [PubMed]
323. Taguchi, Y.V.; Liu, J.; Ruan, J.; Pacheco, J.; Zhang, X.; Abbasi, J.; Keutzer, J.; Mistry, P.K.; Chandra, S.S. Glucosylsphingosine Promotes α-Synuclein Pathology in Mutant GBA-Associated Parkinson's Disease. *J. Neurosci.* **2017**, *37*, 9617–9631. [CrossRef] [PubMed]
324. Zhang, Y.; Shu, L.; Sun, Q.; Zhou, X.; Pan, H.; Guo, J.; Tang, B. Integrated Genetic Analysis of Racial Differences of Common GBA Variants in Parkinson's Disease: A Meta-Analysis. *Front. Mol. Neurosci.* **2018**, *11*, 43. [CrossRef] [PubMed]
325. Xu, Y.H.; Sun, Y.; Ran, H.; Quinn, B.; Witte, D.; Grabowski, G.A. Accumulation and distribution of α-synuclein and ubiquitin in the CNS of Gaucher disease mouse models. *Mol. Genet. Metab.* **2011**, *102*, 436–447. [CrossRef] [PubMed]
326. Kim, D.; Hwang, H.; Choi, S.; Kwon, S.H.; Lee, S.; Park, J.H.; Kim, S.; Ko, H.S. D409H GBA1 mutation accelerates the progression of pathology in A53T α-synuclein transgenic mouse model. *Acta Neuropathol. Commun.* **2018**, *6*, 32. [CrossRef] [PubMed]
327. Farfel-Becker, T.; Vitner, E.B.; Pressey, S.N.; Eilam, R.; Cooper, J.D.; Futerman, A.H. Spatial and temporal correlation between neuron loss and neuroinflammation in a mouse model of neuronopathic Gaucher disease. *Hum. Mol. Genet.* **2011**, *20*, 1375–1386. [CrossRef]
328. Fishbein, I.; Kuo, Y.M.; Giasson, B.I.; Nussbaum, R.L. Augmentation of phenotype in a transgenic Parkinson mouse heterozygous for a Gaucher mutation. *Brain* **2014**, *137*, 3235–3247. [CrossRef]
329. Silvestri, L.; Caputo, V.; Bellacchio, E.; Atorino, L.; Dallapiccola, B.; Valente, E.M.; Casari, G. Mitochondrial import and enzymatic activity of PINK1 mutants associated to recessive parkinsonism. *Hum. Mol. Genet.* **2005**, *14*, 3477–3492. [CrossRef]
330. Yamano, K.; Youle, R.J. PINK1 is degraded through the N-end rule pathway. *Autophagy* **2013**, *9*, 1758–1769. [CrossRef]
331. Kondapalli, C.; Kazlauskaite, A.; Zhang, N.; Woodroof, H.I.; Campbell, D.G.; Gourlay, R.; Burchell, L.; Walden, H.; Macartney, T.J.; Deak, M.; et al. PINK1 is activated by mitochondrial membrane potential depolarization and stimulates Parkin E3 ligase activity by phosphorylating Serine 65. *Open Biol.* **2012**, *2*, 120080. [CrossRef]
332. Shiba-Fukushima, K.; Imai, Y.; Yoshida, S.; Ishihama, Y.; Kanao, T.; Sato, S.; Hattori, N. PINK1-mediated phosphorylation of the Parkin ubiquitin-like domain primes mitochondrial translocation of Parkin and regulates mitophagy. *Sci. Rep.* **2012**, *2*, 1002. [CrossRef]

333. Sarraf, S.A.; Raman, M.; Guarani-Pereira, V.; Sowa, M.E.; Huttlin, E.L.; Gygi, S.P.; Harper, J.W. Landscape of the PARKIN-dependent ubiquitylome in response to mitochondrial depolarization. *Nature* **2013**, *496*, 372–376. [CrossRef]
334. Wauer, T.; Simicek, M.; Schubert, A.; Komander, D. Mechanism of phospho-ubiquitin-induced PARKIN activation. *Nature* **2015**, *524*, 370–374. [CrossRef]
335. Tanaka, A.; Cleland, M.M.; Xu, S.; Narendra, D.P.; Suen, D.F.; Karbowski, M.; Youle, R.J. Proteasome and p97 mediate mitophagy and degradation of mitofusins induced by Parkin. *J. Cell. Biol.* **2010**, *191*, 1367–1380. [CrossRef]
336. Valente, E.M.; Abou-Sleiman, P.M.; Caputo, V.; Muqit, M.M.; Harvey, K.; Gispert, S.; Ali, Z.; del Turco, D.; Bentivoglio, A.R.; Healy, D.G.; et al. Hereditary early-onset Parkinson's disease caused by mutations in PINK1. *Science* **2004**, *304*, 1158–1160. [CrossRef]
337. Plun-Favreau, H.; Klupsch, K.; Moisoi, N.; Gandhi, S.; Kjaer, S.; Frith, D.; Harvey, K.; Deas, E.; Harvey, R.J.; McDonald, N.; et al. The mitochondrial protease HtrA2 is regulated by Parkinson's disease-associated kinase PINK1. *Nat. Cell. Biol.* **2007**, *9*, 1243–1252. [CrossRef]
338. Pridgeon, J.W.; Olzmann, J.A.; Chin, L.S.; Li, L. PINK1 protects against oxidative stress by phosphorylating mitochondrial chaperone TRAP1. *PLoS Biol.* **2007**, *5*, e172. [CrossRef]
339. Kim, Y.; Park, J.; Kim, S.; Song, S.; Kwon, S.K.; Lee, S.H.; Kitada, T.; Kim, J.M.; Chung, J. PINK1 controls mitochondrial localization of Parkin through direct phosphorylation. *Biochem. Biophys. Res. Commun.* **2008**, *377*, 975–980. [CrossRef]
340. Shimura, H.; Schlossmacher, M.G.; Hattori, N.; Frosch, M.P.; Trockenbacher, A.; Schneider, R.; Mizuno, Y.; Kosik, K.S.; Selkoe, D.J. Ubiquitination of a new form of α-synuclein by parkin from human brain: Implications for Parkinson's disease. *Science* **2001**, *293*, 263–269. [CrossRef]
341. Beilina, A.; van der Brug, M.; Ahmad, R.; Kesavapany, S.; Miller, D.W.; Petsko, G.A.; Cookson, M.R. Mutations in PTEN-induced putative kinase 1 associated with recessive parkinsonism have differential effects on protein stability. *Proc. Natl. Acad. Sci. USA* **2005**, *102*, 5703–5708. [CrossRef]
342. Steinlechner, S.; Stahlberg, J.; Volkel, B.; Djarmati, A.; Hagenah, J.; Hiller, A.; Hedrich, K.; Konig, I.; Klein, C.; Lencer, R. Co-occurrence of affective and schizophrenia spectrum disorders with PINK1 mutations. *J. Neurol. Neurosurg. Psychiatry* **2007**, *78*, 532–535. [CrossRef] [PubMed]
343. Samaranch, L.; Lorenzo-Betancor, O.; Arbelo, J.M.; Ferrer, I.; Lorenzo, E.; Irigoyen, J.; Pastor, M.A.; Marrero, C.; Isla, C.; Herrera-Henriquez, J.; et al. PINK1-linked parkinsonism is associated with Lewy body pathology. *Brain* **2010**, *133*, 1128–1142. [CrossRef] [PubMed]
344. Puschmann, A.; Fiesel, F.C.; Caulfield, T.R.; Hudec, R.; Ando, M.; Truban, D.; Hou, X.; Ogaki, K.; Heckman, M.G.; James, E.D.; et al. Heterozygous PINK1 p.G411S increases risk of Parkinson's disease via a dominant-negative mechanism. *Brain* **2017**, *140*, 98–117. [CrossRef] [PubMed]
345. Morais, V.A.; Haddad, D.; Craessaerts, K.; De Bock, P.J.; Swerts, J.; Vilain, S.; Aerts, L.; Overbergh, L.; Grunewald, A.; Seibler, P.; et al. PINK1 loss-of-function mutations affect mitochondrial complex I activity via NdufA10 ubiquinone uncoupling. *Science* **2014**, *344*, 203–207. [CrossRef] [PubMed]
346. Rakovic, A.; Shurkewitsch, K.; Seibler, P.; Grunewald, A.; Zanon, A.; Hagenah, J.; Krainc, D.; Klein, C. Phosphatase and tensin homolog (PTEN)-induced putative kinase 1 (PINK1)-dependent ubiquitination of endogenous Parkin attenuates mitophagy: Study in human primary fibroblasts and induced pluripotent stem cell-derived neurons. *J. Biol. Chem.* **2013**, *288*, 2223–2237. [CrossRef]
347. Seibler, P.; Graziotto, J.; Jeong, H.; Simunovic, F.; Klein, C.; Krainc, D. Mitochondrial Parkin recruitment is impaired in neurons derived from mutant PINK1 induced pluripotent stem cells. *J. Neurosci.* **2011**, *31*, 5970–5976. [CrossRef]
348. Vos, M.; Geens, A.; Bohm, C.; Deaulmerie, L.; Swerts, J.; Rossi, M.; Craessaerts, K.; Leites, E.P.; Seibler, P.; Rakovic, A.; et al. Cardiolipin promotes electron transport between ubiquinone and complex I to rescue PINK1 deficiency. *J. Cell. Biol.* **2017**, *216*, 695–708. [CrossRef]
349. Cooper, O.; Seo, H.; Andrabi, S.; Guardia-Laguarta, C.; Graziotto, J.; Sundberg, M.; McLean, J.R.; Carrillo-Reid, L.; Xie, Z.; Osborn, T.; et al. Pharmacological rescue of mitochondrial deficits in iPSC-derived neural cells from patients with familial Parkinson's disease. *Sci. Transl. Med.* **2012**, *4*, 1–13. [CrossRef]
350. Chung, S.Y.; Kishinevsky, S.; Mazzulli, J.R.; Graziotto, J.; Mrejeru, A.; Mosharov, E.V.; Puspita, L.; Valiulahi, P.; Sulzer, D.; Milner, T.A.; et al. Parkin and PINK1 Patient iPSC-Derived Midbrain Dopamine Neurons Exhibit Mitochondrial Dysfunction and α-Synuclein Accumulation. *Stem. Cell. Rep.* **2016**, *7*, 664–677. [CrossRef]
351. Azkona, G.; Lopez de Maturana, R.; del Rio, P.; Sousa, A.; Vazquez, N.; Zubiarrain, A.; Jimenez-Blasco, D.; Bolanos, J.P.; Morales, B.; Auburger, G.; et al. LRRK2 Expression Is Deregulated in Fibroblasts and Neurons from Parkinson Patients with Mutations in PINK1. *Mol. Neurobiol.* **2018**, *55*, 506–516. [CrossRef]
352. Gautier, C.A.; Kitada, T.; Shen, J. Loss of PINK1 causes mitochondrial functional defects and increased sensitivity to oxidative stress. *Proc. Natl. Acad. Sci. USA* **2008**, *105*, 11364–11369. [CrossRef]
353. Gispert, S.; Ricciardi, F.; Kurz, A.; Azizov, M.; Hoepken, H.H.; Becker, D.; Voos, W.; Leuner, K.; Muller, W.E.; Kudin, A.P.; et al. Parkinson phenotype in aged PINK1-deficient mice is accompanied by progressive mitochondrial dysfunction in absence of neurodegeneration. *PLoS ONE* **2009**, *4*, e5777. [CrossRef]
354. Glasl, L.; Kloos, K.; Giesert, F.; Roethig, A.; di Benedetto, B.; Kuhn, R.; Zhang, J.; Hafen, U.; Zerle, J.; Hofmann, A.; et al. Pink1-deficiency in mice impairs gait, olfaction and serotonergic innervation of the olfactory bulb. *Exp. Neurol.* **2012**, *235*, 214–227. [CrossRef]

355. Kitada, T.; Pisani, A.; Porter, D.R.; Yamaguchi, H.; Tscherter, A.; Martella, G.; Bonsi, P.; Zhang, C.; Pothos, E.N.; Shen, J. Impaired dopamine release and synaptic plasticity in the striatum of PINK1-deficient mice. *Proc. Natl. Acad. Sci. USA* **2007**, *104*, 11441–11446. [CrossRef]
356. Morais, V.A.; Verstreken, P.; Roethig, A.; Smet, J.; Snellinx, A.; Vanbrabant, M.; Haddad, D.; Frezza, C.; Mandemakers, W.; Vogt-Weisenhorn, D.; et al. Parkinson's disease mutations in PINK1 result in decreased Complex I activity and deficient synaptic function. *EMBO Mol. Med.* **2009**, *1*, 99–111. [CrossRef]
357. Smith, G.A.; Isacson, O.; Dunnett, S.B. The search for genetic mouse models of prodromal Parkinson's disease. *Exp. Neurol.* **2012**, *237*, 267–273. [CrossRef]
358. Jiang, P.; Dickson, D.W. Parkinson's disease: Experimental models and reality. *Acta Neuropathol.* **2018**, *135*, 13–32. [CrossRef]
359. Akundi, R.S.; Huang, Z.; Eason, J.; Pandya, J.D.; Zhi, L.; Cass, W.A.; Sullivan, P.G.; Bueler, H. Increased mitochondrial calcium sensitivity and abnormal expression of innate immunity genes precede dopaminergic defects in Pink1-deficient mice. *PLoS ONE* **2011**, *6*, e16038. [CrossRef]
360. Dave, K.D.; De Silva, S.; Sheth, N.P.; Ramboz, S.; Beck, M.J.; Quang, C.; Switzer, R.C., 3rd; Ahmad, S.O.; Sunkin, S.M.; Walker, D.; et al. Phenotypic characterization of recessive gene knockout rat models of Parkinson's disease. *Neurobiol. Dis.* **2014**, *70*, 190–203. [CrossRef]
361. Villeneuve, L.M.; Purnell, P.R.; Boska, M.D.; Fox, H.S. Early Expression of Parkinson's Disease-Related Mitochondrial Abnormalities in PINK1 Knockout Rats. *Mol. Neurobiol.* **2016**, *53*, 171–186. [CrossRef]
362. Grant, L.M.; Kelm-Nelson, C.A.; Hilby, B.L.; Blue, K.V.; Paul Rajamanickam, E.S.; Pultorak, J.D.; Fleming, S.M.; Ciucci, M.R. Evidence for early and progressive ultrasonic vocalization and oromotor deficits in a PINK1 gene knockout rat model of Parkinson's disease. *J. Neurosci. Res.* **2015**, *93*, 1713–1727. [CrossRef] [PubMed]
363. Narendra, D.; Tanaka, A.; Suen, D.F.; Youle, R.J. Parkin is recruited selectively to impaired mitochondria and promotes their autophagy. *J. Cell. Biol.* **2008**, *183*, 795–803. [CrossRef] [PubMed]
364. Matsumine, H.; Saito, M.; Shimoda-Matsubayashi, S.; Tanaka, H.; Ishikawa, A.; Nakagawa-Hattori, Y.; Yokochi, M.; Kobayashi, T.; Igarashi, S.; Takano, H.; et al. Localization of a gene for an autosomal recessive form of juvenile Parkinsonism to chromosome 6q25.2-27. *Am. J. Hum. Genet.* **1997**, *60*, 588–596. [PubMed]
365. Mata, I.F.; Lockhart, P.J.; Farrer, M.J. Parkin genetics: One model for Parkinson's disease. *Hum. Mol. Genet.* **2004**, *13*, R127–R133. [CrossRef] [PubMed]
366. Kitada, T.; Asakawa, S.; Hattori, N.; Matsumine, H.; Yamamura, Y.; Minoshima, S.; Yokochi, M.; Mizuno, Y.; Shimizu, N. Mutations in the parkin gene cause autosomal recessive juvenile parkinsonism. *Nature* **1998**, *392*, 605–608. [CrossRef]
367. Charan, R.A.; LaVoie, M.J. Pathologic and therapeutic implications for the cell biology of parkin. *Mol. Cell. Neurosci.* **2015**, *66*, 62–71. [CrossRef]
368. Conceicao, I.C.; Rama, M.M.; Oliveira, B.; Cafe, C.; Almeida, J.; Mouga, S.; Duque, F.; Oliveira, G.; Vicente, A.M. Definition of a putative pathological region in PARK2 associated with autism spectrum disorder through in silico analysis of its functional structure. *Psychiatr. Genet.* **2017**, *27*, 54–61. [CrossRef]
369. Lucking, C.B.; Durr, A.; Bonifati, V.; Vaughan, J.; de Michele, G.; Gasser, T.; Harhangi, B.S.; Meco, G.; Denefle, P.; Wood, N.W.; et al. Association between early-onset Parkinson's disease and mutations in the parkin gene. *N. Engl. J. Med.* **2000**, *342*, 1560–1567. [CrossRef]
370. Kann, M.; Jacobs, H.; Mohrmann, K.; Schumacher, K.; Hedrich, K.; Garrels, J.; Wiegers, K.; Schwinger, E.; Pramstaller, P.P.; Breakefield, X.O.; et al. Role of parkin mutations in 111 community-based patients with early-onset parkinsonism. *Ann. Neurol.* **2002**, *51*, 621–625. [CrossRef]
371. Foroud, T.; Uniacke, S.K.; Liu, L.; Pankratz, N.; Rudolph, A.; Halter, C.; Shults, C.; Marder, K.; Conneally, P.M.; Nichols, W.C.; et al. Heterozygosity for a mutation in the parkin gene leads to later onset Parkinson disease. *Neurology* **2003**, *60*, 796–801. [CrossRef]
372. Periquet, M.; Latouche, M.; Lohmann, E.; Rawal, N.; de Michele, G.; Ricard, S.; Teive, H.; Fraix, V.; Vidailhet, M.; Nicholl, D.; et al. Parkin mutations are frequent in patients with isolated early-onset parkinsonism. *Brain* **2003**, *126*, 1271–1278. [CrossRef]
373. Kachidian, P.; Gubellini, P. Genetic Models of Parkinson's Disease. In *Clinical Trials In Parkinson's Disease*; Perez-Lloret, S., Ed.; Springer: New York, NY, USA, 2021; pp. 37–84.
374. Abbas, N.; Lucking, C.B.; Ricard, S.; Durr, A.; Bonifati, V.; de Michele, G.; Bouley, S.; Vaughan, J.R.; Gasser, T.; Marconi, R.; et al. A wide variety of mutations in the parkin gene are responsible for autosomal recessive parkinsonism in Europe. French Parkinson's Disease Genetics Study Group and the European Consortium on Genetic Susceptibility in Parkinson's Disease. *Hum. Mol. Genet.* **1999**, *8*, 567–574. [CrossRef]
375. Lohmann, E.; Periquet, M.; Bonifati, V.; Wood, N.W.; de Michele, G.; Bonnet, A.M.; Fraix, V.; Broussolle, E.; Horstink, M.W.; Vidailhet, M.; et al. How much phenotypic variation can be attributed to parkin genotype? *Ann. Neurol.* **2003**, *54*, 176–185. [CrossRef]
376. Benbunan, B.R.; Korczyn, A.D.; Giladi, N. Parkin mutation associated parkinsonism and cognitive decline, comparison to early onset Parkinson's disease. *J. Neural Transm.* **2004**, *111*, 47–57. [CrossRef]
377. Srivastava, A.; Tang, M.X.; Mejia-Santana, H.; Rosado, L.; Louis, E.D.; Caccappolo, E.; Comella, C.; Colcher, A.; Siderowf, A.; Jennings, D.; et al. The relation between depression and parkin genotype: The CORE-PD study. *Park. Relat. Disord.* **2011**, *17*, 740–744. [CrossRef]

378. Farrer, M.; Chan, P.; Chen, R.; Tan, L.; Lincoln, S.; Hernandez, D.; Forno, L.; Gwinn-Hardy, K.; Petrucelli, L.; Hussey, J.; et al. Lewy bodies and parkinsonism in families with parkin mutations. *Ann. Neurol.* **2001**, *50*, 293–300. [CrossRef]
379. Oczkowska, A.; Kozubski, W.; Lianeri, M.; Dorszewska, J. Mutations in PRKN and SNCA Genes Important for the Progress of Parkinson's Disease. *Curr. Genom.* **2013**, *14*, 502–517. [CrossRef]
380. Imaizumi, Y.; Okada, Y.; Akamatsu, W.; Koike, M.; Kuzumaki, N.; Hayakawa, H.; Nihira, T.; Kobayashi, T.; Ohyama, M.; Sato, S.; et al. Mitochondrial dysfunction associated with increased oxidative stress and α-synuclein accumulation in PARK2 iPSC-derived neurons and postmortem brain tissue. *Mol. Brain* **2012**, *5*, 35. [CrossRef]
381. Bogetofte, H.; Jensen, P.; Ryding, M.; Schmidt, S.I.; Okarmus, J.; Ritter, L.; Worm, C.S.; Hohnholt, M.C.; Azevedo, C.; Roybon, L.; et al. PARK2 Mutation Causes Metabolic Disturbances and Impaired Survival of Human iPSC-Derived Neurons. *Front. Cell. Neurosci.* **2019**, *13*, 297. [CrossRef]
382. Shaltouki, A.; Sivapatham, R.; Pei, Y.; Gerencser, A.A.; Momcilovic, O.; Rao, M.S.; Zeng, X. Mitochondrial alterations by PARKIN in dopaminergic neurons using PARK2 patient-specific and PARK2 knockout isogenic iPSC lines. *Stem. Cell. Rep.* **2015**, *4*, 847–859. [CrossRef]
383. Jiang, H.; Ren, Y.; Yuen, E.Y.; Zhong, P.; Ghaedi, M.; Hu, Z.; Azabdaftari, G.; Nakaso, K.; Yan, Z.; Feng, J. Parkin controls dopamine utilization in human midbrain dopaminergic neurons derived from induced pluripotent stem cells. *Nat. Commun.* **2012**, *3*, 668. [CrossRef]
384. Yokota, M.; Kakuta, S.; Shiga, T.; Ishikawa, K.I.; Okano, H.; Hattori, N.; Akamatsu, W.; Koike, M. Establishment of an in vitro model for analyzing mitochondrial ultrastructure in PRKN-mutated patient iPSC-derived dopaminergic neurons. *Mol. Brain* **2021**, *14*, 58. [CrossRef] [PubMed]
385. Aboud, A.A.; Tidball, A.M.; Kumar, K.K.; Neely, M.D.; Han, B.; Ess, K.C.; Hong, C.C.; Erikson, K.M.; Hedera, P.; Bowman, A.B. PARK2 patient neuroprogenitors show increased mitochondrial sensitivity to copper. *Neurobiol. Dis.* **2015**, *73*, 204–212. [CrossRef] [PubMed]
386. Pickrell, A.M.; Youle, R.J. The roles of PINK1, parkin, and mitochondrial fidelity in Parkinson's disease. *Neuron* **2015**, *85*, 257–273. [CrossRef] [PubMed]
387. Scarffe, L.A.; Stevens, D.A.; Dawson, V.L.; Dawson, T.M. Parkin and PINK1: Much more than mitophagy. *Trends. Neurosci.* **2014**, *37*, 315–324. [CrossRef] [PubMed]
388. Grenier, K.; McLelland, G.L.; Fon, E.A. Parkin- and PINK1-Dependent Mitophagy in Neurons: Will the Real Pathway Please Stand Up? *Front. Neurol.* **2013**, *4*, 100. [CrossRef]
389. Sterky, F.H.; Lee, S.; Wibom, R.; Olson, L.; Larsson, N.G. Impaired mitochondrial transport and Parkin-independent degeneration of respiratory chain-deficient dopamine neurons in vivo. *Proc. Natl. Acad. Sci. USA* **2011**, *108*, 12937–12942. [CrossRef]
390. Kageyama, Y.; Hoshijima, M.; Seo, K.; Bedja, D.; Sysa-Shah, P.; Andrabi, S.A.; Chen, W.; Hoke, A.; Dawson, V.L.; Dawson, T.M.; et al. Parkin-independent mitophagy requires Drp1 and maintains the integrity of mammalian heart and brain. *EMBO J.* **2014**, *33*, 2798–2813. [CrossRef]
391. Zhong, P.; Hu, Z.; Jiang, H.; Yan, Z.; Feng, J. Dopamine Induces Oscillatory Activities in Human Midbrain Neurons with Parkin Mutations. *Cell. Rep.* **2017**, *19*, 1033–1044. [CrossRef]
392. Brittain, J.S.; Brown, P. Oscillations and the basal ganglia: Motor control and beyond. *Neuroimage* **2014**, *85*, 637–647. [CrossRef]
393. Ren, Y.; Jiang, H.; Hu, Z.; Fan, K.; Wang, J.; Janoschka, S.; Wang, X.; Ge, S.; Feng, J. Parkin mutations reduce the complexity of neuronal processes in iPSC-derived human neurons. *Stem. Cells* **2015**, *33*, 68–78. [CrossRef]
394. Cartelli, D.; Amadeo, A.; Calogero, A.M.; Casagrande, F.V.M.; de Gregorio, C.; Gioria, M.; Kuzumaki, N.; Costa, I.; Sassone, J.; Ciammola, A.; et al. Parkin absence accelerates microtubule aging in dopaminergic neurons. *Neurobiol. Aging* **2018**, *61*, 66–74. [CrossRef]
395. Vercammen, L.; van der Perren, A.; Vaudano, E.; Gijsbers, R.; Debyser, Z.; van den Haute, C.; Baekelandt, V. Parkin protects against neurotoxicity in the 6-hydroxydopamine rat model for Parkinson's disease. *Mol. Ther.* **2006**, *14*, 716–723. [CrossRef]
396. Yamada, M.; Mizuno, Y.; Mochizuki, H. Parkin gene therapy for α-synucleinopathy: A rat model of Parkinson's disease. *Hum. Gene Ther.* **2005**, *16*, 262–270. [CrossRef]
397. Lo Bianco, C.; Schneider, B.L.; Bauer, M.; Sajadi, A.; Brice, A.; Iwatsubo, T.; Aebischer, P. Lentiviral vector delivery of parkin prevents dopaminergic degeneration in an α-synuclein rat model of Parkinson's disease. *Proc. Natl. Acad. Sci. USA* **2004**, *101*, 17510–17515. [CrossRef]
398. Manfredsson, F.P.; Burger, C.; Sullivan, L.F.; Muzyczka, N.; Lewin, A.S.; Mandel, R.J. rAAV-mediated nigral human parkin over-expression partially ameliorates motor deficits via enhanced dopamine neurotransmission in a rat model of Parkinson's disease. *Exp. Neurol.* **2007**, *207*, 289–301. [CrossRef]
399. Goldberg, M.S.; Fleming, S.M.; Palacino, J.J.; Cepeda, C.; Lam, H.A.; Bhatnagar, A.; Meloni, E.G.; Wu, N.; Ackerson, L.C.; Klapstein, G.J.; et al. Parkin-deficient mice exhibit nigrostriatal deficits but not loss of dopaminergic neurons. *J. Biol. Chem.* **2003**, *278*, 43628–43635. [CrossRef] [PubMed]
400. Hennis, M.R.; Seamans, K.W.; Marvin, M.A.; Casey, B.H.; Goldberg, M.S. Behavioral and neurotransmitter abnormalities in mice deficient for Parkin, DJ-1 and superoxide dismutase. *PLoS ONE* **2013**, *8*, e84894. [CrossRef]
401. Hennis, M.R.; Marvin, M.A.; Taylor, C.M., 2nd; Goldberg, M.S. Surprising behavioral and neurochemical enhancements in mice with combined mutations linked to Parkinson's disease. *Neurobiol. Dis.* **2014**, *62*, 113–123. [CrossRef] [PubMed]

402. Palacino, J.J.; Sagi, D.; Goldberg, M.S.; Krauss, S.; Motz, C.; Wacker, M.; Klose, J.; Shen, J. Mitochondrial dysfunction and oxidative damage in parkin-deficient mice. *J. Biol. Chem.* **2004**, *279*, 18614–18622. [CrossRef] [PubMed]
403. Perez, F.A.; Palmiter, R.D. Parkin-deficient mice are not a robust model of parkinsonism. *Proc. Natl. Acad. Sci. USA* **2005**, *102*, 2174–2179. [CrossRef]
404. Von Coelln, R.; Thomas, B.; Savitt, J.M.; Lim, K.L.; Sasaki, M.; Hess, E.J.; Dawson, V.L.; Dawson, T.M. Loss of locus coeruleus neurons and reduced startle in parkin null mice. *Proc. Natl. Acad. Sci. USA* **2004**, *101*, 10744–10749. [CrossRef]
405. Zhu, X.R.; Maskri, L.; Herold, C.; Bader, V.; Stichel, C.C.; Gunturkun, O.; Lubbert, H. Non-motor behavioural impairments in parkin-deficient mice. *Eur. J. Neurosci.* **2007**, *26*, 1902–1911. [CrossRef]
406. Itier, J.M.; Ibanez, P.; Mena, M.A.; Abbas, N.; Cohen-Salmon, C.; Bohme, G.A.; Laville, M.; Pratt, J.; Corti, O.; Pradier, L.; et al. Parkin gene inactivation alters behaviour and dopamine neurotransmission in the mouse. *Hum. Mol. Genet.* **2003**, *12*, 2277–2291. [CrossRef]
407. Kitada, T.; Tong, Y.; Gautier, C.A.; Shen, J. Absence of nigral degeneration in aged parkin/DJ-1/PINK1 triple knockout mice. *J. Neurochem.* **2009**, *111*, 696–702. [CrossRef]
408. Doherty, K.M.; Hardy, J. Parkin disease and the Lewy body conundrum. *Mov. Disord.* **2013**, *28*, 702–704. [CrossRef]
409. Lu, X.H.; Fleming, S.M.; Meurers, B.; Ackerson, L.C.; Mortazavi, F.; Lo, V.; Hernandez, D.; Sulzer, D.; Jackson, G.R.; Maidment, N.T.; et al. Bacterial artificial chromosome transgenic mice expressing a truncated mutant parkin exhibit age-dependent hypokinetic motor deficits, dopaminergic neuron degeneration, and accumulation of proteinase K-resistant α-synuclein. *J. Neurosci.* **2009**, *29*, 1962–1976. [CrossRef]
410. Van Rompuy, A.S.; Lobbestael, E.; van der Perren, A.; van den Haute, C.; Baekelandt, V. Long-term overexpression of human wild-type and T240R mutant Parkin in rat substantia nigra induces progressive dopaminergic neurodegeneration. *J. Neuropathol. Exp. Neurol.* **2014**, *73*, 159–174. [CrossRef]
411. Wilson, M.A. The role of cysteine oxidation in DJ-1 function and dysfunction. *Antioxid. Redox. Signal* **2011**, *15*, 111–122. [CrossRef]
412. Bandopadhyay, R.; Kingsbury, A.E.; Cookson, M.R.; Reid, A.R.; Evans, I.M.; Hope, A.D.; Pittman, A.M.; Lashley, T.; Canet-Aviles, R.; Miller, D.W.; et al. The expression of DJ-1 (PARK7) in normal human CNS and idiopathic Parkinson's disease. *Brain* **2004**, *127*, 420–430. [CrossRef]
413. Rizzu, P.; Hinkle, D.A.; Zhukareva, V.; Bonifati, V.; Severijnen, L.A.; Martinez, D.; Ravid, R.; Kamphorst, W.; Eberwine, J.H.; Lee, V.M.; et al. DJ-1 colocalizes with tau inclusions: A link between parkinsonism and dementia. *Ann. Neurol.* **2004**, *55*, 113–118. [CrossRef] [PubMed]
414. Oh, S.E.; Mouradian, M.M. Regulation of Signal Transduction by DJ-1. *Adv. Exp. Med. Biol.* **2017**, *1037*, 97–131. [CrossRef] [PubMed]
415. Kahle, P.J.; Waak, J.; Gasser, T. DJ-1 and prevention of oxidative stress in Parkinson's disease and other age-related disorders. *Free Radic. Biol. Med.* **2009**, *47*, 1354–1361. [CrossRef] [PubMed]
416. Batelli, S.; Invernizzi, R.W.; Negro, A.; Calcagno, E.; Rodilossi, S.; Forloni, G.; Albani, D. The Parkinson's disease-related protein DJ-1 protects dopaminergic neurons in vivo and cultured cells from alpha-synuclein and 6-hydroxydopamine toxicity. *Neurodegener. Dis.* **2015**, *15*, 13–23. [CrossRef] [PubMed]
417. Kim, R.H.; Smith, P.D.; Aleyasin, H.; Hayley, S.; Mount, M.P.; Pownall, S.; Wakeham, A.; You-Ten, A.J.; Kalia, S.K.; Horne, P.; et al. Hypersensitivity of DJ-1-deficient mice to 1-methyl-4-phenyl-1,2,3,6-tetrahydropyrindine (MPTP) and oxidative stress. *Proc. Natl. Acad. Sci. USA* **2005**, *102*, 5215–5220. [CrossRef]
418. Meulener, M.; Whitworth, A.J.; Armstrong-Gold, C.E.; Rizzu, P.; Heutink, P.; Wes, P.D.; Pallanck, L.J.; Bonini, N.M. Drosophila DJ-1 mutants are selectively sensitive to environmental toxins associated with Parkinson's disease. *Curr. Biol.* **2005**, *15*, 1572–1577. [CrossRef]
419. Ottolini, D.; Cali, T.; Negro, A.; Brini, M. The Parkinson disease-related protein DJ-1 counteracts mitochondrial impairment induced by the tumour suppressor protein p53 by enhancing endoplasmic reticulum-mitochondria tethering. *Hum. Mol. Genet.* **2013**, *22*, 2152–2168. [CrossRef]
420. Taira, T.; Saito, Y.; Niki, T.; Iguchi-Ariga, S.M.; Takahashi, K.; Ariga, H. DJ-1 has a role in antioxidative stress to prevent cell death. *EMBO Rep.* **2004**, *5*, 213–218. [CrossRef]
421. Thomas, K.J.; McCoy, M.K.; Blackinton, J.; Beilina, A.; van der Brug, M.; Sandebring, A.; Miller, D.; Maric, D.; Cedazo-Minguez, A.; Cookson, M.R. DJ-1 acts in parallel to the PINK1/parkin pathway to control mitochondrial function and autophagy. *Hum. Mol. Genet.* **2011**, *20*, 40–50. [CrossRef]
422. Kinumi, T.; Kimata, J.; Taira, T.; Ariga, H.; Niki, E. Cysteine-106 of DJ-1 is the most sensitive cysteine residue to hydrogen peroxide-mediated oxidation in vivo in human umbilical vein endothelial cells. *Biochem. Biophys. Res. Commun.* **2004**, *317*, 722–728. [CrossRef]
423. Andres-Mateos, E.; Perier, C.; Zhang, L.; Blanchard-Fillion, B.; Greco, T.M.; Thomas, B.; Ko, H.S.; Sasaki, M.; Ischiropoulos, H.; Przedborski, S.; et al. DJ-1 gene deletion reveals that DJ-1 is an atypical peroxiredoxin-like peroxidase. *Proc. Natl. Acad. Sci. USA* **2007**, *104*, 14807–14812. [CrossRef] [PubMed]
424. Smith, N.; Wilson, M.A. Structural Biology of the DJ-1 Superfamily. *Adv. Exp. Med. Biol.* **2017**, *1037*, 5–24. [CrossRef] [PubMed]
425. Lev, N.; Barhum, Y.; Pilosof, N.S.; Ickowicz, D.; Cohen, H.Y.; Melamed, E.; Offen, D. DJ-1 protects against dopamine toxicity: Implications for Parkinson's disease and aging. *J. Gerontol. Biol. Sci. Med. Sci.* **2013**, *68*, 215–225. [CrossRef] [PubMed]

426. Shendelman, S.; Jonason, A.; Martinat, C.; Leete, T.; Abeliovich, A. DJ-1 is a redox-dependent molecular chaperone that inhibits α-synuclein aggregate formation. *PLoS Biol.* **2004**, *2*, e362. [CrossRef] [PubMed]
427. Atieh, T.B.; Roth, J.; Yang, X.; Hoop, C.L.; Baum, J. DJ-1 Acts as a Scavenger of α-Synuclein Oligomers and Restores Monomeric Glycated α-Synuclein. *Biomolecules* **2021**, *11*, 1466. [CrossRef] [PubMed]
428. Zondler, L.; Miller-Fleming, L.; Repici, M.; Goncalves, S.; Tenreiro, S.; Rosado-Ramos, R.; Betzer, C.; Straatman, K.R.; Jensen, P.H.; Giorgini, F.; et al. DJ-1 interactions with α-synuclein attenuate aggregation and cellular toxicity in models of Parkinson's disease. *Cell. Death Dis.* **2014**, *5*, e1350. [CrossRef]
429. Xu, C.Y.; Kang, W.Y.; Chen, Y.M.; Jiang, T.F.; Zhang, J.; Zhang, L.N.; Ding, J.Q.; Liu, J.; Chen, S.D. DJ-1 Inhibits α-Synuclein Aggregation by Regulating Chaperone-Mediated Autophagy. *Front. Aging Neurosci.* **2017**, *9*, 308. [CrossRef]
430. Lee, J.Y.; Song, J.; Kwon, K.; Jang, S.; Kim, C.; Baek, K.; Kim, J.; Park, C. Human DJ-1 and its homologs are novel glyoxalases. *Hum. Mol. Genet.* **2012**, *21*, 3215–3225. [CrossRef]
431. Bonifati, V.; Rizzu, P.; van Baren, M.J.; Schaap, O.; Breedveld, G.J.; Krieger, E.; Dekker, M.C.; Squitieri, F.; Ibanez, P.; Joosse, M.; et al. Mutations in the DJ-1 gene associated with autosomal recessive early-onset parkinsonism. *Science* **2003**, *299*, 256–259. [CrossRef]
432. Sironi, F.; Primignani, P.; Ricca, S.; Tunesi, S.; Zini, M.; Tesei, S.; Cilia, R.; Pezzoli, G.; Seia, M.; Goldwurm, S. DJ1 analysis in a large cohort of Italian early onset Parkinson Disease patients. *Neurosci. Lett.* **2013**, *557*, 165–170. [CrossRef]
433. Weissbach, A.; Wittke, C.; Kasten, M.; Klein, C. 'Atypical' Parkinson's disease—Genetic. *Int. Rev. Neurobiol.* **2019**, *149*, 207–235. [CrossRef]
434. Abou-Sleiman, P.M.; Healy, D.G.; Quinn, N.; Lees, A.J.; Wood, N.W. The role of pathogenic DJ-1 mutations in Parkinson's disease. *Ann. Neurol.* **2003**, *54*, 283–286. [CrossRef]
435. Hague, S.; Rogaeva, E.; Hernandez, D.; Gulick, C.; Singleton, A.; Hanson, M.; Johnson, J.; Weiser, R.; Gallardo, M.; Ravina, B.; et al. Early-onset Parkinson's disease caused by a compound heterozygous DJ-1 mutation. *Ann. Neurol.* **2003**, *54*, 271–274. [CrossRef]
436. Annesi, G.; Savettieri, G.; Pugliese, P.; D'Amelio, M.; Tarantino, P.; Ragonese, P.; La Bella, V.; Piccoli, T.; Civitelli, D.; Annesi, F.; et al. DJ-1 mutations and parkinsonism-dementia-amyotrophic lateral sclerosis complex. *Ann. Neurol.* **2005**, *58*, 803–807. [CrossRef]
437. Olzmann, J.A.; Brown, K.; Wilkinson, K.D.; Rees, H.D.; Huai, Q.; Ke, H.; Levey, A.I.; Li, L.; Chin, L.S. Familial Parkinson's disease-associated L166P mutation disrupts DJ-1 protein folding and function. *J. Biol. Chem.* **2004**, *279*, 8506–8515. [CrossRef]
438. Taipa, R.; Pereira, C.; Reis, I.; Alonso, I.; Bastos-Lima, A.; Melo-Pires, M.; Magalhaes, M. DJ-1 linked parkinsonism (PARK7) is associated with Lewy body pathology. *Brain* **2016**, *139*, 1680–1687. [CrossRef]
439. Kilarski, L.L.; Pearson, J.P.; Newsway, V.; Majounie, E.; Knipe, M.D.; Misbahuddin, A.; Chinnery, P.F.; Burn, D.J.; Clarke, C.E.; Marion, M.H.; et al. Systematic review and UK-based study of PARK2 (parkin), PINK1, PARK7 (DJ-1) and LRRK2 in early-onset Parkinson's disease. *Mov. Disord.* **2012**, *27*, 1522–1529. [CrossRef]
440. Kasten, M.; Hartmann, C.; Hampf, J.; Schaake, S.; Westenberger, A.; Vollstedt, E.J.; Balck, A.; Domingo, A.; Vulinovic, F.; Dulovic, M.; et al. Genotype-Phenotype Relations for the Parkinson's Disease Genes Parkin, PINK1, DJ1: MDSGene Systematic Review. *Mov. Disord.* **2018**, *33*, 730–741. [CrossRef]
441. Kumaran, R.; Vandrovcova, J.; Luk, C.; Sharma, S.; Renton, A.; Wood, N.W.; Hardy, J.A.; Lees, A.J.; Bandopadhyay, R. Differential DJ-1 gene expression in Parkinson's disease. *Neurobiol. Dis.* **2009**, *36*, 393–400. [CrossRef]
442. Choi, J.; Sullards, M.C.; Olzmann, J.A.; Rees, H.D.; Weintraub, S.T.; Bostwick, D.E.; Gearing, M.; Levey, A.I.; Chin, L.S.; Li, L. Oxidative damage of DJ-1 is linked to sporadic Parkinson and Alzheimer diseases. *J. Biol. Chem.* **2006**, *281*, 10816–10824. [CrossRef]
443. Tanudjojo, B.; Shaikh, S.S.; Fenyi, A.; Bousset, L.; Agarwal, D.; Marsh, J.; Zois, C.; Heman-Ackah, S.; Fischer, R.; Sims, D.; et al. Phenotypic manifestation of α-synuclein strains derived from Parkinson's disease and multiple system atrophy in human dopaminergic neurons. *Nat. Commun.* **2021**, *12*, 3817. [CrossRef] [PubMed]
444. Mencke, P.; Boussaad, I.; Onal, G.; Kievit, A.J.A.; Boon, A.J.W.; Mandemakers, W.; Bonifati, V.; Kruger, R. Generation and characterization of a genetic Parkinson's-disease-patient derived iPSC line DJ-1-delP (LCSBi008-A). *Stem. Cell. Res.* **2022**, *62*, 102792. [CrossRef] [PubMed]
445. Mazza, M.C.; Beilina, A.; Roosen, D.A.; Hauser, D.; Cookson, M.R. Generation of iPSC line from a Parkinson patient with PARK7 mutation and CRISPR-edited Gibco human episomal iPSC line to mimic PARK7 mutation. *Stem. Cell. Res.* **2021**, *55*, 102506. [CrossRef] [PubMed]
446. Li, Y.; Ibanez, D.P.; Fan, W.; Zhao, P.; Chen, S.; Md Abdul, M.; Jiang, Y.; Fu, L.; Luo, Z.; Liu, Z.; et al. Generation of an induced pluripotent stem cell line (GIBHi004-A) from a Parkinson's disease patient with mutant DJ-1/PARK7 (p.L10P). *Stem. Cell. Res.* **2020**, *46*, 101845. [CrossRef] [PubMed]
447. Chen, L.; Cagniard, B.; Mathews, T.; Jones, S.; Koh, H.C.; Ding, Y.; Carvey, P.M.; Ling, Z.; Kang, U.J.; Zhuang, X. Age-dependent motor deficits and dopaminergic dysfunction in DJ-1 null mice. *J. Biol. Chem.* **2005**, *280*, 21418–21426. [CrossRef] [PubMed]
448. Goldberg, M.S.; Pisani, A.; Haburcak, M.; Vortherms, T.A.; Kitada, T.; Costa, C.; Tong, Y.; Martella, G.; Tscherter, A.; Martins, A.; et al. Nigrostriatal dopaminergic deficits and hypokinesia caused by inactivation of the familial Parkinsonism-linked gene DJ-1. *Neuron* **2005**, *45*, 489–496. [CrossRef]
449. Pham, T.T.; Giesert, F.; Rothig, A.; Floss, T.; Kallnik, M.; Weindl, K.; Holter, S.M.; Ahting, U.; Prokisch, H.; Becker, L.; et al. DJ-1-deficient mice show less TH-positive neurons in the ventral tegmental area and exhibit non-motoric behavioural impairments. *Genes Brain Behav.* **2010**, *9*, 305–317. [CrossRef]

450. Raman, A.V.; Chou, V.P.; Atienza-Duyanen, J.; Di Monte, D.A.; Bellinger, F.P.; Manning-Bog, A.B. Evidence of oxidative stress in young and aged DJ-1-deficient mice. *FEBS Lett.* **2013**, *587*, 1562–1570. [CrossRef]
451. Rousseaux, M.W.; Marcogliese, P.C.; Qu, D.; Hewitt, S.J.; Seang, S.; Kim, R.H.; Slack, R.S.; Schlossmacher, M.G.; Lagace, D.C.; Mak, T.W.; et al. Progressive dopaminergic cell loss with unilateral-to-bilateral progression in a genetic model of Parkinson disease. *Proc. Natl. Acad. Sci. USA* **2012**, *109*, 15918–15923. [CrossRef]
452. Lin, Z.; Chen, C.; Yang, D.; Ding, J.; Wang, G.; Ren, H. DJ-1 inhibits microglial activation and protects dopaminergic neurons in vitro and in vivo through interacting with microglial p65. *Cell. Death Dis.* **2021**, *12*, 715. [CrossRef]
453. Stauch, K.L.; Villeneuve, L.M.; Purnell, P.R.; Pandey, S.; Guda, C.; Fox, H.S. SWATH-MS proteome profiling data comparison of DJ-1, Parkin, and PINK1 knockout rat striatal mitochondria. *Data Brief* **2016**, *9*, 589–593. [CrossRef]

Article

Relationship between Substantia Nigra Neuromelanin Imaging and Dual Alpha-Synuclein Labeling of Labial Minor in Salivary Glands in Isolated Rapid Eye Movement Sleep Behavior Disorder and Parkinson's Disease

Graziella Mangone [1,2,*], Marion Houot [1,3,4], Rahul Gaurav [1,5,6], Susana Boluda [1,7], Nadya Pyatigorskaya [1,5,6], Alizé Chalancon [1,2], Danielle Seilhean [1,7], Annick Prigent [8], Stéphane Lehéricy [1,5,6], Isabelle Arnulf [1,9], Jean-Christophe Corvol [1,2], Marie Vidailhet [1,2,6], Charles Duyckaerts [1,7] and Bertrand Degos [10,11]

1. Centre National de la Recherche Scientifique CNRS, Institut National de la Santé Et de la Recherche Médicale INSERM, Institut du Cerveau ICM, Sorbonne Université, 47-83 Bd de l'Hôpital, 75013 Paris, France
2. Clinical Investigation Center for Neurosciences, Department of Neurology, Assistance Publique Hôpitaux de Paris, Pitié-Salpêtrière Hospital, 47-83 Bd de l'Hôpital, 75013 Paris, France
3. Centre of Excellence of Neurodegenerative Disease (CoEN), AP-HP, Pitié-Salpêtrière Hospital, 47-83 Bd de l'Hôpital, 75013 Paris, France
4. Department of Neurology, Institute of Memory and Alzheimer's Disease (IM2A), AP-HP, Pitié-Salpêtrière Hospital, 47-83 Bd de l'Hôpital, 75013 Paris, France
5. Centre de NeuroImagerie de Recherche CENIR, Institut du Cerveau ICM, 47-83 Bd de l'Hôpital, 75013 Paris, France
6. Institut du Cerveau-ICM, Team "Movement Investigations and Therapeutics" (MOV'IT), 47-83 Bd de l'Hôpital, 75013 Paris, France
7. Department of Neuropathology, Assistance Publique Hôpitaux de Paris, Pitié-Salpêtrière Hospital, 47-83 Bd de l'Hôpital, 75013 Paris, France
8. HYSTOMICS Platform, Centre National de la Recherche Scientifique-CNRS, Institut National de la Santé Et de la Recherche Médicale-INSERM, Institut du Cerveau-ICM, Sorbonne Université, 47-83 Bd de l'Hôpital, 75013 Paris, France
9. Sleep Disorders Unit, Assistance Publique Hôpitaux de Paris, Pitié-Salpêtrière Hospital, 47-83 Bd de l'Hôpital, 75013 Paris, France
10. Neurology Unit, Avicenne University Hospital, Hôpitaux Universitaires de Paris-Seine Saint Denis, Sorbonne Paris Nord, AP-HP, NS-PARK/FCRIN Network, 125 Route de Stalingrad, 93009 Bobigny, France
11. Dynamics and Pathophysiology of Neuronal Networks Team, Center for Interdisciplinary Research in Biology, Collège de France, PSL University, 11 Place Marcelin Berthelot, 75005 Paris, France
* Correspondence: graziella.mangone@icm-institute.org; Tel.: +33-142165766; Fax: +33-142165767

Abstract: We investigated the presence of misfolded alpha-Synuclein (α-Syn) in minor salivary gland biopsies in relation to substantia nigra pars compacta (SNc) damage measured using magnetic resonance imaging in patients with isolated rapid eye movement sleep behavior disorder (iRBD) and Parkinson's disease (PD) as compared to healthy controls. Sixty-one participants (27 PD, 16 iRBD, and 18 controls) underwent a minor salivary gland biopsy and were scanned using a 3 Tesla MRI. Deposits of α-Syn were found in 15 (55.6%) PD, 7 (43.8%) iRBD, and 7 (38.9%) controls using the anti-aggregated α-Syn clone 5G4 antibody and in 4 (14.8%) PD, 3 (18.8%) iRBD and no control using the purified mouse anti-α-Syn clone 42 antibody. The SNc damages obtained using neuromelanin-sensitive imaging did not differ between the participants with versus without α-Syn deposits (irrespective of the antibodies and the disease group). Our study indicated that the α-Syn detection in minor salivary gland biopsies lacks sensitivity and specificity and does not correlate with the SNc damage, suggesting that it cannot be used as a predictive or effective biomarker for PD.

Keywords: Parkinson's disease; alpha-Synuclein; minor salivary gland biopsy; immunostaining; microscopy; neuromelanin

1. Introduction

Parkinson's disease (PD) is the second most common progressive neurodegenerative disorder after Alzheimer's disease, affecting 1–2% of individuals over 60 years of age, and increasing to 4–5% of the population by the age of 85 years [1]. According to the clinical criteria of the Movement Disorders Society (MDS), the diagnosis of PD is based on typical motor symptoms such as bradykinesia, rigidity, and tremor. However, non-motor symptoms such as hyposmia, isolated rapid eye movement sleep behavior disorder (iRBD), oral and gastrointestinal dysfunction, and depression can precede the onset of motor symptoms by a decade. The accuracy of the clinical diagnosis of PD is estimated between 46% and 90% [2] and depends on prolonged clinical observation and clinical response to Levodopa. Therefore, there is a critical need for reliable diagnostic biomarkers.

The pathological hallmark of PD is the degeneration of dopaminergic neurons in the substantia nigra pars compacta (SNc) and the presence of intra-neuronal inclusions (Lewy bodies) enriched in misfolded alpha-Synuclein (α-Syn) [3]. The accumulation of α-Syn is not limited to the brain but also occurs in the peripheral autonomic nervous system that innervates the skin, olfactory mucosa, gastrointestinal tract, salivary glands, retina, adrenal gland, and heart in PD patients [3].

We investigated whether minor salivary gland biopsies (MSGBs) could be used as an early predictive biomarker for the diagnosis of PD by investigating the presence of misfolded α-Syn immunostaining in patients with early idiopathic PD and iRBD, a prodromal form of α-synucleinopathies (PD, dementia with Lewy bodies, and multiple systemic atrophy) compared to HC. In addition, we investigated whether the presence of α-Syn immunostaining in MSGB was associated with more severe motor or cognitive disorders or with the severity of lesions in the SNc using MRI measures. We explored the neurodegenerative changes in the SNc using neuromelanin-sensitive MRI in PD [4] and iRBD [5].

2. Materials and Methods

2.1. Standard Protocol Approval, Patient Consent, Funding

The study was sponsored by INSERM, approved by French regulatory authorities and by the local Committee on Ethics and Human Research, and conducted in accordance with the Declaration of Helsinki. All patients and HCs provided written informed consent (CPP Paris VI, RCB 2014-A00725-42) for inclusion. All participants were fully informed of assured anonymity, why the research was being conducted, how their data would be used, and the associated risk. The study was funded by a grant from the French State "Investissements d'Avenir" (ANR-10-IAIHU-06). Additional funding was obtained from a grant from the France Parkinson Association and EDF foundation.

2.2. Subjects

Participants were included in the ICEBERG study (ClinicalTrials.gov Identifier: NCT02305147), an ongoing observational, prospective, monocentric, four-year natural history study including patients with PD, iRBD, and HCs conducted at the Paris Brain Institute (Institut du Cerveau—ICM, Pitié-Salpêtrière Hospital, Paris, France). PD patients were defined according to International Movement Disorders Society (MDS) criteria and had less than 4 years disease duration at inclusion. All patients with iRBD were defined by sleep neurologists (I.A. and S.L.S.) using polysomnography (i.e., tonic muscle atonia greater than twice the minimal level during >50% of epochs, during >18% of REM sleep, and/or behaviors on video during REM sleep) following the criteria of the International Classification of Sleep Disorders, third edition (ICSD-3, 2014) [6]. All subjects are comprehensively assessed at baseline and every year thereafter. Subjects underwent clinical (motor, neuropsychiatric, sleep, ocular, and cognitive evaluations) and imaging assessments. In all participants, clinical, MRI, MSGB, and polysomnography evaluations were performed during the inclusion visit. All participants signed informed consent.

2.3. Clinical Assessment

The (part I–IV) Movement Disorders Society Unified PD Rating Scale (MDS-UPDRS) [7] was carried out in all groups by movement disorder specialists (M.V., G.M., and J.-C.C.). In the PD patients, the motor examination was carried out in the OFF-drug conditions, after a 12 h interruption of the antiparkinsonian medication, and in the ON-drug conditions after the administration of a single suprathreshold dose of antiparkinsonian medications. Dopaminergic drugs and doses were converted to levodopa-equivalent daily doses (LEDD) [8]. Cognition was assessed using the Montreal Cognitive Assessment (MoCA) (range 0–30) [9].

2.4. MRI Data Acquisition and Analysis

All subjects were scanned on a 3 Tesla Prisma MRI (Siemens, Erlangen, Germany) using a 64-channel head coil for signal reception. The MRI protocol included a whole-brain T_1-weighted three-dimensional (3D) anatomical image acquired using magnetization-prepared two rapid gradient echoes (MP2RAGE) and a T_1-weighted two-dimensional (2D) turbo spin echo (TSE) protocol for neuromelanin-sensitive imaging [4]. For the TSE acquisition, the transverse slices were oriented perpendicular to the long axis of the brainstem, and the field of view included both the SNc and the locus coeruleus. Briefly, two experienced raters blind to the subject's clinical status manually segmented the SNc (S.L., R.G.). A background region comprising the tegmentum and superior cerebral peduncles was also manually traced in order to obtain the signal-to-noise ratio (SNR) by normalizing the mean signal in the SNc relative to the background signal as previously described [4]. A total of six scans were not analyzed either due to image artifacts (one PD patient) or the unavailability of MRI (three PD, one iRBD, and one HC). All analyses were performed using software programs written with an in-house algorithm in MATLAB 9.3 (version R2017b; MathWorks Inc, Natick, MA, USA), Statistical Parametric Mapping (SPM) (version SPM12, London, UK, http://www.fil.ion.ucl.ac.uk/spm, accessed on 20 September 2022), FreeSurfer (v5.3.0; MGH, Harvard, MA, USA, http://surfer.nmr.mgh.harvard.edu/, accessed on 20 September 2022), and FSL (Version 5.0; FMRIB Software Library, Oxford, UK).

2.5. Labial Salivary Gland Biopsy

After an anesthetic injection of lidocaine 2% into the inner side of the lower lip, a 0.5–1 cm horizontal incision was performed along the long axis of the lower lip mucosa just lateral to the midline while stretching the lip, allowing the removal of one–two accessory salivary glands with a scalpel, which were placed into a saline solution. Hemostasis was achieved by maintaining pressure with gauze. No antibiotics or pain medications were needed, and no adverse event was reported.

2.6. Immunohistochemistry

Small salivary gland biopsies were immersed in saline solution in a 4% buffered formalin solution for less than 72 h and embedded in paraffin. Serial 5 µm-thick sections were performed. To identify nerve fibers and α-Syn aggregates, immunohistochemistry was performed with the following antibodies: anti-phosphorylated high and medium molecular weight neurofilament (anti-NF) (clone SMI 310, mouse monoclonal, Biolegend, dilution 1/4000), anti-aggregated-α-Syn (clone 5G4, amino acids 46–53, mouse monoclonal, Merck, dilution 1/4000) [10], and anti-α-Syn (clone 42/Synuclein, amino acids 15–123, mouse monoclonal, BD Transduction Laboratories, dilution 1/4000) [11]. Immunohistochemistry was performed in a Nexes station automated system (Ventana Medical System, Inc., Roche, Basel, Switzerland). A pre-treatment was performed using a proprietary high pH buffer (pH8) (CC1). The detection of the anti-NF and the anti-α-Syn was performed with the UltraView Universal DAB Detection Kit (Ventana Medical System, Inc., Roche, Basel, Switzerland). The detection of the anti-aggregated-α-Syn was carried out with the OptiView DAB Immunohistochemistry Detection Kit (Ventana Medical System, Inc., Roche, Basel, Switzerland). The concordance between the two neuropathologists (C.D., S.B) who examined the slides blindly was 93%. The following variables were collected: lymphocytes infiltrate, α-Syn aggregates, and neurofilaments (Supplementary Figure S1).

2.7. Statistical Analyses

In order to compare the SNc measurements and the presence of α-Syn deposits between PD, iRBD, and HCs, a general linear model (GLM) adjusted for age and gender was performed, with identity link and normal distribution for the former and with logit link and Bernoulli distribution for the latter. Post hoc Tukey HSD tests were performed when the group effect was significant. Using Wilcoxon–Mann–Whitney tests, we conducted separate analyses in the PD, iRBD, and HC groups to compare demographics, clinical score, and SNc measurements between subjects who showed α-Syn deposits either using α-Syn clone 5G4 or clone 42 antibodies to those who did not. Differences were considered significant at $p < 0.05$. Statistical analyses were performed using R 4.1.2. (R Foundation for Statistical Computing, Vienna, Austria. https://www.R-project.org/, accessed on 20 September 2022).

3. Results

The study data were collected and managed using REDCap electronic data capture tools hosted at the ICM [12]. There was no age difference between the PD, iRBD, and HC groups, with a mean age of 64.5 ± 8.3 years old in the cohort (Table 1). There were more men in the iRBD group than in the PD and HC groups. The PD patients had a lower educational level than the iRBD patients. The PD patients had a mean age at disease onset of 61.5 ± 9 years old, disease duration of 21.7 ± 12.8 months, and took an LEDD of 342.3 ± 176.8 mg/d, including 77.4 ± 88.6 mg/d of dopamine agonists and 169.4 ± 178.7 mg/d of levodopa. Patients had iRBD for 88.3 ± 93.3 months and a mean age at iRBD onset of 59.6 ± 10.7 years old. The cognitive scores did not differ between the groups. The motor scores were worse in the PD than in the iRBD and HC groups. The non-motor symptom scores of PD and iRBD patients differed from those of HC and were similar between PD and iRBD for the Scales for Outcomes in Parkinson's disease—Autonomic (SCOPA-AUT), Non-Motor Symptoms Scale (NMSS), and the University of Pennsylvania Odor Identification Test (UPSIT)—and differed from the HC group as reported in Table 1. Deposits of α-Syn were detected in nerve fibers of the minor salivary glands in 15 (55.6%) PD patients, (43.8%) iRBD patients, and 7 (38.9%) HCs using the anti-aggregated α-Syn clone 5G4 antibody and in 4 (14.8%) PD patients, 3 (18.8%) iRBD patients, and no HC using the purified mouse anti-α-Syn, clone 42 antibody (Supplementary Figure S1). No difference was found between the three groups (α-Syn clone 5G4 antibody, $p = 0.49$; anti-α-Syn, clone 42 antibody $p = 0.12$). The SNRs were lower in the PD than in the other groups (mean difference estimate (MDE) $\pm$ standard error (SE), vs. controls: -1.90 ± 0.61, post hoc $p = 0.009$; vs. iRBD patients: -1.73 ± 0.68, post hoc $p = 0.037$) (Table 1). The volumes of the SNc were also lower in PD and iRBD than in the controls but the difference did not reach significance (Supplementary Figure S2). The comparison between subjects with α-Syn deposits and subjects without α-Syn deposits in each group did not differ for disease course, as well as clinical motor and non-motor symptoms scores, except for the olfaction scores, which were higher in PD patients with misfolded α-Syn deposits (Table 2). In addition, no difference in SNc volume and SNR were found in PD and iRBD, suggesting that there was no relationship between neuromelanin loss and α-Syn deposition as evaluated by MSGB (Table 2).

Table 1. Demographic, clinical, biopsies, and MRI measures in patients with Parkinson's disease (PD) and with isolated rapid eye movement sleep disorder (iRBD) as well as in healthy controls.

	PD (a)	iRBD (b)	Healthy Controls (c)	p ‡
Number	27	16	18	0.003 *
Demographics				
Gender (Male)	17 (63%) [b]	15 (94%) [a,c]	7 (39%) [b]	0.003 *
Age (years)	63.3 ± 8.9	68.1 ± 5.7	62.93 ± 9.0	0.124
Educational level	5.7 ± 1.6 [b]	6.8 ± 0.5 [a]	6.6 ± 0.8	0.024 *
Clinical data				
Age at onset (years)	61.5 ± 9.0	59.6 ± 10.7	NA	NA
Disease duration (months)	21.7 ± 12.8	88.3 ± 93.3	NA	NA
Dopaminergic agonists (mg)	77.4 ± 88.6	NA	NA	NA
Levodopa (mg)	169.4 ± 178.7	NA	NA	NA
LEDD (mg)	342.3 ± 176.8	NA	NA	NA
Clinical scores				
Montreal Cognitive Assessment ƒ	27.6 ± 2.1	27.8 ± 2.3	27.8 ± 1.5	0.986
MDS-UPDRS I	9.7 ± 3.7 [c]	7.50 ± 3.1 [c]	5.06 ± 2.7 [a,b]	<0.001 *
MDS-UPDRS II	8.5 ± 3.4 [b,c]	1.4 ± 1.1 [a]	0.8 ± 1.3 [a]	<0.001 *
MDS-UPDRS III Off Med	31.6 ± 9.2 [b,c]	9.9 ± 4.1 [a]	6.1 ± 7.1 [a]	<0.001 *
MDS-UPDRS III On Med	27.6 ± 7.3	NA	NA	NA
MDS-UPDRS IV	0.2 ± 0.8	0.00 ± 0.00	0.00 ± 0.00	0.365
Hoehn and Yahr	2.0 ± 0.2 [b,c]	0.4 ± 0.8 [a]	0.2 ± 0.7 [a]	<0.001 *
SCOPA-AUT	11.7 ± 6.3 [c]	10.0 ± 5.6	5.6 ± 4.2 [a]	0.002 *
NMSS scale	8.0 ± 3.9 [c]	8.9 ± 3.7 [c]	3.2 ± 3.1 [a,b]	<0.001 *
UPSIT	22.0 ± 5.3 [c]	19.6 ± 5.0 [c]	32.6 ± 6.4 [a,b]	<0.001 *

Table 1. Cont.

	PD (a)	iRBD (b)	Healthy Controls (c)	p ‡
α-Syn deposits, No with (%)				
α-Syn clone 5G4 antibodies	15 (55.6%)	7 (43.8%)	7 (38.9%)	0.491
α-Syn clone 42 antibodies	4 (14.8%)	3 (18.8%)	0 (0.00%)	0.115
either α-Syn clone 5G4 or 42 antibodies	17 (63.0%)	8 (50.0%)	7 (38.9%)	0.286
SNc measurements				
Volume (mm³)	234.4 ± 45.4	244.8 ± 18.4	257.7 ± 56.1	0.228
Signal-to-Noise Ratio (SNR)	109.8 ± 1.5 [b,c]	111.6 ± 2.1 [a]	111.7 ± 2.1 [a]	0.004 *

Data are given as mean ± standard deviation for continuous variables and as numbers (percentages) for categorical variables. * The level of significance was set at $p < 0.05$. ‡ For demographics, the Kruskal-Wallis test was used to compare the groups for continuous variables and Fisher's exact test for categorical variables. Post hoc comparisons were performed using pairwise Wilcoxon–Mann–Whitney tests for continuous variables and pairwise Fisher's exact tests for qualitative variables. Benjamini Hochberg correction was applied to correct for multiple testing. For clinical scores and SNc measurements, linear models adjusted for age and sex were performed. For α-Syn deposits, logistic regressions adjusted for age and sex were performed. Post hoc Tukey HSD tests were performed for pairwise comparison. ∫ Linear model with MoCA as the dependent variable was adjusted for age, sex, and educational level. [a] differs from the PD group; [b] differs from the iRBD group; [c] differs from controls. Abbreviations: LEDD Levodopa Equivalent Daily Dose; MoCA = Montreal Cognitive Assessment; MDS-UPDRS = Movement Disorders Society Unified Parkinson's Disease Rating Scale; SCOPA-AUT = Scales for Outcomes in Parkinson's disease—Autonomic; NMSS = Non-Motor Symptoms Scale; UPSIT University of Pennsylvania Smell Identification Test. Sociocultural level score ranges from 0–7. Cognition score ranges: MoCA total score 0–30; (high scores = better cognitive performances). Motor score ranges: MDS-UPDRS score ranges: MDS-UPDRS I 0–52, MDS-UPDRS II 0–52, MDS-UPDRS III 0–132, MDS-UPDRS IV 0–24 (high scores = worse clinical assessment); Hoehn and Yahr staging 0–5 (low scores = better clinical assessment); Non-motor score ranges: SCOPA-AUT 0–69; NMSS 0–30 (high scores = worse autonomic functions); UPSIT 0–40 (low scores = worse smell assessment).

Table 2. Comparison between subjects with α-Syn deposits and subjects without α-Syn deposits in each clinical group.

	PD			iRBD			Healthy Controls		
	With α-Syn Deposits	Without α-Syn Deposits	p ‡	With α-Syn Deposits	Without α-Syn Deposits	p ‡	With α-Syn Deposits	Without α-Syn Deposits	p ‡
Demography									
Number (%)	17 (63%)	10 (37%)		8 (50%)	8 (50%)		7 (39%)	11 (61.1%)	
Gender (Male)	10 (58.8%)	7 (70.0%)	0.69	8 (100.0%)	7 (87.5%)	1.000	3 (42.9%)	4 (36.4%)	1.000
Age (years)	64. ± 9.4	61.8 ± 8.4	0.547	69.1 ± 4.8	67.2 ± 6.6	0.600	59.8 ± 10.3	64.9 ± 7.9	0.342
Educational level	5.6 ± 1.8	5.9 ± 1.4	0.643	6.7 ± 0.7	6.9 ± 0.3	0.927	6.71 ± 0.49	6.5 ± 0.9	0.954
Clinical data									
Age at onset (years)	62.5 ± 9.4	59.7 ± 8.5	0.482	59.8 ± 13.0	59.4 ± 8.8	0.565	NA	NA	NA
Disease duration (months)	19.7 ± 12.6	24.9 ± 13.2	0.248	100.0 ± 118.8		NA	NA	NA	NA
Dopaminergic agonists	50.2 ± 88.7	118.1 ± 74.8	0.019 *	NA	NA	NA	NA	NA	NA
Levodopa	139.7 ± 180.1	219.9 ± 173.5	0.237	NA	NA	NA	NA	NA	NA
LEDD	291.8 ± 175.3	418.0 ± 157.8	0.211	NA	NA	NA	NA	NA	NA
Clinical scores									
MoCA (/30)	27.6 ± 2.2	27.5 ± 1.9	0.738	28.2 ± 1.6	27.2 ± 3.0	0.591	27.6 ± 1.7	28.0 ± 1.4	0.577
MDS-UPDRS I	9.6 ± 4.2	9.9 ± 2.7	0.631	6.2 ± 3.3	8.7 ± 2.4	0.064	5.6 ± 3.1	4.7 ± 2.4	0.291
MDS-UPDRS II	8.3 ± 3.6	8.8 ± 3.3	0.577	1.9 ± 1.2	0.9 ± 0.8	0.091	0.6 ± 1.0	0.9 ± 1.4	0.665
MDS-UPDRS III Off Med	30.8 ± 9.9	32.9 ± 8.2	0.258	8.0 ± 4.2	11.7 ± 3.1	0.071	2.57 ± 2.1	8.3 ± 8.4	0.102
MDS-UPDRS III On Med	27.2 ± 7.1	28.2 ± 8.0	0.731						
MDS-UPDRS IV	0.3 ± 1.0	0.0 ± 0.0	0.269	0.0 ± 0.0	0.0 ± 0.0		0.0 ± 0.0	0.0 ± 0.0	
Hoehn and Yahr	2.1 ± 0.2	2.0 ± 0.0	0.443	0.6 ± 0.9	0.2 ± 0.7	0.298	0.0 ± 0.0	0.3 ± 0.9	0.425
SCOPA-AUT	13.1 ± 7.1	9.4 ± 3.9	0.158	10.4 ± 5.4	9.9 ± 5.9	0.791	6.3 ± 5.4	5.2 ± 3.5	0.964
NMSS	8.1 ± 4.0	7.8 ± 3.8	0.899	8.1 ± 3.5	9.7 ± 4.0	0.337	3.3 ± 3.4	3.2 ± 3.0	0.927
UPSIT	24.3 ± 4.3	18.2 ± 4.7	0.003 *	19.1 ± 2.8	20.1 ± 6.7	0.833	34.0 ± 3.5	31.7 ± 7.8	0.728
SNc measurements									
Volume (mm³)	242.3 ± 47.6	224.0 ± 42.5	0.535	247.1 ± 19.54	242.99 ± 18.64	0.699	255.4 ± 53.9	259.6 ± 61.01	0.958
Signal-to-Noise Ratio (SNR)	109.9 ± 1.6	109.7 ± 1.5	0.770	112.2 ± 2.0	110.9 ± 2.1	0.418	111.4 ± 1.3	111.9 ± 2.6	0.380

Notes. Data are given as mean ± standard deviation for continuous variables and as count (percentages) for categorical variables. * The level of significance was set at $p < 0.05$. ‡ Wilcoxon–Mann–Whitney tests were used to compare groups for continuous variables and Fisher's exact test for categorical variables. Abbreviations: LEDD Levodopa Equivalent Daily Dose; SCOPA-AUT = Scales for Outcomes in Parkinson's disease—Autonomic; MoCA = Montreal Cognitive Assessment; MDS-UPDRS = Movement Disorders Society Unified Parkinson's Disease Rating Scale; NMSS = Non-Motor Symptoms Scale; UPSIT = University of Pennsylvania Smell Identification Test. Sociocultural level score ranges from 0–7. Cognition score ranges: MoCA total score 0–30; (high scores = better cognitive performances). Motor score ranges: MDS-UPDRS score ranges: MDS-UPDRS I 0–52, MDS-UPDRS II 0–52, MDS-UPDRS III 0–132, MDS-UPDRS IV 0–24 (high scores = worse clinical assessment); Hoehn and Yahr staging 0–5 (low scores = better clinical assessment); Non-motor score ranges: SCOPA-AUT 0–69; NMSS 0–30 (high scores = worse autonomic functions); UPSIT 0–40 (low scores = worse smell assessment).

4. Discussion

In this study, misfolded α-Syn deposits were found in the salivary glands of two-thirds of the PD patients, half of the patients with iRBD, but surprisingly also in more than one-third of HC, which is higher than in most previous studies [13,14], indicating that these measures lack specificity at this age. These cases may be considered "incidental or preclinical cases" given that post-mortem studies in the brain have shown Lewy body pathology in about 10–20% of people over the age of 60 without parkinsonism or dementia during their lifetime [15].

There were no differences for any SNc measures between subjects with vs. without α-Syn deposits, regardless of the antibody used for the clinical group. The volumes of the SNc were also lower in PD and iRBD than in HCs but the difference did not reach significance. These results indicate an absence of a link between neuromelanin loss in the SNc and α-Syn deposits in the salivary glands.

Although the procedure was safe, our results suggested that MSGB lacks sufficient accuracy to detect α-Syn deposits in salivary glands in PD and in iRBD, thus it cannot be considered a useful and relevant biomarker of synucleinopathy. The sensitivity and specificity of the measures of α-Syn deposits in the MSGB were disappointing, as reported in unlike previous studies.

The abnormal accumulation of α-Syn around gland cells was reported in minor salivary glands in PD patients but in none of the HCs [16,17]. Another study found an abnormal accumulation of α-Syn in 3 out of 16 PD patients, and 2 out of 11 HCs exhibited weak phosph-α-Syn [13]. In another study, the ratio of nerve fibers immunoreactive to α-Syn was slightly decreased in seven PD patients as compared to seven HCs [18]. Conversely, Ser129-phosphorylated-α-Synuclein immunoreactive nerve fibers were identified in five of seven PD cases but no HCs [18]. Phosph-α-Syn was detected in 31 of 62 patients with iRBD, in 7 of 13 patients with PD, in 5 of 10 patients with dementia with Lewy bodies, and in 1 of the 33 HCs [14].

The differences between our results and those of previous studies using MSGB are likely due to different methods used to search peripheral Lewy bodies, including biological specimen collection techniques, biopsy locations, histological methods, observation, and experience criteria.

Many studies have shown that submandibular gland biopsy is more sensitive and specific than the MSGB in the diagnosis of PD. α-Syn aggregates were detected in the nerve fibers of the glandular parenchyma in 8 (89%) of 9 patients with iRBD and 8 (67%) of 12 with PD, but none of the HCs [19]. In an autopsy-based study of submandibular glands, Lewy pathology was present in all PD patients (nine of nine cases) and incidental Lewy body disease (two of three cases) but not in multiple system atrophy or HC [11]. Using sections of large segments (simulating open biopsy) and needle cores of submandibular glands from 128 subjects, immunoreactive phosph-α-Syn-positive nerve fibers were present in all 28 PD patients and three patients with Alzheimer's disease and Lewy bodies, but none in HCs. Cores from frozen submandibular glands were positive for phosph-α-Syn in 17 of 19 PD patients [20].

Based on these results, needle core biopsy of submandibular glands and MSGB was performed in patients with PD showing 9/12 biopsies positive for phosph-α-Syn (75%) while only 1/15 MSGB were positive (6.7%) [21]. Subsequently, a study of submandibular gland needle biopsies from patients with early PD (<5 years of disease duration) showed positive staining in 14 of 19 patients (74%) and only 2 of 9 HCs (22%) [22].

Another factor of variability comes from the wide range of available antibodies that have been used in the different studies depending on the divergent fixation, epitope exposure, and signal development methods [23]. It is noteworthy that some studies looked for α-Syn and others for phosph-α-Syn. However, it became clear that peripheral α-Syn is detectable even in HCs. Therefore, previous studies suggested using antibodies against phosph-α-Syn with proteinase K pretreatment, since phosph-α-Syn is expressed at very low levels in normal controls, and proteinase K is able to digest normal α-Syn. They concluded that phosph-α-Syn is the best hallmark for α-Synucleinopathy [20].

Baseline reductions in neuromelanin-based SN volume and signal intensity were observed in early PD (although not significant), as reported previously in de novo patients [24] and early PD [25] and in line with histological studies [26]. In previous studies, volume reductions correlated positively with disease duration [27,28]. iRBD patients also displayed lower neuromelanin volume and signal intensity than healthy controls [4,5]. The lack of significance in PD and iRBD was probably due to the small number of subjects [4].

5. Limitation of Study

Our study has several limitations including a small sample size of each subgroup and a small number of labial salivary glands analyzed. The short disease duration of PD and the realization of the MSGBs at the time of the inclusion visit are limiting factors in interpreting our results. It would have been interesting to perform again MSGBs at a later stage to investigate whether the presence of peripheral α-Syn changed over time. The choice of antibodies could also explain the differences observed between studies. For our study, we performed immunohistochemistry with two -Syn antibodies: (a) clone 42/-synuclein [11], which is a highly sensitive antibody and has been previously described to detect -Syn pathological aggregates in the submaxillary gland and peripheral nervous system, (b) 5G4 clone [10] which also has high sensitivity and detects the oligomeric and fibrillar forms of -Syn but not the normal soluble monomeric form of the protein. The latter was employed in substitution of the phosph-α-Syn, expecting to observe a higher number of cases with pathology while preserving the specificity of the detection of the pathological aggregates.

6. Conclusions

In recent years, a major research effort has been made to find biomarkers that would allow an accurate and early diagnosis of synucleinopathies. Recent immunohistochemistry studies have demonstrated, with varying success, that accumulation of α-Syn also occurred in the peripheral autonomic nervous system that innervates the skin, olfactory mucosa, gastrointestinal tract, retina, adrenal gland, and heart in PD [3] as well as in human fluid (saliva, red blood cells, and cerebrospinal fluid) in iRBD patients [29,30]. Therefore, further explorations aimed at studying the pathological protein by non-immuno-histochemical techniques and characterizing the inflammatory infiltrate in peripheral tissues from multiple organs in combination with human fluid are needed. Novel methods for pathological α-Syn detection in human tissues, including real-time quaking-induced conversion (RT-QuIC) and protein misfolding cyclic amplification (PMCA), seem to have higher diagnostic sensitivity.

Supplementary Materials: The following supporting information can be downloaded at: https://www.mdpi.com/article/10.3390/genes13101715/s1, Figure S1: Immunohistochemistry of minor salivary glands in a PD patient. Figure S2: Manual tracing of the SNc in neuromelanin-sensitive images: neuromelanin images of a representative PD (left column), an iRBD patient (middle column) and an HC (right column).

Author Contributions: Conceptualization: G.M., M.V., C.D. and B.D.; Data curation: G.M. and A.C.; Formal analysis: M.H., R.G., S.B., N.P. and C.D.; Funding acquisition: M.V. and B.D.; Investigation: G.M., S.L., I.A., J.-C.C., M.V. and B.D.; Methodology: M.H. and B.D.; Project administration: A.C., M.V. and B.D.; Resources: G.M., R.G., S.B., N.P., A.C., D.S., A.P., S.L., I.A., J.-C.C., M.V., C.D. and B.D.; Software: M.H., R.G. and N.P.; Supervision: G.M., A.C., S.L., I.A., J.-C.C., M.V., C.D. and B.D.; Validation: G.M., M.H., R.G., S.B., N.P., A.C., D.S., S.L., I.A., J.-C.C., M.V., C.D. and B.D.; Visualization: G.M., M.H., R.G., S.B., N.P., A.C., D.S., S.L., I.A., J.-C.C., M.V., C.D. and B.D.; Writing—original draft: G.M.; Writing—review and editing: G.M., M.H., R.G., S.B., N.P., A.C., D.S., A.P., S.L., I.A., J.-C.C., M.V., C.D. and B.D. All authors have read and agreed to the published version of the manuscript.

Funding: The research leading to these results received funding from a grant from the France Parkinson Association, from the French State "Investissements d'Avenir" (ANR-10-IAIHU-06), and from the EDF foundation.

Institutional Review Board Statement: The local ethics committee approved this study (CPP Paris VI, RCB: 2014-A00725-42).

Informed Consent Statement: Informed consent was obtained from all subjects involved in the study.

Data Availability Statement: Due to its proprietary nature, supporting data cannot be made openly available. For further information about the data and conditions for access, you can contact by email the Principal Investigator of ICEBERG study (Pr Marie Vidailhet: marie.vidailhet@aphp.fr).

Acknowledgments: The authors gratefully acknowledge all patients who participated in this study, the Clinical Investigation Center for Neurosciences at the ICM, the France Parkinson Association, and the EDF Foundation.

Conflicts of Interest: G.M. declares no conflict of interest. M.H. declares no conflict of interest. R.G. has received research grants from BIOGEN INC, USA outside this work. Susana Boluda declares no conflict of interest. N.P. received honoraria for lecturing for GE Healthcare outside this work. Alizé Chalancon declares no conflict of interest. D.S. declares no conflict of interest. A.P. declares no conflict of interest. S.L. has received research grants from BIOGEN INC, USA outside this work. I.A. has received an ICRIN grant from the Paris Brain Institute outside this work. J.-C.C. has received research grants from Sanofi and the Michael Joe Fox Foundation outside this work; scientific advisory board or speaker fees from Air Liquide, Biogen, Biophytis, Denali, Ever Pharma, Idorsia, Prevail Therapeutics, Theranexus, and UCB. M.V. declares no conflict of interest. C.D. declares no conflict of interest. B.D. has received research support from Orkyn, Merz-Pharma, and a grant from Contrat de Recherche Clinique 2021 (CRC 2021) outside this work.

References

1. Thomas, B.; Beal, M.F. Parkinson's disease. *Hum. Mol. Genet.* **2007**, *6*, R183-94. [CrossRef] [PubMed]
2. Hughes, A.J.; Daniel, S.E.; Kilford, L.; Lees, A.J. Accuracy of clinical diagnosis of idiopathic Parkinson's disease: A clinico-pathological study of 100 cases. *J. Neurol. Neurosurg. Psychiatry* **1992**, *55*, 181–184. [CrossRef] [PubMed]
3. Goedert, M.; Spillantini, M.G.; Del Tredici, K.; Braak, H. 100 years of Lewy pathology. *Nat. Rev. Neurol.* **2012**, *9*, 13–24. [CrossRef] [PubMed]
4. Gaurav, R.; Yahia-Cherif, L.; Pyatigorskaya, N.; Mangone, G.; Biondetti, E.; Valabrègue, R.; Ewenczyk, C.; Hutchison, R.M.; Cedarbaum, J.M.; Corvol, J.; et al. Longitudinal Changes in Neuromelanin MRI Signal in Parkinson's Disease: A Progression Marker. *Mov. Disord.* **2021**, *36*, 1592–1602. [CrossRef]
5. Pyatigorskaya, N.; Gaurav, R.; Arnaldi, D.; Leu-Semenescu, S.; Yahia-Cherif, L.; Valabrègue, R.; Vidailhet, M.; Arnulf, I.; Lehéricy, S. Magnetic resonance Imaging Biomarkers to assess Substantia Nigra Damage in Idiopathic Rapid Eye Movement Sleep Be-havior Disorders. *Sleep* **2017**, *40*, 1–8. [CrossRef]
6. Sateja, M.J. International classification of sleep disorders-third edition: Highlights and modifications. *Chest* **2014**, *146*, 1387–1394. [CrossRef]
7. Goetz, C.G.; Tilley, B.C.; Shaftman, S.R.; Stebbins, G.T.; Fahn, S.; Martinez-Martin, P.; Poewe, W.; Sampaio, C.; Stern, M.B.; Dodel, R.; et al. Movement Disorder Society-sponsored revision of the Unified Parkinson's Disease Rating Scale (MDS-UPDRS): Scale presentation and clinimetric testing results. *Mov. Disord.* **2008**, *23*, 2129–2170. [CrossRef]
8. Tomlinson, C.L.; Stowe, R.; Patel, S.; Rick, C.; Gray, R.; Clarke, C.E. Systematic review of levodopa dose equivalency reporting in Parkinson's disease. *Mov. Disord.* **2010**, *25*, 2649–2653. [CrossRef]
9. Nasreddine, Z.S.; Phillips, N.A.; Bédirian, V.; Charbonneau, S.; Whitehead, V.; Collin, I.; Cummings, J.L.; Chertkow, H. The Montreal Cognitive Assessment, MoCA: A brief screening tool for mild cognitive impairment. *J. Am. Geriatr. Soc.* **2005**, *53*, 695–699. [CrossRef]
10. Kovacs, G.G.; Wagner, U.; Dumont, B.; Pikkarainen, M.; Osman, A.A.; Streichenberger, N.; Leisser, I.; Verchère, J.; Baron, T.; Alafuzoff, I.; et al. An antibody with high reactivity for disease-associated α-synuclein reveals extensive brain pathology. *Acta Neuropathol.* **2012**, *124*, 37–50. [CrossRef]
11. Del Tredici, K.; Hawkes, C.H.; Ghebremedhin, E.; Braak, H. Lewy pathology in the submandibular gland of individuals with incidental Lewy body disease and sporadic Parkinson's disease. *Acta Neuropathol.* **2010**, *119*, 703–713. [CrossRef] [PubMed]
12. Harris, P.A.; Taylor, R.; Thielke, R.; Payne, J.; Gonzalez, N.; Conde, J.G. Research electronic data capture (REDCap)—A metadata-driven methodology and workflow process for providing translational research informatics support. *J. Biomed. Inform.* **2009**, *42*, 377–381. [CrossRef] [PubMed]
13. Folgoas, E.; Lebouvier, T.; Leclair-Visonneau, L.; Cersosimo, M.-G.; Barthelaix, A.; Derkinderen, P.; Letournel, F. Diagnostic value of minor salivary glands biopsy for the detection of Lewy pathology. *Neurosci. Lett.* **2013**, *551*, 62–64. [CrossRef] [PubMed]
14. Iranzo, A.; Borrego, S.; Vilaseca, I.; Marti, C.; Serradell, M.; Sanchez-Valle, R.; Kovacs, G.G.; Valldeoriola, F.; Gaig, C.; Santamaria, J.; et al. α-Synuclein aggregates in labial salivary glands of idiopathic rapid eye movement sleep behavior dis-order. *Sleep* **2018**, *41*, zsy101. [CrossRef]

15. Gibb, W.R.; Lees, A.J. The relevance of the Lewy body to the pathogenesis of idiopathic Parkinson's disease. *J. Neurol. Neurosurg. Psychiatry* **1988**, *51*, 745–752. [CrossRef]
16. Cersosimo, M.G.; Perandones, C.; Micheli, F.E.; Raina, G.B.; Beron, A.M.; Nasswetter, G.; Radrizzani, M.; Benarroch, E.E. Al-pha-synuclein immunoreactivity in minor salivary gland biopsies of Parkinson's disease patients. *Mov. Disord.* **2011**, *26*, 188–190. [CrossRef] [PubMed]
17. Gao, L.; Chen, H.; Li, X.; Li, F.; Ou-Yang, Q.; Feng, T. The diagnostic value of minor salivary gland biopsy in clinically diagnosed patients with Parkinson's disease: Comparison with DAT PET scans. *Neurol. Sci.* **2015**, *36*, 1575–1580. [CrossRef]
18. Carletti, R.; Campo, F.; Fusconi, M.; Pellicano, C.; De Vincentiis, M.; Pontieri, F.E.; Di Gioia, C.R. Phosphorylated α-synuclein immunoreactivity in nerve fibers from minor salivary glands in Parkinson's disease. *Park. Relat. Disord.* **2017**, *38*, 99–101. [CrossRef]
19. Vilas, D.; Iranzo, A.; Tolosa, E.; Aldecoa, I.; Berenguer, J.; Vilaseca, I.; Martí, C.; Serradell, M.; Lomeña, F.; Alós, L.; et al. Assessment of α-synuclein in submandibular glands of patients with idiopathic rapid-eye movement sleep behaviour disorder: A case-control study. *Lancet Neurol.* **2016**, *15*, 708–718. [CrossRef]
20. Beach, T.G.; Adler, C.H.; Dugger, B.N.; Serrano, G.; Hidalgo, J.; Henry-Watson, J.; Shill, H.A.; Sue, L.I.; Sabbagh, M.N.; Akiyama, H.; et al. Submandibular Gland Biopsy for the Diagnosis of Parkinson Disease. *J. Neuropathol. Exp. Neurol.* **2013**, *72*, 130–136. [CrossRef]
21. Adler, C.H.; Dugger, B.N.; Hinni, M.L.; Lott, D.G.; Driver-Dunckley, E.; Hidalgo, J.; Henry-Watson, J.; Serrano, G.; Sue, L.I.; Nagel, T.; et al. Submandibular gland needle biopsy for the diagnosis of Parkinson disease. *Neurology* **2014**, *82*, 858–864. [CrossRef] [PubMed]
22. Adler, C.H.; Dugger, B.N.; Hentz, J.G.; Hinni, M.L.; Lott, D.G.; Driver-Dunckley, E.; Mehta, S.; Serrano, G.; Sue, L.I.; Duffy, A.; et al. Peripheral synucleinopathy in early Parkinson's disease: Subman-dibular gland needle biopsy findings. *Mov. Disord.* **2016**, *31*, 250–256. [CrossRef] [PubMed]
23. Schneider, S.A.; Boettner, M.; Alexoudi, A.; Zorenkov, D.; Deuschl, G.; Wedel, T. Can we use peripheral tissue biopsies to diagnose Parkinson's disease? A review of the literature. *Eur. J. Neurol.* **2015**, *23*, 247–261. [CrossRef] [PubMed]
24. Wang, J.; Li, Y.; Huang, Z.; Wan, W.; Zhang, Y.; Wang, C.; Cheng, X.; Ye, F.; Liu, K.; Fei, G.; et al. Neuro-melanin-sensitive magnetic resonance imaging features of the substantia nigra and locus coeruleus in de novo Parkinson's disease and its phenotypes. *Eur. J. Neurol.* **2018**, *25*, 949–955. [CrossRef]
25. Reimão, S.; Ferreira, S.; Nunes, R.; Lobo, P.P.; Neutel, D.; Abreu, D.; Gonçalves, N.; Campos, J.; Ferreira, J.J. Magnetic resonance correlation of iron content with neuromelanin in the substantia nigra of early-stage Parkinson's disease. *Eur. J. Neurol.* **2015**, *23*, 368–374. [CrossRef] [PubMed]
26. Fearnley, J.M.; Lees, A.J. Ageing and Parkinson's disease: Substantia nigra regional selectivity. *Brain* **1991**, *114 Pt 5*, 2283–2301. [CrossRef]
27. Pyatigorskaya, N.; Magnin, B.; Mongin, M.; Yahia-Cherif, L.; Valabregue, R.; Arnaldi, D.; Ewenczyk, C.; Poupon, C.; Vidailhet, M.; Lehéricy, S. Comparative Study of MRI Biomarkers in the Substantia Nigra to Discriminate Idiopathic Parkinson Disease. *Am. J. Neuroradiol.* **2018**, *39*, 1460–1467. [CrossRef]
28. Schwarz, S.T.; Rittman, T.; Gontu, V.; Morgan, P.S.; Bajaj, N.; Auer, D.P. T1-Weighted MRI shows stage-dependent substantia nigra signal loss in Parkinson's disease. *Mov. Disord.* **2011**, *26*, 1633–1638. [CrossRef]
29. Miglis, M.; Adler, C.H.; Antelmi, E.; Arnaldi, D.; Baldelli, L.; Boeve, B.F.; Cesari, M.; Dall'Antonia, I.; Diederich, N.J.; Doppler, K.; et al. Biomarkers of conversion to α-synucleinopathy in isolated rapid-eye-movement sleep behavior disorder. *Lancet Neurol.* **2020**, *20*, 671–684. [CrossRef]
30. Zitser, J.; Gibbons, C.; Miglis, M.G. The role of tissue biopsy as a biomarker in REM sleep behavior disorder. *Sleep Med. Rev.* **2020**, *51*, 101283. [CrossRef]

Article

A Machine Learning Approach to Parkinson's Disease Blood Transcriptomics

Ester Pantaleo [1,2,3,†], Alfonso Monaco [1,†], Nicola Amoroso [1,4], Angela Lombardi [1,3,*], Loredana Bellantuono [1,2], Daniele Urso [5,6], Claudio Lo Giudice [7], Ernesto Picardi [7,8], Benedetta Tafuri [5], Salvatore Nigro [5,9], Graziano Pesole [7,8], Sabina Tangaro [1,10], Giancarlo Logroscino [2,5,‡] and Roberto Bellotti [1,3,‡]

1. Istituto Nazionale di Fisica Nucleare (INFN), Sezione di Bari, Via A. Orabona 4, 70125 Bari, Italy; ester.pantaleo@uniba.it (E.P.); alfonso.monaco@ba.infn.it (A.M.); nicola.amoroso@uniba.it (N.A.); loredana.bellantuono@uniba.it (L.B.); sabina.tangaro@uniba.it (S.T.); roberto.bellotti@uniba.it (R.B.)
2. Dipartimento di Scienze Mediche di Base, Neuroscienze e Organi di Senso, Università degli Studi di Bari Aldo Moro, Piazza G. Cesare 11, 70124 Bari, Italy; giancarlo.logroscino@uniba.it
3. Dipartimento Interateneo di Fisica M. Merlin, Università degli Studi di Bari Aldo Moro, Via G. Amendola 173, 70125 Bari, Italy
4. Dipartimento di Farmacia-Scienze del Farmaco, Università degli Studi di Bari Aldo Moro, Via A. Orabona 4, 70125 Bari, Italy
5. Centro per le Malattie Neurodegenerative e l'Invecchiamento Cerebrale, Dipartimento di Ricerca Clinica in Neurologia, Università degli Studi di Bari Aldo Moro, Pia Fondazione Cardinale G. Panico, 73039 Tricase, Italy; daniele.urso@kcl.ac.uk (D.U.); benedetta.tafuri@gmail.com (B.T.); salvatoreangelo.nigro@gmail.com (S.N.)
6. Institute of Psychiatry, Psychology and Neuroscience, King's College London, De Crespigny Park, London SE5 8AF, UK
7. Dipartimento di Bioscienze, Biotecnologie e Biofarmaceutica, Università degli Studi di Bari Aldo Moro, Via A. Orabona 4, 70125 Bari, Italy; claudio.logiudice@uniba.it (C.L.G.); ernesto.picardi@uniba.it (E.P.); graziano.pesole@uniba.it (G.P.)
8. Istituto di Biomembrane, Bioenergetica e Biotecnologie Molecolari, Consiglio Nazionale delle Ricerche, Via G. Amendola 122/O, 70126 Bari, Italy
9. Istituto di Nanotecnologia (NANOTEC), Consiglio Nazionale delle Ricerche, Via Monteroni, 73100 Lecce, Italy
10. Dipartimento di Scienze del Suolo, della Pianta e degli Alimenti, Università degli Studi di Bari Aldo Moro, Via A. Orabona 4, 70125 Bari, Italy
* Correspondence: angela.lombardi@uniba.it
† These authors contributed equally to this work.
‡ These authors contributed equally to this work.

Abstract: The increased incidence and the significant health burden associated with Parkinson's disease (PD) have stimulated substantial research efforts towards the identification of effective treatments and diagnostic procedures. Despite technological advancements, a cure is still not available and PD is often diagnosed a long time after onset when irreversible damage has already occurred. Blood transcriptomics represents a potentially disruptive technology for the early diagnosis of PD. We used transcriptome data from the PPMI study, a large cohort study with early PD subjects and age matched controls (HC), to perform the classification of PD vs. HC in around 550 samples. Using a nested feature selection procedure based on Random Forests and XGBoost we reached an AUC of 72% and found 493 candidate genes. We further discussed the importance of the selected genes through a functional analysis based on GOs and KEGG pathways.

Keywords: blood transcriptomics; Parkinson's disease; machine learning; xgboost; feature selection; oxidative stress; inflammation; mitochondrial dysfunction

1. Introduction

Parkinson's disease (PD) is a chronic, degenerative disease of the central nervous system with a pattern of incidence that increases with age; as the population ages, its

burden is poised to increase [1]. Despite considerable research efforts, PD is incurable; available treatments can only help manage the symptoms, and its diagnosis often occurs a long time after onset after substantial loss of function of substantia nigra dopamine neurons [2].

Massively parallel analysis of cellular RNAs can provide an unbiased set of biomarkers of PD and can generate hypotheses about disease mechanisms. It may be particularly useful for decoding a disease with considerable environmental and epigenetic contributions not readily explained by variations in the genomic fingerprint such as PD [3]. Brain transcriptomics has already shown its potential to uncover the functional mechanisms at the basis of this disease although its signal is confounded by underlying differences in cell type composition and it can only be performed after death [4]. Whole blood transcriptomics represents a convenient and less invasive alternative to brain transcriptomics for early PD diagnosis, as blood is a readily accessible peripheral biofluid and blood and brain share significant transcriptional profile similarities [5,6] although more investigations are needed in this field. A number of experimental observations have shown molecular and biochemical changes in the blood cells of PD subjects [7,8] and RNA-sequencing experiments on blood leukocytes have revealed the diagnostic potential of long non coding RNAs (lncRNAs) [9]. Some studies have identified biomarkers from blood that are robust and have great potential for helping reduce misdiagnosis [10–12].

As high throughput technologies such as transcriptome sequencing can now generate huge amounts of biological data at relatively low costs, the processing and extraction of relevant signal requires the adoption of artificial intelligence methodologies. A number of Machine Learning (ML) approaches have been undertaken for PD classification that use as input vocal and gait [13] or neuroimaging [14] features, or genetic risk scores from Genome Wide Association Studies (GWAS) studies [15] and microarray transcriptional profiles [16,17]. We used advanced Machine Learning techniques for feature selection and classification of early (drug-naive) PD patients and healthy controls (HCs) using gene expression data from blood RNA sequencing.

For blood transcriptomics, experience suggests that large cohorts are needed, and that drug-naive patients should be used, as medications certainly affect gene expression [18]. Microarray assays for whole blood transcriptomics have been used to classify early stage drug-naive PDs vs. HCs [19,20] with a small number of samples (less than 50 PD subjects), while previous experiments with a large number of samples used PD subjects on dopaminergic medication [17].

Given the importance of using large cohorts of drug-naive patients, we used open access gene expression data from the Parkinson's Progression Markers Initiative (PPMI), an international study that has enrolled the largest to date cohort of untreated PD patients (around 430 subjects) across multiple sites (www.ppmi-info.org/data accessed on 11 March 2022) [21].

2. Materials and Methods

2.1. PPMI Data

We downloaded PPMI whole blood transcriptome data from the LONI Image and Data Archive (IDA) (data dowloaded in July 2021). From the available set of sequenced samples, we selected 579 samples collected from different individuals, namely 390 subjects in the early PD cohort and 189 age-matched subjects in the HC cohort. Therefore the dataset consisted of twice as many PD cases as HCs. Each sample had expression values (read counts) for a total of around 60,000 transcripts. The early PD cohort included subjects with PD that were not treated with dopaminergic medications, that were not carriers of 'LRRK2', 'GBA' or 'SNCA' mutations, and that did not have a first relative with one or more mutations. Sequence data had been aligned to GRCh37(hs37d5) by STAR (v2.4K) [22] using exon-exon junctions from GENCODE v19 and gene count data had been obtained via featureCounts [23] by the same GENCODE annotations. Samples that failed quality control were excluded [24].

Subject metadata that we downloaded from the PPMI website included biological variables such as age, sex, clinical site and clinical measures of motor symptoms such as indicators of tremor dominant (TD) or postural instability gait difficulty (PIGD), of non motor symptoms such as categorical REM sleep behavior disorder (RBD), of cognitive impairment (CI) such as the Montreal Cognitive Assessment or MoCA index (adjusted for education), and of olfactory function (UPSIT or University of Pennsylvania Smell Identification Test score). Additional metadata included technical variables such as, for instance, RIN (RNA integrity number), percent usable bases, total number of reads, sequencing plate. Table 1 reports some statistics on the metadata.

Table 1. Relevant clinical, pathological and technical metadata of the cohort divided by disease status.

Variable	PD	HC
Gender (male %)	252/390 (64%)	123/189 (65%)
Age at enrollment	62 ± 10	61 ± 11
Disease duration	2 ± 2	-
RBD	37%	20%
TD	70%	13%
Number of sites	25	23
MoCA ≤ 26 (CI-adjusted for education)	33%	0.5%
RIN	8 ± 1.7	8 ± 1.7

For up-to-date information on the study and for access to the data, visit www.ppmi-info.org accessed on 11 March 2022.

2.2. Overview of the Methodology

Our computational workflow consists of three main phases: (i) a first preprocessing phase, which was essential to manage the informative content of highly heterogeneous and computationally demanding data such as transcriptomes; (ii) a second learning phase, which exploited a feature importance evaluation embedded in a Random Forest (RF) classification procedure [25,26] and whose best features were used to feed an eXtreme Gradient Boosting (XGBoost) algorithm [27]; (iii) finally, an unbiased evaluation of classification performances and of the set of important features obtained through a nested cross-validation scheme. A schematic overview of our workflow is presented in Figure 1. A detailed description of the previously mentioned processing steps is presented in the following methodological subsections.

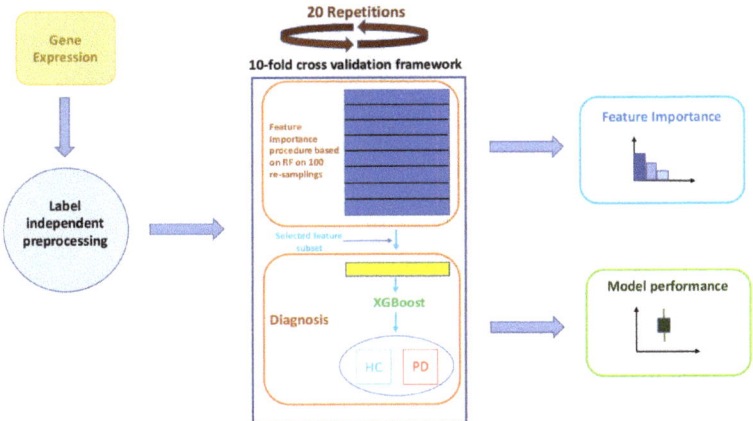

Figure 1. Schematic workflow of the performed analyses. The main phases are: (i) preprocessing, (ii) learning and (iii) performance evaluation.

For all analyses we used R version 4.0.3, packages xgboost v1.6.0.1, caret v6.0-90, and Bioconductor packages DESeq2 v1.30.1, limma v4.46.0, enrichR v3.0, AnnotationDbi v1.52.0, and org.Hs.eg.db v3.12.0. The code used to conduct this research is available upon request.

2.3. Empowering Informative Content of Gene Expression Values

The first phase of our workflow consists of multiple preprocessing procedures. This phase is essential given the large number of features, namely gene expression values based on the GENCODE v19 comprehensive annotation. A number of label independent filtering steps, where the labels are "PD" and "HC", were required to extract informative content.

First, we selected only transcripts corresponding to protein coding genes and long intergenic non-coding RNAs (lincRNAs). Second, we discarded 2667 transcripts driving technical variance [24], which left us with 18,727 protein coding genes and 7444 lincRNAs. Third, we removed lowly expressed genes, by keeping only genes that had more than five counts in at least 10% of the individuals, which left us with 21,273 genes. Fourth, we estimated size factors, normalized the library size bias using these factors, performed independent filtering to remove lowly expressed genes using the mean of normalized counts as a filter statistic. This left us with 12,612 genes. Finally, we applied a variance stabilizing transformation to accommodate the problem of unequal variance across the range of mean values. We used DESeq2 to perform theses steps [28].

Afterwards, we used control samples to estimate the batch effect of the site, that we subsequently removed in both controls and cases [29] using limma [30]. To perform this step we removed subjects from sites with no control samples or with only one control sample, i.e., sites "14" (1 sample), "26" (16 samples), "55" (4 samples), and "59" (10 samples), see Figure 2.

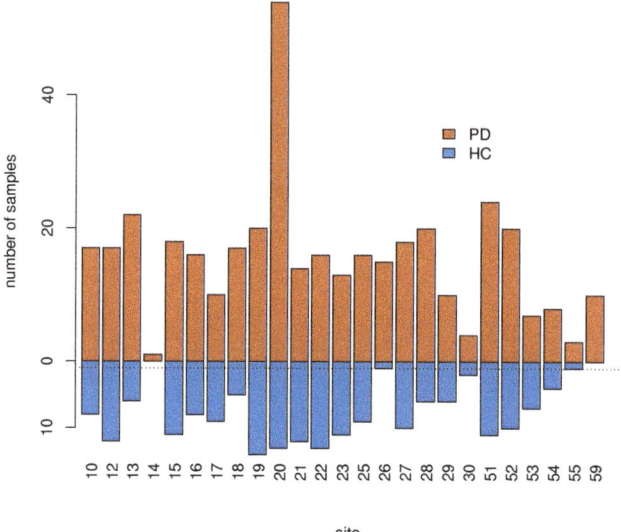

Figure 2. Samples were collected across 25 different sites labeled with an integer number. Sites "14", "26", "55", and "59" had 0 or 1 control sample only (horizontal dotted line) and were excluded from the classification analysis as batch effects due to site could not be estimated and therefore corrected for.

After this step, we were left with a total of 548 samples. Then we removed further confounding effects due to sex and RIN value, again with limma. Thus, for the subsequent analyses we considered a database including 548 subjects described by 12,612 genes.

2.4. Differential Expression Analysis

Before moving to the second phase of our workflow, namely the learning phase (see next section), we performed differential expression (DE) analysis, which is a classical and univariate approach towards the identification of biomarkers from RNASeq data. We will also test the performance of our ML approach (XGBoost) when we used as input the set of DE genes obtained with DESeq2 instead of the set of genes selected with RF. In the discussion we will contrast the results of this univariate approach with results from our machine learning multivariate approach. For DE analysis we used DESeq2 [28], a popular tool. As it is standard procedure, we used as input to the algorithm counts prior to independent filtering, batch correction and variance stabilization and defined a design matrix with four variables: the normalized RIN value, factor site, factor gender and the disease label. For the comparison between PD and HC, DESeq2 returns a positive fold change value to indicate an increase of expression of a gene in PD subjects vs. HC, and a negative fold change to indicate a decrease in expression. It also uses a shrinkage procedure to combine information from multiple genes, but its approach is univariate as it tests each gene individually for DE using a beta binomial generalized linear model. DESeq2 corrects for multiple testing using a Benjamini–Hochberg adjusted p-value. Genes with adjusted p-value < 0.05 are called significantly differentially expressed in the two classes. We will evaluate the fold change of genes with its associated error and adjusted p-value and compare results with a multivariate analysis that uses Machine Learning algorithms.

2.5. A Robust Learning Scheme

After performing DE analysis, we moved to the second phase of our workflow, the learning phase.

Our filtering procedure described in Section 2.3 had already significantly reduced the amount of gene expression to consider. Nonetheless, we designed and implemented an additional feature selection procedure (nested within the learning phase) to further reduce the number of genes with the two-fold goal of enhancing classification performances and optimizing model interpretability.

Within a repeated stratified (to tackle the control-patient mismatch) 10-fold cross-validation framework (20 iterations), we trained multiple RF models (100 repetitions, where each repetition used a different seed of the random generation process) to evaluate permutation feature importance measures. We chose RF for two main reasons: on the one hand, RF is easy to tune as it only depends on two parameters, namely the number of trees to grow and the number of features randomly selected at each split; on the other hand, RF is an extremely efficient algorithm on high dimensional data. Each forest was grown using 1000 trees, a sufficient value to allow the algorithm to reach a stable plateau of the out-of-bag internal error. The features selected at each split were $\sqrt{f}$ with f being the overall number of genes, which is the default value for this parameter. As already mentioned, another important advantage of the RF classifier is its embedded feature importance evaluation; during the training phase, the algorithm can assess how much each feature decreases the impurity of a tree, or the likelihood of incorrect classification of a new instance of a random variable and then can make an average over all trees [26]. Using this embedded feature importance procedure, we determined the overall feature importance ranking by averaging over the 100 repetitions. Then, a subset of size C of the most important features was used to train an XGBoost model; the XGBoost classification performance was evaluated on the validation set, for the twenty 10-fold cross validation iterations, in order to obtain an unbiased performance evaluation. As with RF, the XGBoost algorithm belongs to the set of learning approaches called *ensemble*, which combines and manages the predictions of several weak models to obtain a more robust model. While RF relies on bagging (Bootstrap aggregation), XGBoost exploits the Gradient Boosting framework. In the Gradient Boosting method, new models are applied to predicting residuals or errors of previous models and then added together to obtain the final predictive model. This approach implements a gradient descent algorithm to minimize the loss when including new models [27].

Overall, our procedure is very robust because, in addition to the high number of iterations implemented, we also use two different classification algorithms in the training and test phases which makes the results independent from the model. Then, to compare the performance of the ML approach to the performance of a simpler XGBoost classification algorithm that uses as input features the set of DE genes obtained with DESeq2, we trained the algorithm on 90% of the data and tested it on 10%.

Finally, we tested if the predicted probability of the algorithm was different between PD subjects with different endo-phenotypes: (i) MoCA $\leq$ 26 and MoCA higher than 26; (ii) PDs with RBD an PDs without RBD; (iii) PDs with TD and PDs with PIGD or undetermined; (iv) PDs with Normosmia and PDs with Hyposmia or Anosmia; (v) PD subjects belonging to different age categories, namely age $\geq$ 56 and age $<$ 56.

2.6. Performance Evaluation

The last phase of our workflow is performance evaluation. A binary classification problem has only two class labels; therefore, the resulting model decisions can fall into four categories: true positives (TP) when the model correctly predicts the positive class, erroneous positive predictions (false positives, FP) and, analogously, true negatives (TN) and false negatives (FN).

Given these four cases, one can define several metrics; in particular, we considered here [31]:

- Accuracy
$$\frac{TP + TN}{TP + TN + FP + FN};$$

- Sensitivity
$$\frac{TP}{TP + FN};$$

- Specificity
$$\frac{TN}{TN + FP};$$

- Balanced Accuracy
$$\frac{Sensitivity + Specificity}{2};$$

- F1
$$\frac{2TP}{2TP + FP + FN};$$

- Area Under the Receiver Operating Characteristics (ROC) Curve (AUC), which plots sensitivity against specificity by varying the decision threshold.

Sensitivity and specificity evaluate how well the model performs on the positive and the negative class, respectively. The other metrics provide an overall performance evaluation. Although these "overall" metrics are roughly equivalent, their values can ease the comparison of our results with the state-of-the-art.

3. Results

3.1. Evaluating the Informative Content of Transcriptomic Data

The first research question addressed by this work concerned the evaluation of the informative content provided by blood transcriptomic data. We first assessed the informative content through a univariate DE analysis and we found a total of 1368 up-regulated genes and 911 down-regulated genes with an adjusted p-value less than 0.05. Of the DE genes, however, only one gene, namely 'RAP1GAP', had a log fold change (lfc) higher than 0.5 in absolute value (lfc = -0.65 ± 0.15, adjusted p-value$\sim 10^{-5}$). In general, the DE signal, except for this gene, was very low.

We then evaluated the informative content of blood transcriptomic data using a multivariate ML procedure and the classification AUC as a performance measure, see Figure 3.

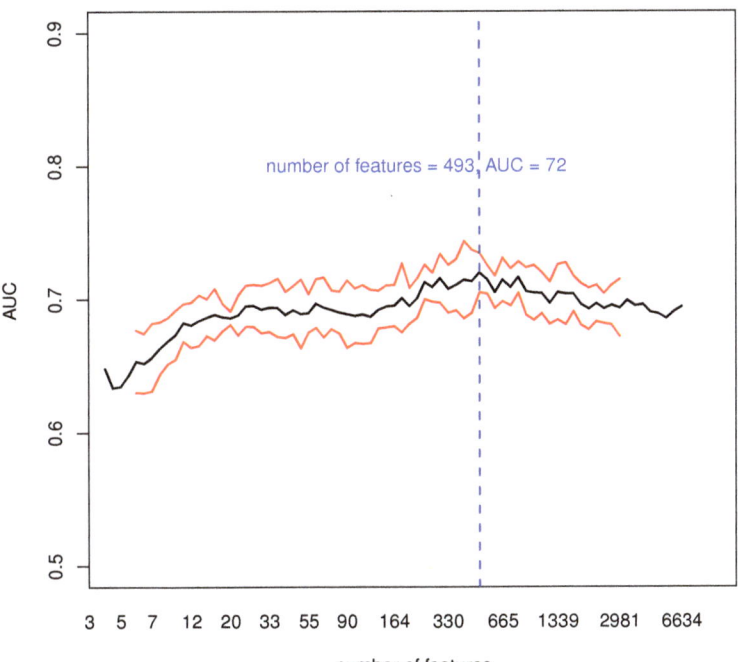

Figure 3. In black, the median AUC over 20 runs of 10-fold cross validation; in red, the median AUC ± its mean absolute deviation; in blue, the number of features (genes) where the maximum median AUC (72%) was reached. For each run, we collected the AUC values obtained at different thresholds C (or equivalently a different number of genes) and we interpolated these values to build a curve. Then we obtained the black curve as the median of 20 curves, one for each 10-fold Cross-Validation (CV) run.

Figure 3 shows the cross-validation median AUC with its mean absolute deviation for a different number C of input features. The maximum median AUC of 72% with a mean absolute deviation of 1.5% is reached with a number of input features equal to $C = 493$. Despite classification results should depend on the number of features (genes) used to learn the model, this analysis shows that over an extremely broad range of features the informative content remains stable and accurate. For what concerns the other classification metrics obtained using the previously mentioned 493 features, a detailed overview is presented in Table 2.

Table 2. Average performances of XGBoost over 20 runs of 10-fold cross validation.

	Mean	Standard Deviation
AUC	71.3	1.2
Accuracy	69.3	1.2
Sensitivity	81.7	1.6
Specificity	45.5	2.3
Balanced Accuracy	63.6	1.3
F1	77.8	0.9

The model is generally accurate, as shown by "global" metrics (AUC, F1, accuracy); it is worth noting the performance drop revealed by the balanced accuracy, which reflects the data imbalance. The same consideration holds for the performance gap in terms of sensitivity and specificity.

We tested the performance of an XGBoost classification algorithm that used as input features the set of DE genes obtained with DESeq2. We obtained an AUC of 64%, which is considerably lower than the performance of our ML approach based on RF and XGBoost, which proves how a multivariate ML model can be more effective on this type of data compared to classical DE approaches.

A final note on the performance of the algorithm with respect to PD endo-phenotypes. The predicted probability of the algorithm was higher for PD subjects in different age categories: the algorithm had an average predicted probability higher for PD individuals with age ≥ 56 (p-value 0.004, Wilcoxon test, average predicted probability = 0.77 for age < 56 and 0.84 for age ≥ 56), while there was no significant difference between PD subjects belonging to the other considered endo-phenotypic classes.

3.2. Evaluating Gene Importance

As the RF feature importance procedure in principle returns a different feature ranking at each iteration (both because of the different cross-validation splits and intrinsic RF variability), we designed an experiment to investigate which were the most important genes for classification. Provided that the highest performance value was obtained with 493 features, within the cross-validation scheme, we evaluated the probability that an input feature (gene) is one of the top 493 genes, see Figure 4.

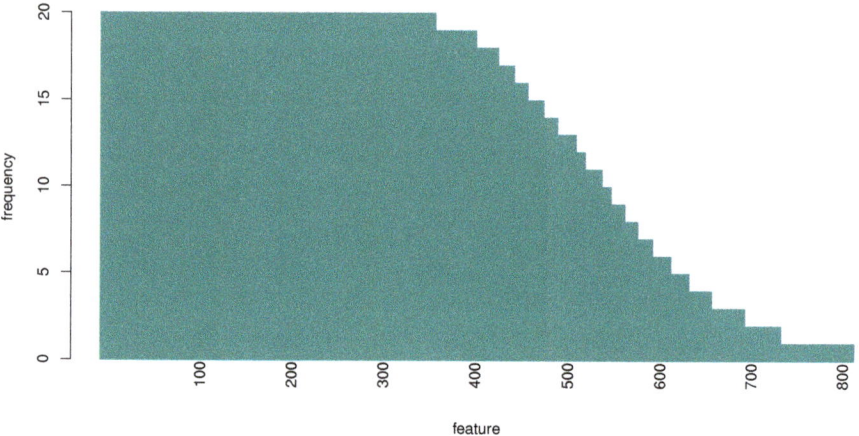

Figure 4. Histogram of the frequency of occurrence of the top 493 genes over 20 repetitions. At each repetition we collected the 493 most important genes; over 20 repetitions we gathered in total around 800 genes, many of which (365) appeared in all 20 repetitions.

Among the most frequently selected genes, the 20 most important genes (according to the average importance ranking) are listed in Table 3; a list with the genes that have been selected in at least 70% of the iterations is presented in Table A1. This list includes 434 protein coding genes and 61 are lincRNAs (lincRNAs are marked with an asterisk).

Table 3. List of the 20 most important protein coding genes and lincRNAs, ordered by importance. LincRNAs are marked with an asterisk. For each gene, four attributes are listed: (i) Up-arrow/Down arrow: significant over/under-expression in PD subjects compared to HC; (ii) HGCN HUGO Gene Nomenclature Committee symbol (or Ensembl ID when missing); (iii) Average XGBoost importance over 20 runs of 5-fold cross validation; (iv) Number of times that a gene is selected over 20 runs of feature selection. For a more complete list, including the genes that are selected 70% of the times, see Table A1.

e	Symbol	imp	f	e	Symbol	imp	f
↑	MYOM1	82.1	20	↑	SLC25A20	62.7	20
	NRM	46.4	20	↓	PHF7	45.9	20
↑	ENSG00000277763 *	39.4	20		ICA1	36	20
↑	CPT1A	33.8	20		LINC02422 *	33.3	20
	GSTM1	32.4	20		PCDHGA6	31.6	20
	AK5	31.5	20	↓	GCNT2	29.9	20
	CERS4	29.7	20	↓	YJU2	29.4	20
	SURF6	27.7	20		ENSG00000281181 *	26.7	20
	ENSG00000285774 *	26.7	20	↑	ENSG00000272688 *	26.2	20
	SERF1B	25.8	20		ENSG00000284773	25.7	20

LincRNAs are marked with an asterisk.

3.3. Gene Set Enrichment Analysis

We performed KEGG (Kyoto Encyclopedia of Genes and Genomes) pathway and GO (Gene Ontology) functional annotation enrichment analysis with respect to biological processes, cellular components and molecular functions using enrichR [32] on the list of most frequent genes (Table A1). Figures 5–7 report all the resulting significant groups at a False Discovery Rate (FDR) < 0.05; no GO molecular function was significant.

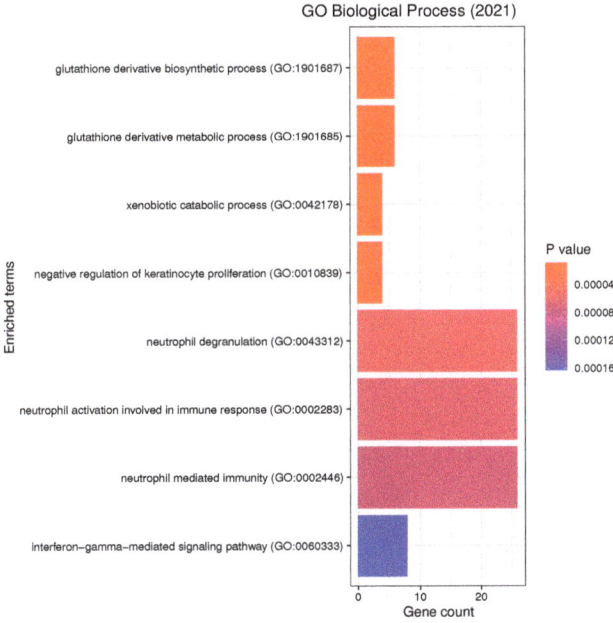

Figure 5. List of all the GO Biological Processes that are enriched in the selected genes, with the respective number of genes belonging to each term. The analysis was performed with enrichR at an FDR < 0.05.

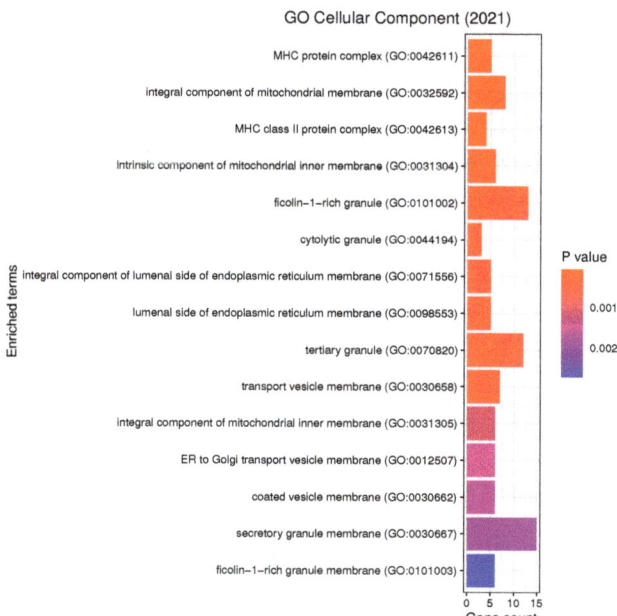

Figure 6. List of all the GO Cellular Components that are enriched in the selected genes with the respective number of genes belonging to each term. The analysis was performed with enrichR at an FDR < 0.05.

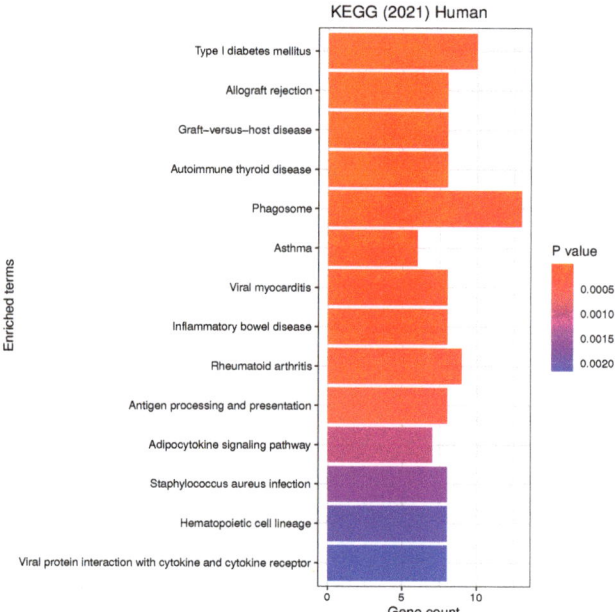

Figure 7. List of all the KEGG pathways that are enriched in the selected genes with the respective number of genes belonging to each term. The analysis was performed with enrichR at an FDR < 0.05.

4. Discussion

4.1. A Robust Machine Learning Model

With the robust methodology we implemented, we identified a set of around 500 genes that could discriminate between PD and HC with an AUC of 72%. Over 20 runs of cross validation (Figure 3) the AUC had a slightly increasing pattern for increasing values of C, and reached a maximum at a number of features C = 493, then slowly decreased. This behavior showed that the informative content of the selected genes was stable and accurate. While there was an imbalance between sensitivity and specificity, it was moderate and, if needed, this discrepancy could be mitigated with additional under-sampling of over-sampling techniques that could be embedded in the described methodology.

Comparing our performance with the state-of-the-art is not straightforward because of the nature of the data and because of ongoing research in the area. A comparable study on a large cohort of 523 individuals performed on blood microarray gene expression data and using Support Vector Machines reports an AUC of 79% on the validation set and of 74% on the test set [17]. However in that study PD subjects with a positive family history were not excluded and most importantly PD patients were treated with dopaminergic medication. Dopaminergic medication alters gene expression and thus confounds the underlying signal: higher discriminative performances are to be expected but are misleading.

A multivariate study not yet published [33] and performed on the same PPMI cohort as ours used a multi-modal approach that combines the informative content of transcriptomics, clinico-demographic data, genome sequencing data, and poligenic risk scores (PRS). To compare our results with theirs, we considered their transcriptomics-only model. They used Support Vector Machines (but they tested and tuned 12 different ML algorithms) and divided the PPMI cohort at baseline into a training (70%) and a validation set (30%), then they tested the resulting model on independent data from the PDBP (Parkinson's Disease Biomarkers Program) cohort, although performances of the uni-modal model were not reported for this test set. After careful preprocessing, where they used limma to adjust for additional covariates of sex, plate, age, ten principal components, and percentage usable bases and then normalized counts, they used significantly over- or under-expressed protein coding genes as determined through logistic regression (p-value < 0.01) on the training set as input features to a Support Vector Machine classification algorithm. Using only transcriptome data they reached an AUC of 79.73% on the validation set, 73.89% accuracy, 54.60% balanced accuracy, 97% sensitivity, and 12% specificity. When they combined transcriptomics with the other multi-modal data using a union of the features as input features; after tuning, they reached an AUC of 85.03%, 75% accuracy, 68.09% balanced accuracy, 93% sensitivity and 43% specificity on the independent test set and determined, by comparing the relative importances of the input features, that the UPSIT score, as well as PRS, contributed most to the predictive power of the model, but the accuracy of these were supplemented by many smaller effect transcripts and risk SNPs.

The strength of our work is its high balanced accuracy in delineating cases and controls and its robustness. Our feature selection procedure identified a robust set of around 500 genes listed in Table A1 that may have some impact on PD biology.

4.2. Candidate Genes, GOs and KEGG Pathways

Accurate characterization of the selected genes and of significantly enriched gene sets is beyond the scope of this paper; however, we report a few comments on the enrichment analysis and a few notes on the selected genes.

Our analyses revealed a number of significant functions and pathways, some of which have already been linked to the pathogenesis of PD, such as oxidative stress, inflammation, mitochondrial and vesicular dysfunction, as well as associations between PD and diseases such as diabetes mellitus or inflammatory bowel disease (IBD) (see Figures 5–7). *Oxidative stress* plays an important role in the degeneration of dopaminergic neurons [34]; its involvement in PD is further substantiated by Reactive Oxygen Species (ROS) induced Parkinsonian models and elevated oxidative markers in clinical PD samples [35]. Glutathione (GSH) is a ubiquitous thiol tripeptide that protects against oxidative stress-induced damage by neutralizing reactive oxygen species; its deficiency has been identified as an early event in the progression of PD [36]. *Inflammation* is another important contributor to the pathogenesis of the disease [37]. Interestingly, our GO analysis has identified biological processes that involve neutrophils. A very recent meta-analysis studying the association between the neutrophil-to-lymphocyte ratio (NLR), a well-established indicator of the overall inflammatory status of the organism, and clinical characteristics in PD has demonstrated that PD patients have an altered peripheral immune profile [38]. Neuronal expression of major histocompatibility complex I (MHC-I) and II (MHC-II) also play a neuroinflammatory role in PD [39,40]. The MHC gene family encodes molecules on the surface of cells that enable the immune system to recognize presented self- and foreign-derived peptides. MHC class II-positive microglia are a sensitive index of neuropathological change and are actively associated with damaged neurons and neurites in PD [41]. *Mitochondrial dysfunction* is another pathway that has been implicated in the pathophysiology of PD through both environmental exposure and genetic factors. The discovery of the role of the PD familial genes 'PTEN'-induced putative kinase 1 (PINK1) and parkin (PRKN) in mediating mitochondrial degradation reaffirmed the importance of this process in PD aetiology [42]. *Vesicular dysfunction* is another known contributor of PD [43]. Finally, *diabetes mellitus* and *inflammatory bowel diseases* (IBD) are known PD risk factors. In fact, population-based cohort studies indicate that diabetes and IBD are associated with increased PD risk by about 38% [44] and 22% [45], respectively.

A few notes on the set of genes selected follow. In Table 3 we reported the first 20 most important protein coding genes and lincRNAs in our analysis. We included lincRNAs because long non coding RNAs in general assume various roles, which include regulatory roles, and can thus modulate gene expression of protein coding genes; also they are very relevant in neurobiology, as many are associated with neurological pathologies [9].

'MYOM1', Myomesin1, the most important gene, is a protein coding gene and is up-regulated in PD subjects. Noticeably, 'ENSG00000272688' (Lnc-MYOM1-4) falls within an intron of MYOM1 and is the fifth most important lincRNA; gene 'MYOM2' is also in the list of selected genes and was selected in all the 20 repeated runs. Gene 'MYOM1' is significantly up-regulated in human substantia nigra pars compacta from PD patients [46] and is also one of the most important genes in [33], together with 'SQLE', 'LGALS2', and 'NCR1'. The intersection between our and their set might be larger as in that paper only 29 out of a much larger set of genes selected are reported. Gene 'SLC25A20', Solute Carrier Family 25 Member 20, the second most important gene, was up-regulated in PD, and was one of the nine PD biomarkers identified by Jiang et al. [47], which used a meta-analysis of microarray gene expression data from [17,48,49]. 'PTGDS', another gene in our set, was also one of these nine biomarkers. In our set of genes, 6 other genes, 'SLC18B1', 'SLC25A3', 'SLC11A2', 'SLC25A25', 'SLC25A43', 'SLC38A11' belong to the solute carrier (SLC) superfamily, one of the major sub-groups of membrane proteins in mammalian cells. Their role in neurodegenerative disorders is described thoroughly in [50]. 'NRM', the third most important gene, the integral nuclear membrane protein Nurim, plays a role in the suppression of apoptosis [51], and apoptosis is the main mechanism of neuronal loss in Parkinson's disease [52]. 'PHF7', PHD Finger Protein 7, is a candidate gene for a PD risk locus identified with a meta-analysis of genome-wide association studies [53]. Both protein coding gene 'NUP50' (Nucleoporin 50) and lincRNA 'NUP50-DT' ('NUP50' divergent transcript) are in our gene list. 'CERS4', Ceramide Synthase 4, is involved in Sphingolipid metabolism and its relation to PD is described in [54]. Dysregulation of metabolic pathways by carnitine palmitoyl-transferase 1 'CPT1A' plays a key role in central nervous system disorders [55].

Gene 'RAP1GAP' has been identified by both the DE analysis and the ML methodology (it is selected 20 times over 20 repetitions) (see Table A1). This gene is under-expressed in PD subjects and has a role in orchestrating the development and maintenance of different populations of central and peripheral neurons [56].

4.3. Final Considerations

Two final comments. First, the performance of a classification algorithm that used as input features DE genes, as found by DESeq2, showed much lower performances compared to those obtained with the set of features selected with the ML algorithm, thus confirming the validity of our methodology and the importance of using ML models with gene expression data from RNA sequencing of whole blood where the signal is significantly low. Furthermore, we notice how some of the genes selected by the ML algorithm are not DE between the class of PD and the class of HC subjects (see Table A1) but nonetheless contain a relevant signal.

Last, the different average predicted probability between subjects that falls in different *age of onset* classes (early-onset and late-onset PD subtypes) could reflect the heterogeneity of PD at different ages. In fact, it has been observed that PD patients with older age onset have more severe motor and non-motor burdens and a more widespread involvement of striatal structures [57].

5. Conclusions

We used a robust ML approach to make predictions on PD from whole blood expression data. The studied cohort included 390 early stage drug-naive PD subjects and 189 age-matched HCs. After careful preprocessing, including batch correction and independent gene filtering, we used a feature selection procedure based on RF and re-sampling and an XGBoost algorithm to evaluate PD vs. HC classification performances within a nested 10-fold cross validation scheme. We explored classification performances for different values of C, the number of features selected, and identified a set of around 500 genes listed in Table A1 that corresponded to maximum discriminative power. We also performed an enrichment analysis on this set of genes and identified significant GO terms and KEGG pathways, many of which are in line with the current literature on PD, although further analysis of these sets is needed and is outside the scope of our work. A strength of our methodology is its robustness. The balanced accuracy of our algorithm compares favorably with the state-of-the-art.

This area of research is cutting edge and requires further investigation. A possible extension of our work could be the evaluation of the predictive power of the selected set of genes on an independent dataset. We are also working on a multi-modal approach that combines transcriptome data with epigenomic data (and other data possibly) with the final aim of increasing the predictive performances of our model.

Author Contributions: Conceptualization, E.P. (Ester Pantaleo), A.M., A.L., N.A., S.T. and D.U.; methodology, E.P. (Ester Pantaleo), A.M., A.L., N.A. and S.T.; software, E.P. (Ester Pantaleo); validation, E.P. (Ernesto Picardi), N.A. and A.L.; formal analysis, E.P. (Ester Pantaleo), A.M., A.L., S.T., N.A, C.L.G. and E.P. (Ernesto Picardi); investigation, E.P. (Ernesto Picardi); resources, E.P. (Ester Pantaleo) and D.U.; data curation, E.P. (Ester Pantaleo); writing—original draft preparation, E.P. (Ester Pantaleo), N.A. and A.M.; writing—review and editing, E.P. (Ester Pantaleo), A.M., N.A., A.L., L.B., D.U., B.T., S.N., C.L.G. and E.P. (Ernesto Picardi); visualization, E.P. (Ester Pantaleo), A.M.; supervision, R.B., G.L. and G.P. All authors have read and agreed to the published version of the manuscript.

Funding: This work has been supported by Regione Puglia and CNR funds to "Tecnopolo per la Medicina di Precisione", D.G.R. n. 2117 of 21.11.2018 (B84I18000540002). A.L.s position is funded by the Program Research for Innovation—REFIN funded by Regione Puglia (Italy) in the framework of the POR Puglia FESR FSE 2014-2020 Asse X-Azione 10.4, project code 928A7C98 (Biomarcatori di connettività cerebrale da imaging multimodale per la diagnosi precoce e stadiazione personalizzata di malattie neurodegenerative con metodi avanzati di intelligenza artificiale in ambiente di calcolo distribuito).

Institutional Review Board Statement: Not applicable.

Informed Consent Statement: Not applicable

Data Availability Statement: The pseudocode used for the analysis is reported in Table A1. R codes used to perform the preprocessing and the analysis are available upon request.

Acknowledgments: PPMI—a public-private partnership—is funded by the Michael J. Fox Foundation for Parkinson's Research and funding partners, including Abbvie, Acurex Therapeutics, Allergan, Amathus Therapeutics, Asap Aligning Science Across Parkinson's, Avid Radiopharmaceuticals, Bial Biotech, Biogen, Biolegend, Bristol Myers Squibb, Calico, Celgene, Dacapo Brainscience, Denali, Edmond J. Safra Philantropic Foundation, 4d Pharma plc, GE Heathcare, Genentech, gsk GlaxoSmithKline, Golub Capital, Handl Therapeutics, Insitro, Janssen Neuroscience, Lilly, Lundbeck, Merck, msd Meso Scale Discovery, Neurocrine Biosciences, Pfizer, Piramal, Prevail Therapeutics, Roche, Sanofi Genzyme, Servier, Takeda, TEVA, ucb, VanquaBbio, verily, Voyager Therapeutics, Yumanity Therapeutics.

Conflicts of Interest: The authors declare no conflict of interest.

Appendix A

Algorithm A1 Pseudocode.

1: Let F be the total number of features
2: **for** $r = 1$ to 20 **do**
3:     Divide data into 10 stratified folds using random seed r
4:     **for** fold $k = 1$ to 10 **do**
5:         Set fold k as *validation_set* and the remaining 9 folds as *training_set*
6:         **for** $s = 1$ to 100 **do**
7:             Divide *training_set* into 5 stratified folds using random seed s
8:             Take 4 of the folds as the new training set
9:             Train a RF on this training set with 1000 trees
10:             **for** $f = 1$ to F **do**
11:                 Set $is\_outlier_{r,s,f} = 0$
12:                 Estimate $importance_{r,s,f}$
13:             **end for**
14:             Evaluate $th_{r,s} = \text{MEDIAN}^{(f)}(importance_{r,s,f}) + 1.5 * \text{IQR}^{(f)}(importance_{r,s,f})$ where $\text{MEDIAN}^{(f)}$ means median over the F values of f
15:             **for** $f = 1$ to F **do**
16:                 $is\_outlier_{r,s,f} = \text{IFELSE}(importance_{r,s,f} > th_{r,s}, 1, 0)$
17:             **end for**
18:         **end for**
19:         **for** $f = 1$ to F **do**
20:             Set $percentage\_outlier_{r,f} = 0$
21:             **for** $s = 1$ to 100 **do**
22:                 $percentage\_outlier_{r,f} \mathrel{+}= is\_outlier_{r,s,f}$
23:             **end for**
24:         **end for**
25:         **for** $C = 1$ to 100 **do**
26:             Evaluate $is\_selected_{r,f,C} = \text{IFELSE}(percentage\_outlier_{r,f} > C, 1, 0)$
27:             Train XGBoost on the *training_set* using only features f with $is\_selected_{r,f,C} = 1$
28:             Estimate performance $\text{ROCAUC}_{r,k,C}$ on the *validation_set*
29:         **end for**
30:     **end for**
31: **end for**
32: **for** $C = 1$ to 100 **do**
33:     Evaluate $m\_\text{ROCAUC}_{r,C} = \text{MEDIAN}^{(k)}(\text{ROCAUC}_{r,k,C})$ over the 10 values of k
34: **end for**
35: **for** $C = 1$ to 100 **do**
36:     Evaluate $m\_\text{ROCAUC}_C = \text{MEDIAN}^{(r)}(\text{ROCAUC}_{r,C})$ over the 20 values of r
37: **end for**
38: Let $C^* = \text{ARGMAX}_C(m\_\text{ROCAUC}_C)$
39: **for** $f = 1$ to F **do**
40:     Set $count\_selected_f = 0$
41:     **for** $r = 1$ to 20 **do**
42:         $count\_selected_f \mathrel{+}= is\_selected_{r,f,C^*}$
43:     **end for**
44: **end for**

Table A1. Complete list of the most frequent protein coding genes and lincRNAs, ordered by importance. LincRNAs are marked with an asterisk. For each gene, four attributes are listed: (i) No arrow, an upward pointing arrow, a downward pointing arrow indicate no significant DE between PD and HC, significant over-expression in PD subjects, significant under-expression in PD subjects, respectively; (ii) HGCN symbol (or Ensembl ID when missing); (iii) Average importance over 20 runs of 5-fold cross validation; (iv) Frequency of occurrence over 20 repetitions.

e	Symbol	imp	f	e	Symbol	imp	f	e	Symbol	imp	f
↑	MYOM1	82.1	20		SLC25A20	62.7	20		NRM	46.4	20
↑	PHF7	45.9	20	↑	ENSG00000277763 *	39.4	20		ICA1	36	20
↓	CPT1A	33.8	20	↑	LINC02422 *	33.3	20		GSTM1	32.4	20
	PCDHGA6	31.6	20		AK5	31.5	20	↑	GCNT2	29.9	20
	CERS4	29.7	20		YJU2	29.4	20		SURF6	27.7	20
	ENSG00000281181 *	26.7	20		ENSG00000285774 *	26.7	20		ENSG00000272688 *	26.2	20
	SERF1B	25.8	20		ENSG00000284773	25.7	20	↑	TREML4	25.7	20
	ANKRD34B	25.1	14	↑	NDUFB9	25	20		ERLIN2	24.6	20
	ENSG00000276651 *	24.4	20		CCR4	23.9	20	↑	NFE2L3	23.9	20
	FGGY	23.5	20		ENSG00000234902 *	23.3	20		ARRDC4	23.1	20
	TMTC4	21.6	20	↑	BPHL	20.1	20		C2orf42	20	20
	BTBD19	19.8	15		LOXHD1	19.4	20	↑	DHFR	19.1	20
	LINC02470 *	18.9	20	↑	SHISA4	18.8	20		FKBP5	18.4	20
	ENSG00000234426 *	18.3	20		TKTL1	18.1	20		ATP6V0A2	18	20
	GPR19	17.9	20	↑	ZNF584	17.4	20		FAN1	17.4	20
	MRPS6	17.2	18		TSPAN2	16.9	20		CRAT	16.5	20
	CCRL2	16	20		GTF2IRD2	15.9	20		PUDP	15.8	20
→	NOP16	15.7	20		LINC00243 *	15.6	20		CEP19	15.6	20
	GAB3	15.6	20	↑	ENSG00000269399 *	15.5	20	↑	YOD1	15.3	20
	GET1	15.3	19	↑	NREP	15.2	20	↑	YES1	14.8	15
↑	COL9A3	14.5	20		NSUN4	14.4	20		FARSB	14.3	20
←	GZMB	14.2	20		B4GALNT3	14.1	20		TBL2	14.1	20
←	RAP1GAP	13.7	20	↑	BASP1	13.5	20		PRUNE2	13.5	19
	FBN2	13.3	20		VNN1	13.2	20	↑	LSMEM1	13.2	20
	ZSCAN21	13.1	20		CLEC12A	13	20		COA4	13	20
↑	DPY19L2	12.9	20	↑	RNASET2	12.8	20		DCXR	12.4	20
→	WDR49	12.4	20		CRYZ	12.3	15	↑	LINC00623 *	12.2	20
←	ZNF714	12.2	20	↑	TOR1B	12	20	←	ADGRE5	11.9	20
	SULF2	11.7	20		MSH3	11.7	20		PCDHGB3	11.6	20
↑	SPHK1	11.3	20	→	G6PC3	11	20		MASTL	11	20
	LINC01806 *	11	20	↑	SQLE	11	19	→	PWP2	10.9	20
	TXLNB	10.8	20		ZSWIM3	10.7	20	←	SFXN4	10.6	20
↑	RUBCNL	10.6	20		PNO1	10.2	20		SMIM12	10.2	18
	TNFRSF10B	10.2	20		GPR162	10	16		KRT1	10	20
	B3GAT1	9.8	20		PILRB	9.8	20		TAP2	9.8	20
→	MSR1	9.7	18	←	LINC00482 *	9.6	18		OSER1-DT *	9.4	15
	ASCC1	9.4	20	←	ZNF429	9.4	20		SSPN	9.3	20
	GYPA	9.3	20		FAT4	9.3	20		SLC18B1	9.2	20
	TIPIN	9.1	20		IL18RAP	9	15		GYPE	8.9	20

Table A1. Cont.

e	Symbol	imp	f	e	Symbol	imp	f	e	Symbol	imp	f
	HYAL3	8.7	20		PREX1	8.7	19		KLRC2	8.6	20
←	FCER1A	8.6	19		ENDOG	8.6	20		GSTM3	8.6	20
	TMEM252	8.6	20		SRGAP2C	8.5	17		ATP5MC1	8.5	20
→	FIS1	8.3	20		ARRDC3-AS1 *	8.3	19		LINC01948 *	8.3	16
	TPPP3	8.2	20		HDHD3	8.2	19		LINC01560 *	8.2	20
	IFRD2	8.1	20		STK11	8.1	17		TARBP1	7.9	20
→	LINC00299 *	7.9	19		XCL1	7.8	20		ZNF491	7.6	20
	LINC00570 *	7.4	20		PARS2	7.4	20		INPP1	7.4	20
	TMEM245	7.4	20		NECAP2	7.3	19		PER3	7.3	20
→	CDCA4	7.2	20		NUP210L	7.2	19		GTF2H2	7.1	20
	APOLD1	7.1	20		ETFDH	7	20		GPX4	6.9	20
	PPP1R14B	6.9	20		GOLGA6L9	6.9	20		ATP6AP1L	6.9	20
→	GFUS	6.9	20	→	ENSG00000268240 *	6.7	20		MAST4	6.7	20
→	ENSG00000225750 *	6.7	20	←	BCL6	6.5	20		DDO	6.5	20
←	TMEM185B	6.4	20	←	UPB1	6.4	20		CCR3	6.4	20
	PLIN2	6.4	20		RALY-AS1 *	6.3	20		DDIT4	6.3	19
	FGFR2	6.3	20	→	PAICS	6.3	20		ENSG00000278384	6.2	20
	HLA-DRB5	6.2	16	→	VDR	6.1	19		ENSG00000254789 *	6.1	20
	ENSG00000260077 *	6.1	18	←	CLEC18A	6.1	15		LINC02193 *	6.1	20
	SBNO1	6.1	20		VAV1	6.1	17		SCN3A	6	20
←	CCL4L2	5.9	20		ASB3	5.8	18		GSTM2	5.8	20
←	KDM5B	5.8	20	→	GNAL	5.7	18		KCNMB1	5.6	15
	CSGALNACT1	5.6	20		RNASEH1	5.6	20		ENSG00000285476	5.5	20
←	FBXL13	5.4	20	→	VLDLR	5.4	20	→	FPR2	5.4	20
←	PPP1R3B	5.4	20	→	SRSF8	5.4	20	←	APOO	5.3	20
←	TXNIP	5.3	20		MPG	5.3	19	→	TAS2R43	5.2	20
	FLVCR2	5.1	20		SLC25A3	5.1	20		CD36	5.1	19
	CENPK	5	18		C5	5	14		PRRG4	5	20
	DYRK1B	4.9	17	←	APTR *	4.8	20		TMEM14C	4.8	19
	PF4V1	4.8	20	←	ZNF789	4.7	20	←	UBR7	4.7	17
	HMOX2	4.6	19		PID1	4.6	20		LERFS *	4.5	20
	ENSG00000266302	4.5	20		AKAP5	4.5	20	←	DPCD	4.4	20
	TMTC2	4.4	20		NKAP	4.4	17		ENSG00000276476 *	4.4	20
	EDAR	4.4	20		VSTM1	4.4	20		PDK4	4.3	20
←	HIF1A	4.3	20		GRHPR	4.3	20		TUBB2A	4.3	20
	PALLD	4.3	20	→	LINC01303 *	4.3	20		FPR3	4.3	20
←	TMEM45B	4.3	20	←	RGMB	4.3	20		CREM	4.3	20
	LYRM9	4.3	18		VSIG10	4.3	20		TSPAN17	4.2	20
	BBLN	4.1	16		LTA4H	4.1	20		U2AF1	4.1	20
	PPAN	4.1	20		ARL17B	4.1	20		ENSG00000274922 *	4.1	19
	TM9SF1	4	20		EPPK1	4	20		THBD	4	20
	DRAXIN	4	14		USP12	3.9	20		SLC11A2	3.9	19
	ENSG00000259071 *	3.9	20		SPON2	3.8	20		ENSG00000256427 *	3.8	14
	FAM124B	3.8	20		NBDY	3.8	20		MBNL3	3.8	20

Table A1. Cont.

Symbol	imp	f	e	Symbol	imp	f	e	Symbol	imp	f
COMMD9	3.7	20		CTSK	3.7	20		CYREN	3.7	20
LINC00654 *	3.7	20	→	ENSG00000270972 *	3.6	20		SVBP	3.6	20
TMEM185A	3.6	18	←	CDK6	3.6	20	→	MFSD9	3.6	20
CRTAP	3.5	14		CSTB	3.5	20		PTRHD1	3.5	20
PPIE	3.5	20	→	HLA-DMB	3.5	15		DSC1	3.5	20
CEP85	3.4	20		RNF182	3.4	20	→	HSD17B8	3.4	20
NKX3-1	3.4	20	→	F2R	3.4	20		ENSG00000224635 *	3.4	19
HDHD5	3.4	20	←	ZKSCAN4	3.4	20		KPNB1	3.3	20
LAMP1	3.3	20		ENSG00000277369 *	3.3	20	←	SNHG4 *	3.3	20
MYG1	3.3	15	→	SLC25A25	3.3	18		U2AF1L5	3.3	20
ETHE1	3.2	20	→	KAT2B	3.2	20	→	MIR378D2HG *	3.2	16
TLCD4	3.2	20	←	SPTY2D1	3.2	20		MYOM2	3.2	20
IL18R1	3.1	17		UBE2E2	3.1	20	←	KREMEN1	3.1	20
ENSG00000227920 *	3.1	20		COX5A	3	16	←	LINC00920 *	3	20
NRG1	3	20		GPR15	3	20		UROS	3	20
LINC02520 *	3	17		TGM3	3	20		CCZ1B	3	19
S100B	3	20		NR4A2	2.9	20		SULT1A1	2.9	20
TMEM273	2.9	20		LINC00381 *	2.9	18	←	FMN1	2.8	20
CCDC144A	2.8	20		LMTK2	2.8	20	←	HSDL2	2.8	20
BMX	2.8	20		ZNF559	2.8	20		ELL	2.8	17
MIR646HG *	2.8	20	←	CREG1	2.8	20		DACT1	2.8	19
TBC1D30	2.7	20	←	JUN	2.7	20		CLEC4F	2.7	20
ENSG00000259652 *	2.7	19		POMC	2.6	14	→	THAP7	2.6	20
YDJC	2.6	20		NFE4	2.6	20	←	PDZD4	2.6	20
FTCDNL1	2.6	20		GABARAPL1	2.6	20	←	TIMM9	2.6	20
ANKRD9	2.6	19	→	RNF11	2.5	19	→	ATP6V1F	2.5	20
MTCH2	2.5	20	→	SCO1	2.5	19		NOTCH2NLA	2.5	20
GATD3A	2.5	20		MAP3K7CL	2.5	20		NCAM1	2.4	20
LINC02273 *	2.4	19		PI16	2.4	14	←	CLCN4	2.4	20
CTXN2-AS1 *	2.4	20		MECR	2.4	20	←	ENSG00000273243 *	2.4	17
COL18A1	2.3	15	→	TLK2	2.3	19	→	HMBS	2.3	16
CCDC102A	2.3	20		TTF2	2.3	19		C16orf91	2.3	20
HERPUD1	2.3	20		SLA	2.2	20	→	TMEM102	2.2	14
HLA-DQB1	2.2	20	←	DUSP19	2.2	20		KCTD3	2.2	19
FOLR3	2.2	20		C1orf220 *	2.1	15		PRDM8	2.1	20
KIF1B	2.1	19	←	LINC00298 *	2.1	18		LINC01410 *	2.1	20
LINC02218	2.1	20	←	NKAPL	2.1	20		RAB34	2.1	17
GSTZ1	2.1	19	←	ENSG00000267575 *	2.1	16	←	SYNM	2	19
RNF149	2	20	←	CSRNP1	2	17		LSG1	2	20
TOP1	2	20		IRF1	2	14	→	SYTL3	1.9	20
ZNRD2	2	20		ICAM4	2	20		CLEC12B	1.9	20
NDRG3	1.9	20	←	PAQR8	1.9	19		LGALS2	1.9	15
WDR11	1.9	17		HDAC9	1.9	20		RRS1	1.8	20
ANKRD55	1.9	16		NIT2	1.8	14		ENSG00000272908 *	1.8	20

Table A1. *Cont.*

e	Symbol	imp	f	e	Symbol	imp	f	e	Symbol	imp	f
	PMVK	1.8	20		RFC5	1.8	20		PRADC1	1.8	18
↑	HSD17B13	1.8	18		ZNF487	1.8	20	→	NUP50-DT *	1.8	20
↓	TOR3A	1.7	20	↑	ADAM15	1.7	20	↑	ENSG00000285492 *	1.7	20
←	CA4	1.7	20		PARN	1.7	18		AKR1A1	1.7	20
↓	DOCK4	1.7	20	↑	IRS2	1.7	20	←	CHST2	1.7	20
→	C3orf18	1.6	20		ZNF69	1.6	20	→	CCN3	1.6	20
	CLMN	1.6	20		GCAT	1.6	14		TXN2	1.6	15
←	TPST1	1.6	20	↑	MIR3945HG *	1.5	20		PTPRN2	1.5	20
	ADGRB3	1.5	18	↑	ENSG00000281831 *	1.5	15	←	EIF2D	1.5	20
	OAS1	1.5	14	↓	ACSL1	1.5	20		SRP19	1.5	20
←	NUP50	1.5	20	→	XK	1.5	20		COA1	1.4	19
	KRT72	1.4	20	←	ROPN1L	1.4	16	←	SLC25A43	1.4	20
↑	ENSG00000251093 *	1.4	20		ABCA1	1.4	19	→	AFDN	1.4	18
	TMEM176B	1.3	20	↓	SERINC3	1.3	18		CEMIP2	1.3	20
→	NAXD	1.3	20		NFXL1	1.3	20		ALKBH7	1.3	19
	ENSG00000259959 *	1.3	20	↓	ENSG00000275765 *	1.3	15		BSCL2	1.2	18
→	CISD2	1.2	20	→	DCAF4L1	1.2	19	←	CD93	1.2	19
→	APRT	1.2	20		CYBRD1	1.1	16		NBPF26	1.1	20
↓	MRPS27	1.1	18		GIMAP1	1.1	20	←	RRP7A	1.1	20
	ISCA1	1.1	20	←	FADS2	1.1	19		TRANK1	1.1	18
←	PHACTR1	1.1	20	→	VNN3	1	20		HLX	1	20
	JADE1	1	20		KNOP1	1	20		HLA-DQA2	1	19
	XKR3	1	19		P2RX4	0.9	16	←	CPA3	0.9	19
←	C8orf33	0.9	14	←	MS4A4E	0.9	20		ENSG00000274979 *	0.9	20
	RPGRIP1	0.9	19		NCR1	0.9	20	←	PRF1	0.9	20
	PEA15	0.8	16	←	S100A10	0.8	19		ERO1A	0.8	20
←	ADGRG3	0.8	20	→	BTNL8	0.8	20		EMC9	0.8	20
	LONRF3	0.8	15		SLC38A11	0.7	20	←	BAZ1A	0.7	17
	ACAD11	0.7	18	→	C1orf109	0.7	20		SUV39H1	0.7	14
←	PAAF1	0.7	20		MGST3	0.7	20	←	PHTF1	0.7	20
	CD55	0.6	20		MTPAP	0.7	20		ZNF80	0.6	18
←	SIPA1L2	0.6	17	←	PTGDS	0.6	19	↓	SNX3	0.6	20
	KLF9	0.6	20		TGFA	0.6	20		HLA-DQA1	0.5	20
←	AMACR	0.5	20		NCAPG2	0.5	14		CTSH	0.5	15
↑	ENSG00000282988	0.5	17		PANX1	0.5	20	←	HLA-A	0.5	20
↓	CPD	0.4	20	←	NHS	0.4	16		KRT73	0.4	20
←	METRNL	0.3	17	→	PIGW	0.3	16		AVIL	0.3	20
←	ABCG1	0.2	20	↓	RAB27A	0.2	20		DNAJC3	0	20

LincRNAs are marked with an asterisk.

References

1. GBD Disease Incidence, Prevalence Collaborators. Global, regional, and national incidence, prevalence, and years lived with disability for 354 diseases and injuries for 195 countries and territories, 1990–2017: A systematic analysis for the Global Burden of Disease Study 2017. *Lancet* **2018**, *392*, 1789–1858. [CrossRef]
2. Schapira, A.H.V.; Chaudhuri, K.R.; Jenner, P. Non-motor features of Parkinson disease. *Nat. Rev. Neurosci.* **2017**, *18*, 435–450. [CrossRef] [PubMed]
3. Angelopoulou, E.; Paudel, Y.N.; Papageorgiou, S.G.; Piperi, C. Environmental Impact on the Epigenetic Mechanisms Underlying Parkinson's Disease Pathogenesis: A Narrative Review. *Brain Sci.* **2022**, *12*, 175. [CrossRef] [PubMed]
4. Nido, G.S.; Dick, F.; Toker, L.; Petersen, K.; Alves, G.; Tysnes, O.B.; Jonassen, I.; Haugarvoll, K.; Tzoulis, C. Common gene expression signatures in Parkinson's disease are driven by changes in cell composition. *Acta Neuropathol. Commun.* **2020**, *8*, 55. [CrossRef] [PubMed]
5. Sullivan, P.F.; Fan, C.; Perou, C.M. Evaluating the comparability of gene expression in blood and brain. *Am. J. Med. Genet. B Neuropsychiatr. Genet.* **2006**, *141*, 261–268. [CrossRef] [PubMed]
6. Soreq, L.; Salomonis, N.; Bronstein, M.; Greenberg, D.S.; Israel, Z.; Bergman, H.; Soreq, H. Small RNA sequencing-microarray analyses in Parkinson leukocytes reveal deep brain stimulation-induced splicing changes that classify brain region transcriptomes. *Front. Mol. Neurosci.* **2013**, *6*, 10. [CrossRef] [PubMed]
7. Haas, R.H.; Nasirian, F.; Nakano, K.; Ward, D.; Pay, M.; Hill, R.; Shults, C.W. Low platelet mitochondrial complex I and complex II/III activity in early untreated Parkinson's disease. *Ann. Neurol.* **1995**, *37*, 714–722. [CrossRef] [PubMed]
8. Barbanti, P.; Fabbrini, G.; Ricci, A.; Cerbo, R.; Bronzetti, E.; Caronti, B.; Calderaro, C.; Felici, L.; Stocchi, F.; Meco, G.; et al. Increased expression of dopamine receptors on lymphocytes in Parkinson's disease. *Mov. Disord.* **1999**, *14*, 764–771. [CrossRef]
9. Soreq, L.; Guffanti, A.; Salomonis, N.; Simchovitz, A.; Israel, Z.; Bergman, H.; Soreq, H. Long non-coding RNA and alternative splicing modulations in Parkinson's leukocytes identified by RNA sequencing. *PLoS Comput. Biol.* **2014**, *10*, e1003517. [CrossRef] [PubMed]
10. Grünblatt, E.; Zehetmayer, S.; Jacob, C.P.; Müller, T.; Jost, W.H.; Riederer, P. Pilot study: Peripheral biomarkers for diagnosing sporadic Parkinson's disease. *J. Neural Transm.* **2010**, *117*, 1387–1393. [CrossRef]
11. Shehadeh, L.A.; Yu, K.; Wang, L.; Guevara, A.; Singer, C.; Vance, J.; Papapetropoulos, S. SRRM2, a potential blood biomarker revealing high alternative splicing in Parkinson's disease. *PLoS ONE* **2010**, *5*, e9104. [CrossRef] [PubMed]
12. Molochnikov, L.; Rabey, M.R.; Dobronevsky, E.; Bonuccelli, U.; Ceravolo, R.; Frosini, D.; Grünblatt, E.; Riederer, P.; Jacob, C.; Aharon-Peretz, J.; et al. A molecular signature in blood identifies early Parkinson's disease. *Mol. Neurodegener.* **2012**, *7*, 26. [CrossRef] [PubMed]
13. Su, C.; Tong, J.; Wang, F. Mining genetic and transcriptomic data using machine learning approaches in Parkinson's disease. *NPJ Park. Dis.* **2020**, *6*, 24. [CrossRef]
14. Amoroso, N.; La Rocca, M.; Monaco, A.; Bellotti, R.; Tangaro, S. Complex networks reveal early MRI markers of Parkinson's disease. *Med. Image Anal.* **2018**, *48*, 12–24. [CrossRef]
15. Nalls, M.A.; McLean, C.Y.; Rick, J.; Eberly, S.; Hutten, S.J.; Gwinn, K.; Sutherland, M.; Martinez, M.; Heutink, P.; Williams, N.M.; et al. Parkinson's Disease Biomarkers Program and Parkinson's Progression Marker Initiative investigators. Diagnosis of Parkinson's disease on the basis of clinical and genetic classification: A population-based modelling study. *Lancet Neurol.* **2015**, *14*, 1002–1009. [CrossRef]
16. Monaco, A.; Pantaleo, E.; Amoroso, N.; Bellantuono, L.; Lombardi, A.; Tateo, A.; Tangaro, S.; Bellotti, R. Identifying potential gene biomarkers for Parkinson's disease through an information entropy based approach. *Phys. Biol.* **2020**, *18*, 016003. [CrossRef]
17. Shamir, R.; Klein, C.; Amar, D.; Vollstedt, E.J.; Bonin, M.; Usenovic, M.; Wong, Y.C.; Maver, A.; Poths, S.; Safer, H.; et al. Analysis of blood-based gene expression in idiopathic Parkinson disease. *Neurology* **2017**, *89*, 1676–1683. [CrossRef] [PubMed]
18. Chen-Plotkin, A. Blood transcriptomics for Parkinson disease? *Nat. Rev. Neurol.* **2018**, *14*, 5–6. [CrossRef] [PubMed]
19. Babu, G.S.; Suresh, S. Parkinson's disease prediction using gene expression—A projection based learning meta-cognitive neural classifier approach. *Expert Syst. Appl.* **2013**, *40*, 1519–1529. [CrossRef]
20. Karlsson, M.K.; Sharma, P.; Aasly, J.; Toft, M.; Skogar, O.; Sæbø, S.; Lönneborg, A. Found in transcription: Accurate Parkinson's disease classification in peripheral blood. *J. Park. Dis.* **2013**, *3*, 19–29. [CrossRef]
21. Marek, K.; Chowdhury, S.; Siderowf, A.; Lasch, S.; Coffey, C.S.; Caspell-Garcia, C.; Simuni, T.; Jennings, D.; Tanner, C.M.; Trojanowski, J.Q.; et al. Parkinson's Progression Markers Initiative. The Parkinson's progression markers initiative (PPMI)—Establishing a PD biomarker cohort. *Ann. Clin. Transl. Neurol.* **2018**, *5*, 1460–1477. [CrossRef] [PubMed]
22. Dobin, A.; Davis, C.A.; Schlesinger, F.; Drenkow, J.; Zaleski, C.; Jha, S.; Batut, P.; Chaisson, M.; Gingeras, T.R. STAR: Ultrafast universal RNA-seq aligner. *Bioinformatics* **2013**, *29*, 15–21. [CrossRef] [PubMed]
23. Liao, Y.; Smyth, G.K.; Shi, W. featureCounts: An efficient general purpose program for assigning sequence reads to genomic features. *Bioinformatics* **2014**, *30*, 923–930. [CrossRef] [PubMed]
24. Hutchins, E.; Craig, D.; Violich, I.; Alsop, E.; Casey, B.; Hutten, S.; Reimer, A.; Whitsett, T.G.; Crawford, K.L.; Toga, A.W.; et al. Quality Control Metrics for Whole Blood Transcriptome Analysis in the Parkinson's Progression Markers Initiative (PPMI). *medRxiv* **2021**. [CrossRef]
25. Breiman, L. Random forests. *Mach. Learn.* **2001**, *45*, 5–32. [CrossRef]
26. Breiman, L. Bagging predictors. *Mach. Learn.* **1996**, *24*, 123–140. [CrossRef]

27. Chen, T.; Guestrin, C. XGBoost: A Scalable Tree Boosting System; Association for Computing Machinery: New York, NY, USA, 2016; pp. 785–794. [CrossRef]
28. Love, M.I.; Huber, W.; Anders, S. Moderated estimation of fold change and dispersion for RNA-seq data with DESeq2. *Genome Biol.* **2014**, *15*, 550. [CrossRef]
29. Gibbons, S.M.; Duvallet, C.; Alm, E.J. Correcting for batch effects in case-control microbiome studies. *PLoS Comput. Biol.* **2018**, *14*, e1006102. [CrossRef]
30. Ritchie, M.E.; Phipson, B.; Wu, D.; Hu, Y.; Law, C.W.; Shi, W.; Smyth, G.K. Limma powers differential expression analyses for RNA-sequencing and microarray studies. *Nucleic Acids Res.* **2015**, *43*, e47. [CrossRef]
31. Monaco, A.; Pantaleo, E.; Amoroso, N.; Lacalamita, A.; Lo Giudice, C.; Fonzino, A.; Fosso, B.; Picardi, E.; Tangaro, S.; Pesole, G.; et al. A primer on machine learning techniques for genomic applications. *Comput. Struct. Biotechnol. J.* **2021**, *19*, 4345–4359. [CrossRef]
32. Kuleshov, M.V.; Jones, M.R.; Rouillard, A.D.; Fernandez, N.F.; Duan, Q.; Wang, Z.; Koplev, S.; Jenkins, S.L.; Jagodnik, K.M.; Lachmann, A.; et al. Enrichr: A Comprehensive Gene Set Enrichment Analysis Web Server 2016 Update. *Nucleic Acids Res.* **2016**, *44*, W90–W97. [CrossRef] [PubMed]
33. Makarious, M.B.; Leonard, H.L.; Vitale, D.; Iwaki, H.; Sargent, L.; Dadu, A.; Violich, I.; Hutchins, E.; Saffo, D.; Bandres-Ciga, S.; et al. Multi-Modality Machine Learning Predicting Parkinson's Disease. *bioRxiv* [CrossRef] [PubMed]
34. Gaki, G.S.; Papavassiliou, A.G. Oxidative stress-induced signaling pathways implicated in the pathogenesis of Parkinson's disease. *Neuromol. Med.* **2014**, *16*, 217–230. [CrossRef] [PubMed]
35. Wei, Z.; Li, X.; Liu, Q.; Cheng, Y. Oxidative stress in Parkinson's disease: A systematic review and meta-analysis. *Front. Mol. Neurosci.* **2018**, *11*, 236. [CrossRef] [PubMed]
36. Garcia, A.; León-Martinez, R.; Blanco-Lezcano, L.; Pavón-Fuentes, N.; Lorigados-Pedre, L. Transient glutathione depletion in the substantia nigra compacta is associated with neuroinflammation in rats. *Neuroscience* **2016**, *335*, 207–220. [CrossRef]
37. Tufekci, K.U.; Meuwissen, R.; Genc, S.; Genc, K. Inflammation in Parkinson's disease. *Adv. Protein Chem. Struct. Biol.* **2012**, *88*, 69–132. [CrossRef]
38. Muñoz-Delgado, L.; Macías-García, D.; Jesús, S.; Martín-Rodríguez, J.F.; Labrador-Espinosa, M.Á.; Jiménez-Jaraba, M.V.; Adarmes-Gómez, A.; Carrillo, F.; Mir, P. Peripheral Immune Profile and Neutrophil-to-Lymphocyte Ratio in Parkinson's Disease. *Mov. Disord.* **2021**, *36*, 2426–2430. [CrossRef]
39. Sulzer, D.; Alcalay, R.N.; Garretti, F.; Cote, L.; Kanter, E.; Agin-Liebes, J.; Liong, C.; McMurtrey, C.; Hildebr, W.H.; Mao, X.; et al. T cells from patients with Parkinson's disease recognize α-synuclein peptides. *Nature* **2017**, *546*, 656–661. [CrossRef]
40. Tan, J.S.Y.; Chao, Y.X.; Rötzschke, O.; Tan, E.K. New Insights into Immune-Mediated Mechanisms in Parkinson's Disease. *Int. J. Mol. Sci.* **2020**, *21*, 9302. [CrossRef]
41. Imamura, K.; Hishikawa, N.; Sawada, M.; Nagatsu, T.; Yoshida, M.; Hashizume, Y. Distribution of major histocompatibility complex class II-positive microglia and cytokine profile of Parkinson's disease brains. *Acta Neuropathol.* **2003**, *106*, 518–526. [CrossRef]
42. Malpartida, A.B.; Williamson, M.; Narendra, D.P.; Wade-Martins, R.; Ryan, B.J. Mitochondrial Dysfunction and Mitophagy in Parkinson's Disease: From Mechanism to Therapy. *Trends Biochem. Sci.* **2021**, *46*, 329–343. [CrossRef] [PubMed]
43. Ebanks, K.; Lewis, P.A.; Bandopadhyay, R. Vesicular Dysfunction and the Pathogenesis of Parkinson's Disease: Clues From Genetic Studies. *Front. Neurosci.* **2020**, *13*, 1381. [CrossRef] [PubMed]
44. Yue, X.; Li, H.; Yan, H.; Zhang, P.; Chang, L.; Li, T. Risk of Parkinson Disease in Diabetes Mellitus: An Updated Meta-Analysis of Population-Based Cohort Studies. *Medicine* **2016**, *95*, e3549. [CrossRef] [PubMed]
45. Villumsen, M.; Aznar, S.; Pakkenberg, B.; Jess, T.; Brudek, T. Inflammatory bowel disease increases the risk of Parkinson's disease: A Danish nationwide cohort study 1977–2014. *Gut* **2019**, *68*, 18–24. [CrossRef] [PubMed]
46. Grünblatt, E.; Mandel, S.; Jacob-Hirsch, J.; Zeligson, S.; Amariglo, N.; Rechavi, G.; Li, J.; Ravid, R.; Roggendorf, W.; Riederer, P.; et al. Gene expression profiling of parkinsonian substantia nigra pars compacta; alterations in ubiquitin-proteasome, heat shock protein, iron and oxidative stress regulated proteins, cell adhesion/cellular matrix and vesicle trafficking genes. *J. Neural Transm.* **2004**, *111*, 1543–1573. [CrossRef]
47. Jiang, F.; Wu, Q.; Sun, S.; Bi, G.; Guo, L. Identification of potential diagnostic biomarkers for Parkinson's disease. *FEBS Open Bio.* **2019**, *9*, 1460–1468. [CrossRef]
48. Scherzer, C.R.; Eklund, A.C.; Morse, L.J.; Liao, Z.; Locascio, J.L.; Fefer, D.; Schwarzschild, M.A.; Schlossmacher, M.G.; Hauser, M.A.; Vance, J.M.; et al. Molecular markers of early Parkinson's disease based on gene expression in blood. *Proc. Natl. Acad. Sci. USA* **2007**, *104*, 955–960. [CrossRef]
49. Calligaris, R.; Banica, M.; Roncaglia, P.; Robotti, E.; Finaurini, S.; Vlachouli, C.; Antonutti, L.; Iorio, F.; Carissimo, A.; Cattaruzza, T.; et al. Blood transcriptomics of drug-naive sporadic Parkinson's disease patients. *BMC Genom.* **2015**, *16*, 876. [CrossRef]
50. Ayka, A.; Şehirli, A.Ö. The Role of the SLC Transporters Protein in the Neurodegenerative Disorders. *Clin Psychopharmacol. Neurosci.* **2020**, *18*, 174–187. [CrossRef]
51. Chen, H.; Chen, K.; Chen, J.; Cheng, H.; Zhou, R. The integral nuclear membrane protein nurim plays a role in the suppression of apoptosis. *Curr. Mol. Med.* **2012**, *12*, 1372–1382. [CrossRef]
52. Erekat, N.S. Apoptosis and its Role in Parkinson's Disease. In *Parkinson's Disease: Pathogenesis and Clinical Aspects*; Stoker, T.B., Greenl, J.C., Eds.; Codon Publications: Brisbane, Australia, 2018; Chapter 4.

53. Chang, D.; Nalls, M.A.; Hallgrímsdóttir, I.B.; van der Brug, M.; Cai, F.; International Parkinson's Disease Genomics Consortium; 23andMe Research Team; Kerchner, G.A.; Ayalon, G.; Bingol, B.; et al. A meta-analysis of genome-wide association studies identifies 17 new Parkinson's disease risk loci. *Nat. Genet.* **2017**, *49*, 1511–1516. [CrossRef] [PubMed]
54. Custodia, A.; Aramburu-Núñez, M.; Correa-Paz, C.; Posado-Fernández, A.; Gómez-Larrauri, A.; Castillo, J.; Gómez-Muñoz, A.; Sobrino, T.; Ouro, A. Ceramide Metabolism and Parkinson's Disease Therapeutic Targets. *Biomolecules* **2021**, *11*, 945. [CrossRef] [PubMed]
55. Trabjerg, M.S.; Mørkholt, A.S.; Lichota, J.; Oklinski, M.K.E.; Andersen, D.C.; Jønsson, K.; Mørk, K.; Skjønnemand, M.N.; Kroese, L.J.; Pritchard, C.E.J.; et al. Dysregulation of metabolic pathways by carnitine palmitoyl-transferase 1 plays a key role in central nervous system disorders: Experimental evidence based on animal models. *Sci. Rep.* **2020**, *10*, 15583. [CrossRef] [PubMed]
56. Paratcha, G.; Ledda, F. The GTPase-activating protein Rap1GAP: A new player to modulate Ret signaling. *Cell Res.* **2011**, *21*, 217–219. [CrossRef]
57. Pagano, G.; Ferrara, N.; Brooks, D.J.; Pavese, N. Age at onset and Parkinson disease phenotype. *Neurology* **2016**, *86*, 1400–1407. [CrossRef]

Review

Does the Expression and Epigenetics of Genes Involved in Monogenic Forms of Parkinson's Disease Influence Sporadic Forms?

Aymeric Lanore [1,2,*], Suzanne Lesage [1], Louise-Laure Mariani [1,2], Poornima Jayadev Menon [1,2], Philippe Ravassard [1], Helene Cheval [1], Olga Corti [1], Alexis Brice [1,3] and Jean-Christophe Corvol [1,2]

[1] Institut du Cerveau–Paris Brain Institute–ICM, Inserm, CNRS, Sorbonne Université, 75013 Paris, France; suzanne.lesage@upmc.fr (S.L.); louise-laure.mariani@aphp.fr (L.-L.M.); poornima.menon@icm-institute.org (P.J.M.); philippe.ravassard@icm-institute.org (P.R.); helene.cheval@icm-institute.org (H.C.); olga.corti@icm-institute.org (O.C.); alexis.brice@icm-institute.org (A.B.); jean-christophe.corvol@aphp.fr (J.-C.C.)
[2] Assistance Publique Hôpitaux de Paris, Department of Neurology, CIC Neurosciences, Hôpital Pitié-Salpêtrière, 75013 Paris, France
[3] Assistance Publique Hôpitaux de Paris, Department of Genetics, Hôpital Pitié-Salpêtrière, 75013 Paris, France
* Correspondence: aymeric.lanore@aphp.fr; Tel.: +33-648245623

Abstract: Parkinson's disease (PD) is a disorder characterized by a triad of motor symptoms (akinesia, rigidity, resting tremor) related to loss of dopaminergic neurons mainly in the *Substantia nigra pars compacta*. Diagnosis is often made after a substantial loss of neurons has already occurred, and while dopamine replacement therapies improve symptoms, they do not modify the course of the disease. Although some biological mechanisms involved in the disease have been identified, such as oxidative stress and accumulation of misfolded proteins, they do not explain entirely PD pathophysiology, and a need for a better understanding remains. Neurodegenerative diseases, including PD, appear to be the result of complex interactions between genetic and environmental factors. The latter can alter gene expression by causing epigenetic changes, such as DNA methylation, post-translational modification of histones and non-coding RNAs. Regulation of genes responsible for monogenic forms of PD may be involved in sporadic PD. This review will focus on the epigenetic mechanisms regulating their expression, since these are the genes for which we currently have the most information available. Despite technical challenges, epigenetic epidemiology offers new insights on revealing altered biological pathways and identifying predictive biomarkers for the onset and progression of PD.

Keywords: Parkinson's disease; Parkinson's and related diseases; epigenetic; neurodegeneration; DNA methylation; histone modification; genetic; RNA-based gene regulation

1. Introduction

Parkinson's disease (PD) is a neurodegenerative disease, characterized by progressive degeneration of the dopaminergic neurons of the *Substantia nigra pars compacta*. Its pathology is multifactorial; influenced by both environmental and genetic determinants [1]. Several pathogenic mutations have been linked to autosomal dominant or recessive forms of PD. The discovery of these genes allowed a new insight into the pathophysiology of this disorder [2]. Pathological hallmarks of PD include the presence of cytoplasmic inclusions, called Lewy bodies (LB), mainly composed of aggregated αsynuclein (α-Syn) [3]. Multiplication and point mutations of *SNCA* encoding α-Syn are now recognized to cause autosomal dominant PD and they are suspected of promoting α-Syn aggregation. The *PINK1*, *PARKIN* and *DJ1* genes encode for proteins required by mitochondria, which are essential components of neurons for ATP synthesis, calcium storage, lipid metabolism and neuronal survival [2,4]. The fact that mutations in these genes lead to PD is a strong argument that mitochondrial dysfunction is involved in the pathophysiology of PD.

Genetic mutations account for only 10% of patients with PD, and therefore environmental exposure seems to play a major role in PD [2]. Epigenetic modulation of gene expression by environmental factors is increasingly studied. Epigenetic regulation involves different mechanisms such as modification of the histones of chromatin and DNA methylation changes [5]. Chromatin is a dynamic scaffold and modification of its main components, the histones, can modulate gene expression [6]. The effect of histone modification is mediated either by directly affecting the structure of chromatin, by disrupting the binding of proteins that associate with chromatin or by attracting certain effector proteins to chromatin [5]. Histone acetylation decreases the compression of chromatin and promotes gene transcription. Methylation of histone H3 lysine 4 (H3K4), H3K36 and H3K79 are marks of transcriptional activation, whereas methylation of H3K9, H3K27 and H4K20 are repressive modifications of transcription, involving the recruitment of methylating enzymes and HP1 to the gene promoter [7].

Direct DNA methylation is also a key epigenetic mechanism regulating gene expression. It is a reversible modification of DNA, which consists of the addition of a methyl group to the fifth carbon position of a cytosine, converting it to 5-methylcytosine (5mC). The transfer of a methyl group is carried out by DNA methyltransferase (DNMT) enzymes [8]. This epigenetic mark is frequently found within a 5′-Cytosine-phosphate-Guanosine sequence, called a CpG site [9]. DNA methylation is not homogeneously distributed in the genome, CpG sites are clustered in sequences called CpG islands (CGI). Methylation of promoter-associated CGIs can impair transcription factor binding or recruit repressive binding proteins, thus reduce gene expression [10]. Cytosine methylation is mediated by DNMTs, which can be classified as de novo (DNMT3A and DNMT3B) and maintenance (DNMT1) [11].

Epigenetic regulation is also closely linked to non-coding RNAs. Non-coding RNAs are classified according to their size, with small RNAs less than 200 nucleotides long and long non coding (LncRNAs) longer than 200 nucleotides [12]. Among the small non-coding RNAs, microRNAs (miRNAs) are the most studied. Mature miRNAs bind to complementary sequences of the target messenger RNA (mRNA), often in the 3′ untranslated region (3′UTR), and can increase mRNA degradation but also inhibit translation without reducing mRNA expression [13,14]. The regulation of gene expression by Long non-coding RNAs (LncRNAs) is not yet fully understood. However, they appear to be important genomic regulators, from the epigenetic to the post-translational level [15].

Epigenetic mechanisms influencing the development of sporadic PD have yet to be identified. Genes involved in monogenic forms of PD could be over- or under-expressed in the sporadic form compared to the general population without PD (controls). In the context of altered expression of these genes in sporadic forms, it can be assumed that epigenetic mechanisms may be involved in this dysregulation (Table 1).

We will review here the differences in the expression of these genes between sporadic human PD and controls. We will then consider whether histone modification, DNA methylation but also miRNA expression could account for the difference in expression. Finally, we will discuss whether the observed changes are epiphenomena or are an integral part of the pathophysiology of the disease.

Table 1. Expression profile and epigenetic changes observed for genes involved in monogenic forms of PD.

Studies	Tissues Analyzed	Proteins in Controls vs. sPD	mRNA in Controls vs. sPD	DNA Methylation in Controls vs. sPD	MiRNA Expression in Controls vs. sPD	Reference
		SNCA				
Grundemann et al., 2008	Brain: DA neurons of SN	Increase			-	[16]
Jowaed et al., 2010/Matsumoto et al., 2010	Brain	-	Increase	Hypomethylation	-	[17,18]
Pihlstrom et al., 2015/Ai et al., 2014/Tan et al., 2014	Blood immune cells	-	-	Hypomethylation	-	[19–21]
Minones-Moyano et al., 2011	Brain	-	-	-	Decrease miR-34b/c	[22]
		LRRK2				
Cho et al., 2013	Brain: frontal cortex/striatum	Increase	No difference	-	Decrease miR-205	[23]
Cook et al., 2017	Blood	Increase	-	-	-	[24]
Tan et al., 2014	Blood immune cells	-	-	Hypomethylation	-	[21]
		PRKN				
Beyer et al., 2008	Brain	-	Increase in variant TV3 and TV12	-	-	[25]
Cai et al., 2011	Blood immune cells	-	-	No difference	-	[26]
De Mena et al., 2013	Brain	-	-	No difference	-	[27]
Eryilmaz et al., 2017	Blood immune cells	-	-	Hypomethylation	-	[28]
Ding et al., 2016	Plasma	-	-	-	Decrease miR-181a	[29]
Xing et al., 2020	Brain	-	-	-	Decrease miR-218	[30]
Serafin et al., 2015	Plasma	-	-	-	Increase miR-103a-3p	[31]
		PINK1				
Muqit et al., 2006	Brain	Increase Δ1-PINK1	-	-	-	[32]
Blackinton et al., 2007	Brain: SN	-	No difference	-	-	[33]
Navarro-Sanchez et al., 2018	Brain: SN	-	-	No difference	-	[34]
Fazeli et al., 2020/Dos Santos et al., 2018	PBMC/CSF	-	-	-	Decrease miR-27a	[35,36]
		DJ1				
Kumaran et al., 2009	Brain	Decrease	Decrease	-	-	[37]
Tan et al., 2016	Blood immune cells	-	-	No difference	-	[38]
Chen et al., 2017	Plasma	-	-	-	Increase miR-4639-5p	[39]
		GBA				
Murphy et al., 2014	Brain	Decrease	No difference	-	-	[40]
Moors et al., 2019	Brain	-	No difference	-	-	[41]
Eryilmaz et al., 2017	Blood immune cells	-	-	No difference	-	[28]

sPD: sporadic Parkinson's disease; PBMC: peripheral blood mononuclear cell; CSF: cerebrospinal fluid; DA: dopaminergic; SN: *Substantia nigra*; controls: general population without PD.

2. α Synuclein

2.1. Function and Subcellular Distribution of α-Syn in PD

α-Syn, encoded by SNCA, is a small protein expressed abundantly in neurons of the central nervous system and located mainly at the presynaptic level [42]. In addition to the synaptic localization, the protein has been detected in the nucleus, explaining the name "synuclein" [43,44]. α-Syn is predominantly a soluble and highly mobile protein and since its molecular weight is less than the nuclear pore cutoff (~40 kDa), α-Syn can enter the nucleus by simple diffusion [44].

In PD, α-Syn monomers assemble into insoluble β-sheets-rich fibrils that together compose Lewy bodies (LB) [42]. It is assumed that the pathogenicity of α-Syn is caused by its accumulation and oligomerization preceding the formation of aggregates [45]. Post-mortem brain analysis of subjects without synucleopathy revealed that physiological α-Syn expression is lower in brain regions not subjected to LB accumulation [46].

The N-terminal domain of α-Syn allows its association with anionic phospholipids, preferentially binding to small vesicles [47,48]. It has been shown that α-Syn is involved in the regulation of synaptic vesicle trafficking and neurotransmitter release [42]. The association of α-Syn with lipids seems essential to allow the oligomers of this protein to disrupt the membranes and, consequently, to induce a dysfunction of the vesicular systems [49]. Several studies have reported that SNCA point mutations located in the N-terminal region of α-Syn p.A30P, p.A53T and G51D induce an increase in nuclear localization of α-Syn compared to the wild-type protein [42]. This data suggests that the N terminal region is functionally involved in the subcellular distribution of α-Syn. Other mechanisms proposed to be involved in its subcellular distribution include interactions with nuclear or cytoplasmic proteins, e.g., TRIM28 and oxidative stress [18,24,25].

The nuclear function of α-Syn is undetermined, but the presence of α-Syn in the nucleus seems to promote neurotoxicity, whereas cytoplasmic sequestration is protective in both cell culture and Drosophila [50]. This finding has also been supported by other studies involving SIRT2 inhibition [51].

2.2. Overexpression of α-Syn in PD

SNCA associated PD is characterized by duplications or triplications of the SNCA locus with the number of SNCA alleles correlating with the amount of α-Syn overexpression and also the severity of the clinical phenotype [52,53].

Correspondingly, the development of sporadic PD might be associated with increased SNCA expression or impairment of protein clearance mechanisms. However, studies on SNCA mRNA in brain regions of sporadic PD patients and controls are discordant, with some studies revealing even a decrease in SNCA expression in the *Substantia nigra* of PD patients [54]. A limitation of these studies is that the brain examination is performed at a late stage of the pathology, and due to neurodegeneration, the SNCA mRNA level might reflect expression in the remaining cells but not the expression in the affected neurons [54]. A study of surviving dopaminergic neurons laser-captured from the substantia nigra of post-mortem brains revealed an increase in SNCA mRNA in PD subjects compared to controls [16]. A splice variant of the SNCA gene, SNCA-126, has also been shown to be overexpressed in the *Substantia nigra* of PD patients [55]. In a study, the level of α-Syn protein was also modestly increased in the *Substantia nigra* of PD patients compared to controls [56].

Moreover, single nucleotide polymorphisms (SNPs) of SNCA have been identified in genome-wide associations studies (GWAS) as a risk factor that increases susceptibility to developing sporadic PD [57]. Among these variants a singular SNP, rs356168, in a non-coding distal enhancer element of SNCA leads to an increase in expression of α-Syn [58].

2.3. Epigenetic Regulation of SNCA Expression in Sporadic PD

DNA methylation of SNCA was reported to modulate its expression [17]. Hypomethylation in intron 1 of SNCA was observed in multiple brain regions of PD patients and was associated with increased *SNCA* expression in vitro [17,18]. Among the regulators of DNA methylation, DNMT1 appears to play an important role in *SNCA* expression. DNMT1 is mainly located in the nucleus of neurons and α-Syn aggregation leads to cytoplasmic sequestration of DNMT1 in animal models and patient brains [59] (Figure 1). This mechanism might explain the DNA hypomethylation at the *SNCA* gene and increased *SNCA* transcription. Hypomethylation of the *SNCA* promoter and increased *SNCA* expression in methamphetamine-exposed rats appears to be mediated by decreased occupancy of DNMT1 in the *SNCA* promoter region [60]. These observations were further associated with decreased nuclear localization of DNMT1 [60]. However, it is uncertain whether the observed hypomethylation is the response to cytoplasmic sequestration of DNMT1 or whether this mechanism is involved in the pathogenesis of the disease.

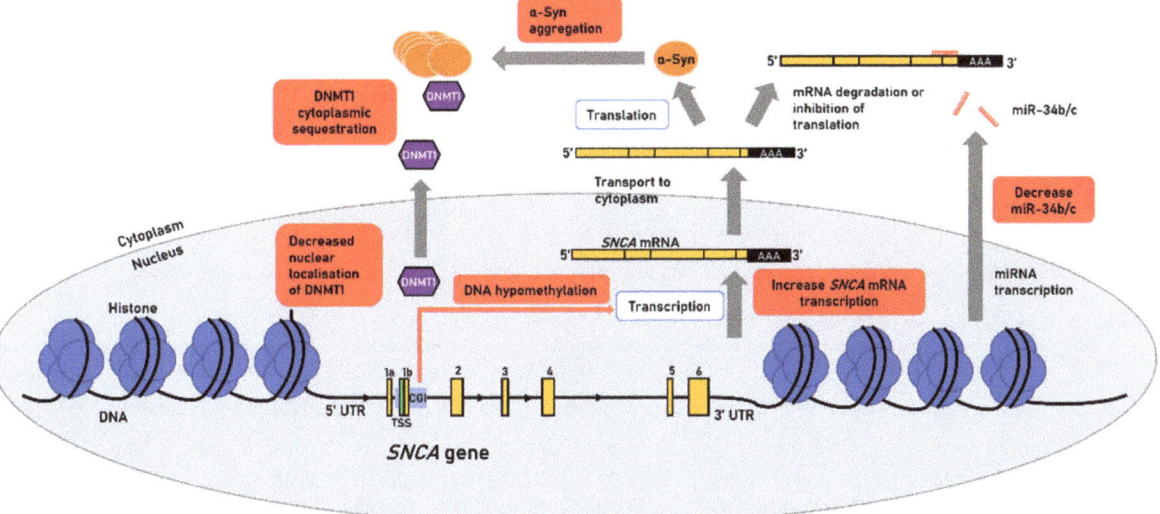

Figure 1. Epigenetic mechanisms and dysregulation of α synuclein. Abbreviations: 5′UTR: 5′ untranslated region; 3′UTR: 3′ untranslated region; 1–6: exon 1 to 6; CGI: CpG island; AAA: polyadenylation; DNMT1: DNA methyltransferase 1; mRNA: messenger RNA; miRNA: microRNA; TSS: transcription start site.

Hypomethylation of *SNCA* intron 1 has also been found in peripheral tissues of PD patients such as blood samples, peripheral blood mononuclear cells (PBMCs) and leukocytes [19–21]. The study of white blood cell *SNCA* methylation in healthy patients revealed a decrease in *SNCA* methylation with age [61]. Among healthy individuals, women had higher methylation of SNCA than men, which may contribute to the lower incidence of sporadic PD in women [61,62]. Furthermore, SNCA methylation in peripheral blood of sporadic PD patients was increased with higher doses of L-dopa and concordantly, L-dopa induced SNCA intron 1 methylation was observed in cultured mononuclear cells from PD patients [61]. These interesting findings highlight how environmental factors are correlated with epigenetic modifications.

An analysis of SNCA histone architecture in post-mortem midbrain samples found that three histone modifications, H3K4 trimethylation (H3K4me3), H3K27 acetylation (H3K27ac) and H3K27me3, were preferentially enriched in SNCA regulatory regions [63] (Figure 2). H3K27ac and H3K4me3 promote transcription and show a peak around the transcription start site [63]. Concordantly, a H3K27ac rich sequence was previously identified within the *SNCA* locus [64]. The involvement of histone modification in *SNCA* expression was first reported in a patient heterozygous for the SNCA p.A53T mutation. The repression of the mutated allele in this subject was not due to DNA methylation but due to histone deacetylation. The use of histone deacetylase (HDAC) inhibitor in cell models reactivated the mutated allele expression [65]. Epidemiological studies have shown that β-2 adrenergic receptor antagonists increase the risk of developing PD. Acetylation of H3K27 *SNCA* histones was proposed as the possible mechanism by which Adrenergic β2 receptor antagonists, potentially via inhibition of the β2-adrenergic receptor pathway, leads to accumulation of α-Syn [66]. However, the analysis of post-mortem midbrain samples found no significant difference for H3K27ac between PD subjects and controls [63]. While the histone mark H3K4me3 was enriched at the SNCA promoter in post-mortem brain samples from PD patients compared to controls. Furthermore, higher levels of H3K4me3 correlated with higher levels of α-Syn [63].

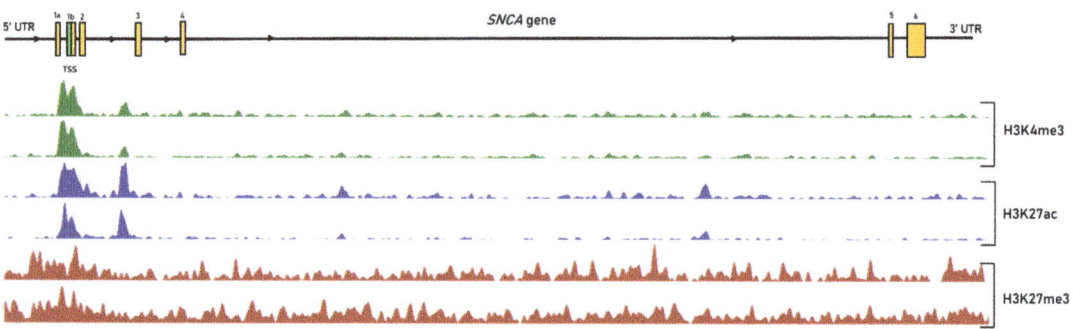

Figure 2. Distribution of histone modifications in the *SNCA* gene. Abbreviations: 5′UTR: 5′ untranslated region; 3′UTR: 3′ untranslated region; 1–6: exon 1 to 6; TSS: transcription start site. Histone modifications, H3K4me3 (green), H3K27ac (blue) and H3K27me3 (red), in the *SNCA* gene from *Substantia nigra* tissues of two healthy adult postmortem brain samples. Adapted from Roadmap Epigenomics Database.

In addition, a-Syn itself is thought to play a role in regulating transcription through histone modification. It has been reported that nuclear α-Syn leads to transcriptional repression of PPARGC1A, encoding PGC-1α, potentially through decreased levels of histone acetylation. PGC-1α is a primary mitochondrial transcription factor involved in the regulation of mitochondrial biogenesis and oxidative metabolism [67,68]. It has been shown in mice that repression of PGC1-a by the PARIS protein leads to progressive loss of dopaminergic neurons [69].

2.4. Non-Coding RNAs Regulating SNCA *Transcription*

Translation of *SNCA* mRNA in the cytoplasm is regulated by specific miRNAs. It has been found that miR-7 and miR-153 binds directly to the 3′-UTR of *SNCA* mRNA, destabilizing the mRNA and significantly reducing its levels [70,71]. Neurotoxin 1-Methyl-4-Phenyl-Pyridinium (MPP+) induced decrease in miR-7 expression possibly contributes to increase *SNCA* expression in vitro and in mice [70]. In contrast, overexpression of miR-7 or miR-153 in cortical neurons by viral transduction showed a protective effect against MPP+ toxicity [72]. In vitro, miR-7 accelerates the clearance of α-Syn; an effect that seems to be mediated by its activation of autophagy [73].

Two other interesting mi-RNAs with multiple potential associations in the pathogenesis of PD are miR-34b/c. Reduced expression of these miRNAs were reported in the amygdala, *Substantia nigra*, frontal cortex and cerebellum of PD patients compared to controls [22]. A decrease in these miRNAs was also observed in the putamen of PD patients [74]. Depletion of miR-34b/c in differentiated SH-SY5Y neurons resulted in reduced cell viability, mitochondrial dysfunction and oxidative stress, and a slight decrease in DJ-1 and Parkin expression [22]. On the other hand, a study found that inhibition of miR-34b/c which targets the 3′-UTR of *SNCA* mRNA is associated with increased expression of mRNA *SNCA* and protein expression [75]. Alternative polyadenylation can lead to different

3′UTR lengths and at least five different 3′UTR lengths of SNCA transcripts have been reported [76]. *SNCA* transcript with longer 3′UTRs may promote protein accumulation and mitochondrial localization [76]. A study using luciferase-SNCA full length 3′UTR reporter vector reported that a miR-34b-mimic induced translation of the long 3′UTR SNCA transcript [76]. In addition, it has been shown that very low frequency magnetic fields decrease the expression of miR-34b/c in vitro [77]. This modulation could be attributed to CpG hyper-methylation within the miR-34b/c promoter observed with exposure to these fields [77].

Other non-coding RNAs might participate in the regulation of *SNCA* expression such as RP11–115D19.1. The expression levels of RP11-115D19.1-003 in the brains of healthy donors and PD patients were strongly and positively correlated with those of *SNCA* [78]. Knockdown of this LncRNA in a cell model led to an increase in *SNCA* expression, suggesting its repressive effect on *SNCA* expression [78].

2.5. Interplay between Epigenetic Mechanisms in the Regulation of SNCA

Different epigenetic mechanisms are involved in the regulation of α-Syn expression, however, the relative weight of each is not determined. Histone modifications such as H3K9me3, H3K27me3 and H4K20me1 cause local chromatin condensation and could lead to an easily reversible repression of gene expression [79,80]. DNA methylation appears to lead to long-term stable gene repression. As H3K4me3 is present in most of CGIs, regardless of whether the associated gene is actively transcribed or not, there appears to be a dependency between these two mechanisms [81]. H3K4me3 is particularly enriched in unmethylated CGIs, which may allow the maintenance of DNA hypomethylation and shape a chromatin environment that favors transcription [81]. This is consistent with the observations of DNA hypomethylation and H3K4me3 enrichment on the *SNCA* gene. However, variability in a-Syn expression was observed despite enrichment of H3K4me3 at the SNCA promoter, raising suspicion of complementary mechanisms. A multivariate analysis considering the level of different epigenetic mechanisms could be interesting to better discern the involvement of these mechanisms.

3. LRRK2

3.1. LRRK2 Protein Function and Localization

LRRK2 mutations induce the autosomal-dominant form of familial PD [82]. *LRRK2* encodes a serine/threonine kinase called dardarin, after the Basque word for tremor. The native protein appears to transit between a monomeric and dimeric form [83,84]. LRRK2 is involved in many cellular functions such as regulation of neurite growth and cytoskeletal dynamics, maintenance of lysosomal function, and synaptic vesicle endocytosis (SVE) [85]. After neurotransmission, the replenishment of synaptic vesicles with neurotransmitters is ensured by the SVE [86]. Some proteins involved in SVE such as synaptojanin 1 (SYNJ1), auxilin (DNAJC6), and endophilin A1 (SH3GL2) are also LRRK2 substrates [86]. Phosphorylation of these proteins by LRRK2 appears to result in SVE dysfunction [87]. The G2019S mutation is the most common LRRK2 mutation. This mutation located in the kinase domain results in an increased kinase activity of the protein and could be toxic by a gain-of-function mechanism [88] (Figure 3). In a Drosophila model of PD, G2019S *LRRK2* mutation suppresses the functions of let-7 miRNA and miR-184*, which regulate the translation of the E2F1/DP complex involved in cell cycle driving [89]. Moreover, LRRK2 also regulates gene transcription through the phosphorylation of HDAC3 by promoting histone deacetylation. In particular, LRRK2 leads to transcriptional repression of MEF2D, a gene associated with neuronal survival [90]. In addition to the G2019S mutation, at least six pathogenic *LRRK2* mutations have been identified as causative for PD that induce autosomal dominant PD [91]. The presence of non-coding variants in LRRK2 in sporadic PD, suggests that altered transcription of this gene is associated with the pathophysiology of sporadic PD [92].

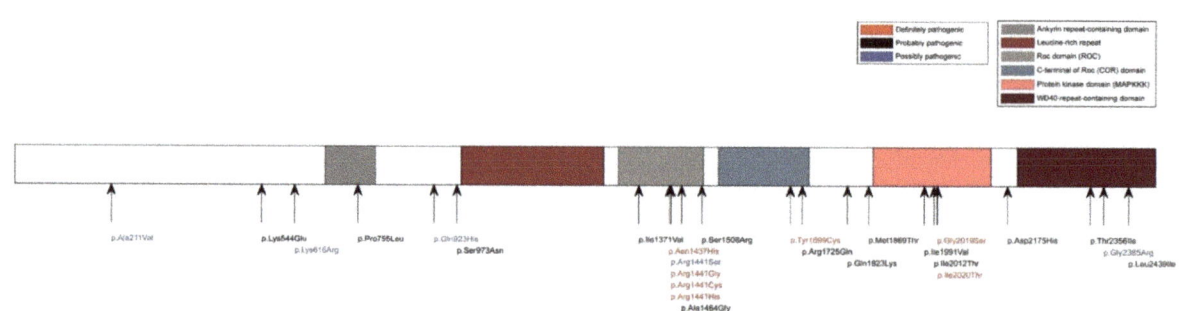

Figure 3. Current known variants in LRRK2 illustrated at the protein level (image obtained from MDS genes). The protein kinase domain of the LRRK2 protein where the G2019S mutation is located is visible in pink in the figure.

It was reported that LRRK2 mRNA and protein levels differ between brain regions, with expression in target areas of the dopaminergic system, such as the striatum and frontal cortex, whereas neurons in the *Substantia nigra* showed very low mRNA and protein expression levels [93,94]. Consistent with this, immunohistochemical analysis of LRRK2 protein in *Substantia nigra* dopamine neuron bodies found no signal in either controls or PD cases [23].

3.2. Regulation of LRRK2 Transcription

LRRK2 protein expression was higher in the frontal cortex and striatal neurons of sporadic PD patients compared to controls, in contrast to mRNA levels which did not vary between patients and controls [23]. In the frontal cortex of sporadic PD patients, an increase in LRRK2 expression was correlated with a decrease in miR-205. miR-205 is able to bind to the 3'UTR of *LRRK2* mRNA and suppress its expression [23]. In neuronal cultures expressing the *LRRK2* R1441G mutation, overexpression of miR-205 protected them from neurite growth defects [23]. The mechanism of miR-205 depletion in PD patients is, however, undetermined. In cancer, epigenetic modifications such as histone modification and DNA methylation, but also microenvironmental changes such as hypoxia, inflammation, and cytokines, contribute to miR-205 dysregulation [95]. It has been suggested that miRNA expression can be regulated by hypoxia in a tissue-specific manner [96]. MiR-205 was found to be induced by hypoxia in cervical and lung cancer cells, potentially through suppression of the apoptosis-stimulating protein p53-2 [97]. Furthermore, miR-205 expression was found upregulated in thymic epithelial cells following inflammatory responses where it helps to preserve the maturation of T cells in response to inflammation [98].

The transcription factor Sp1 promotes LRRK2 expression [99]. However, it has been shown that Sp1 induction also activates miR-205 expression [95], which may be a feedback mechanism for LRRK2 overexpression.

LncRNAs also appear to play a role in the regulation of LRRK2. The LncRNA HOTAIR has been shown to increase *LRRK2* mRNA stability and increase its expression. HOTAIR is upregulated in neurons of MPP+-induced PD mice while its knockdown provides protection against MPP+-induced neuronal apoptosis [100].

3.3. LRRK2 in Immune Cells

LRRK2 is also expressed in immune cells (lymphocytes, monocytes, neutrophils and also microglia) where its expression is tightly regulated by immune stimulation, implicating its potential role as a regulator of immune responses [83,101]. LRRK2 expression is higher in lymphocytes and inflammatory monocytes of late-onset PD patients compared to age-matched individuals, suggesting the role of inflammation in the development of PD [24]. This could explain why hypomethylation of the *LRRK2* gene is observed in leukocytes from PD patients [21]. LRRK2 is also expressed in primary microglia from adult mice and it is upregulated upon IFN-y stimulation or lipopolysaccharide (LPS) treatment [102]. An iPSC study revealed that basal *LRRK2* mRNA expression was lower in sporadic PD microglia, and after treatment of the cells with LPS, sporadic PD microglia had a significantly lower amount of LRRK2 protein than control microglia [103]. However, the mechanisms linking LRRK2 downregulation to microglia dysfunction in PD remains to be elucidated.

4. Dysregulation of Genes Involved in Recessive Forms of PD

Genes involved in recessive forms of PD such as *PRKN*, *PINK1* and *DJ-1* are essential for physiological mitochondrial function.

4.1. Expression Profile of PRKN and Its Regulation

PRKN encodes Parkin, an E3 ubiquitin ligase. Parkin ubiquitinates various proteins, thereby promoting their proteasomal degradation. One of its roles is to control mitochondrial biogenesis, notably mediated by its influence on PGC1-α by ubiquitinating its repressor, PARIS [104]. In addition, Parkin is involved in cell survival signaling pathways [105].

With a genomic sequence above 1.38 Mb, the *PRKN* gene is the second largest in the human genome widely expressed in the brain [106]. More than 170 *PRKN* mutations have been associated with PD, including point mutations and genomic rearrangements [70]. In vitro, it has been reported that some mutations in *PRKN* result in a loss of the ubiquitin–protein ligase activity of Parkin [107]. A deletion in the promoter regions of *PRKN* resulting in the absence of the *PRKN* mRNA transcript has been associated with an early form of PD [108]. Parkin haploinsufficiency has also been identified as a risk factor for familial PD with a tendency towards an earlier age of onset [109]. Inactivation of *PRKN* in mice resulted in motor and cognitive deficits [110]. Overexpression of parkin or restoration of its activity leads to a protective effect against neurodegeneration in cell culture and in animal models [111–113]. These observations suggest that reduced expression of *PRKN* might confer a risk for developing PD. Post-translational modifications of Parkin induced by oxidative and nitrosative stress (sulfonation and S-nitrosylation) are increased in the brain of PD patients. It has been shown that these changes lead to a disruption of the E3 ligase activity of parkin and a dysfunction of the ubiquitin–proteasome system [114,115].

In addition, sulfonation and S-nitrosylation of Parkin alter the solubility of the protein, promoting its intracellular aggregation [115,116]. The decrease in the availability of soluble Parkin through aggregation could be involved in the pathophysiology of the disease.

The study of *PRKN* expression is complicated by the presence of different *PRKN* mRNAs due to alternative splicing of the gene [117]. To date, 26 *PRKN* transcripts have been identified, corresponding to 21 different alternative splice variants [118]. The pattern of *PRKN* expression differs between tissues and cells, with distinct splice variants in human brain regions and leukocytes [117]. Alternative splicing of non-coding sequences can influence the stability, translational activity and subcellular localization of transcripts; [119] while alternative splicing of coding sequences can generate protein isoforms with different biological properties. Alternative splicing could be regulated by LncRNA [120].

Distinct Parkin isoforms have been found to be differentially expressed in specific regions of the rat brain [121] and several isoforms of Parkin have been identified in human blood cells [122]. These observations suggest that the profile of Parkin isoforms may differ between human cells and tissues [118]. Parkin isoforms may have different subcellular locations as well as different functions. Recently it has been shown that intranuclear Parkin could change the transcriptional activity of genes involved in regulating multiple metabolic pathways through interaction with transcription factors [123].

4.2. Role of Parkin and Its Epigenetic Regulation in Sporadic PD

Even though reduction of *PRKN* is potentially involved in the pathophysiology of monogenetic forms of PD, *PRKN* expression does not appear to be decreased in the sporadic form. TV3 and TV12 variants of *PRKN*, resulting from alternative splicing, were overexpressed in the frontal cortex of sporadic PD compared to controls [25], suggesting a change in the expression pattern of *PRKN*. Another study reported that 3, 7, and 11 *PRKN* transcripts were overexpressed in the striatum and cerebellar cortex of PD patients compared to controls [124].

Studies on small cohorts of PD patients revealed no differences in PRKN methylation levels in blood and brain [26,27]. A recent study of blood samples from 91 early-onset PD patients showed hypomethylation of the PRKN promoter in this group compared to healthy controls [28]. However, this hypomethylation may not explain a reduction in PRKN expression and could rather correspond to a compensatory mechanism. Data comparing samples of PD patients and healthy subjects revealed a reduction in miR-181a in the serum [29] and a reduction in miR-218 in the brain of PD patients [30]. In vitro, overexpression of miR-181a and miR-218 each, induced a reduction in PRKN mRNA [125,126]. Among the miRNAs increased in brain or plasma samples from PD patients, some showed in silico binding sequences to PRKN 3'-UTR mRNA. It was confirmed that a selected miRNA, miR-103a-3p [31], directly regulates PRKN mRNA translation leading to a downregulation of Parkin protein level [127]. Furthermore miR-103a-3p inhibition improved mitophagy and had neuroprotective effects in PD models in vitro and in vivo [127].

4.3. PINK1 Expression in PD Patient Brains

The *PINK1* gene encodes a PTEN-induced serine kinase located in mitochondria that protects against mitochondrial dysfunction and regulates the mitochondrial fission/fusion mechanism. PINK1 is imported into the mitochondria and rapidly cleared by the proteasome. However, stress factors can lead PINK1 to accumulate in the outer mitochondrial membrane. PINK1 will then homodimerize, resulting in autophosphorylation, which promotes kinase activation and facilitates binding to its substrates Parkin and ubiquitin [104]. Parkin activation forms ubiquitin chains and this mechanism allows more Parkin to be recruited to the mitochondria, amplifying the damage detection signal. By ubiquitinating various proteins in the outer mitochondrial membrane, Parkin then initiates mitophagy (Figure 4). Alteration of this pathway can lead to the accumulation of dysfunctional mitochondria which may contribute to the loss of dopaminergic cells [104].

Most *PINK1* mutations are point mutations, small insertions or deletions, however, deletions of the entire gene and large complex rearrangements have also been reported [107]. By affecting several mitochondrial phases, including fission, fusion and mitophagy, *PINK1* mutations could lead to respiratory chain dysfunction and impaired ATP production [128].

The full-length form of PINK1 (FL-PINK1) imported into mitochondria undergoes proteolysis to produce Δ1-PINK1 which is then relegated to the cytosol and interacts with Parkin [129]. Binding of Δ1-PINK1 to Parkin impairs the recruitment of Parkin to mitochondria and represses mitophagy [129]. In the brains of PD patients, level of PINK1 mRNA were reported to not differ significantly from those of healthy subjects [33]. Accordingly, a recent study found no difference in *PINK1* methylation in the brains of PD patients compared to controls [34].

NFκB levels are elevated in dopaminergic neurons of Parkinson's disease patients, reflecting an apoptotic and inflammatory state. In vitro PINK1 expression appears to be positively regulated by NFκB. An increase in Δ1-PINK1 level has been reported in the *Substantia nigra* of post-mortem PD patient brains compared to controls [32]. PINK1 translation appears to be critical for the accumulation of the protein during mitochondrial damage. It has been reported that miR-27a/b represses *PINK1* expression by direct binding to the 3'UTR of its mRNA. PINK1 accumulation upon mitochondrial damage was regulated by miR-27a/b expression levels. The latter inhibits *PINK1* translation suppressing autophagic clearance of damaged mitochondria. Furthermore, miR-27a/b expression is increased under chronic mitophagic flux, suggesting a negative feedback regulation between PINK1-mediated mitophagy and these miRNAs [130]. Studies have reported a decrease in miR-27a in PBMCs as well as in cerebrospinal fluid in early-onset PD compared to controls [35,36]. The implication of this observation in the pathophysiology is still unclear.

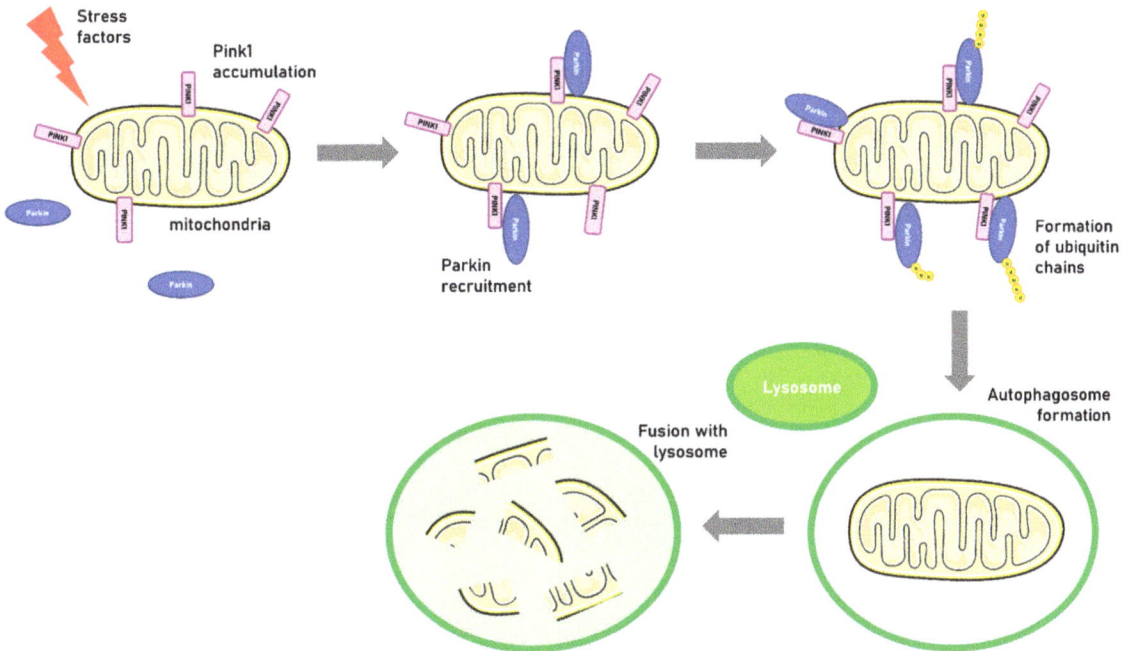

Figure 4. Scheme of mitophagy induced by PINK1 and Parkin. Abbreviation: u: ubiquitin. Under stress factors, PINK1 accumulates and recruits Parkin to the outer membrane of mitochondria. PINK1 and Parkin signaling allows autophagosome formation. Fusion of the autophagosome with the lysosome leads to mitophagy.

4.4. Regulation of DJ-1 Expression

DJ-1 is expressed in many tissues and cells, including neurons and glial cells [131]. DJ-1 has an antioxidant function, notably through the elimination of reactive oxygen species (ROS) [132]. In the presence of oxidative stress, cytoplasmic DJ-1 is translocated to the outer mitochondrial membrane and is thought to play a role in neuroprotection [133]. Different isoforms of *DJ-1* have been identified in brain tissue. Analysis of post-mortem brain samples revealed a decrease in *DJ-1* mRNA and protein, as well as the presence of extra-oxidized *DJ-1* isoforms in subjects with PD compared to controls [37]. Acidic isoforms of the *DJ-1* monomer were selectively accumulated in the brains of sporadic PD patients compared to controls [134]. Depletion of DJ-1 in vitro increases the sensitivity of cells to oxidative stress [135]. However, no significant association was found between polymorphisms within the *DJ-1* gene promoter and the risk of PD [136].

At the epigenetic level, the DNA methylation of *DJ-1* promoter in leukocytes was not different between PD patients and controls [38]. One brain-enriched miRNA, miR-494, was reported to bind to the 3′ UTR of *DJ-1* mRNA and reduce its level [137]. After treating SH-SY5Y cells with MPP+, miR-494-3p expression was increased [138]. In addition, mice overexpressing miR-494 treated with MPTP showed decreased expression of *DJ-1* with exacerbated degeneration of DA neurons and worsened motor impairment [138].

By binding to the 3′UTR of *DJ-1* mRNAs, another miRNA, miR-4639-5p, represses DJ-1 translation. Overexpression of miR-4639-5p in MPP+-treated SH-SY5Y cells results in reduced DJ-1 protein and increased vulnerability to cellular stress [39]. miR-4639-5p expression was higher in the plasma of PD patients than in controls, suggesting its potential role in the pathophysiology of PD [39].

5. GBA a Risk Factor for PD

In PD, clearance of the autophagy–lysosome pathway has been shown to be inefficient, since pathological aSyn can potentially compromise several stages of this pathway [86]. Mutations in the *GBA* gene encoding the lysosomal enzyme, glucocerebrosidase (GCase) involved in the lysosomal storage disorder, Gaucher's disease, have been identified as the most common PD risk factor, highlighting the key role of lysosomal dysfunction in PD [86]. Post-mortem brain analysis revealed a decrease in GCase protein level and activity with increased α-Syn levels in PD patients compared to controls [40]. However, this decrease in the GCase activity does not seem to correlate with a low level of *GBA* mRNA expression [40,41].

To date, whole genome methylation studies have not demonstrated differential methylations of the *GBA* gene [28]. It was shown that miR-127-5p decreases GCase activity and protein levels, this effect being indirectly mediated by a decrease in LIMP-2, a receptor involved in GCase trafficking from the endoplasmic reticulum to the lysosome [139]. In addition, miR-16-5p and miR-195-5p have been shown to increase *GBA* transcript and GCase protein levels. However, their pathophysiological mechanism is not elucidated [139]. Recently, another hypothesis has been explored, based on the observation that specific lncRNAs can limit miRNA activity by sequestration, thus upregulating the expression of target genes. In this context, transcribed pseudogenes, nonfunctional segments of DNA that resemble functional genes, are ideal candidates because they share miRNA binding sites with the transcripts of target genes. In vitro, it has been reported that miR-22-3p directly binds to *GBA* and its pseudogene *GBAP1*, thus downregulating these two genes by decreasing their mRNA levels. Overexpression of *GBAP1* 3' UTR in cell lines resulted in sequestration of mir-22-3p, thereby increasing *GBA* mRNA and GCase levels [140]. It is possible that dysregulation of miR-22-3p or *GBAP1* may participate in the pathophysiology of PD.

6. Discussion

The expression of genes involved in familial forms of PD seems to also be dysregulated in sporadic PD patients. However, it remains unclear whether the change in gene expression corresponds to a pathophysiological mechanism, a marker of degeneration or a protective effect.

For *SNCA*, its overexpression seems to be a prerequisite for aSyn aggregation leading to Lewy body (LB) formation [45]. Hypomethylation of *SNCA* could be involved in its overexpression. A pathophysiological mechanism suggested is the sequestration of cytoplasmic DNMT1 by the aggregates leading to the dysregulation of aSyn homeostasis [59]. The decrease in the brain level of miR-34b/c could also participate in the overexpression of *SNCA* [75].

The increase in LRRK2 kinase activity mediated by the G2019S and other disease-causing mutations raises the question whether *LRRK2* is overexpressed in sporadic PD [88]. Increased expression of *LRRK2* was found in the brains of PD patients, without increased transcripts [23]. A step in the pathophysiology could be a decrease in miR-205 expression, but the causes of this dysregulation is not clear [23]. Furthermore, the regulation of *LRRK2* expression by immune stimulation in blood tissue suggests the implication of neuroinflammatory mechanisms [101] and could represent an accessible epigenetic biomarker.

Due to alternative splicing and multiple splice isoforms, it is more challenging to understand the relationship between Parkin expression and its pathophysiology in sporadic PD [117]. Epigenetic mechanisms could be involved in the modification of the *PRKN* expression profile.

Mitochondrial damage could induce overexpression of PINK1 [130]. However, the role of increased *PINK1* expression is unclear. It could be either protective or participating in neuronal death. Inflammation in PD seems to be most intense at the beginning, just after clinical diagnosis, attenuating in later stages [141,142]. It has been reported that miR-27a is downregulated in macrophages after stimulation [143]. The decrease in miR-27a observed in early PD may be involved in the inflammation-induced upregulation of PINK1.

Decreased transcription and translation of *DJ-1* in sporadic forms of PD may also be involved in the pathophysiology of the disease [37]. Oxidative stress leads to the expression of miR-494-3p [144]. The induction of miR-4639-5p expression by oxidative stress remains to be explored. We can hypothesize that miRs could be sensors of oxidative stress and contribute to the cellular response by downregulating DJ1.

Decreased GCase protein levels and activity may also be involved in the development of sporadic PD [40].

MiRNA and lncRNA mechanisms may be involved in the regulation of *GBA* expression but their change in expression level in sporadic PD patients is not clear. The mechanisms regulating miRNA expression are not well elucidated. An interesting approach would be to explore DNA methylation patterns or histone modifications at the transcription sites of these miRNA-regulating genes involved in monogenic forms of PD. Furthermore, epigenetic modifications involve tissue-specific processes. The observation of these mechanisms in pathologically relevant cell types and access to these cells is complex.

Analysis of native brain tissue allows us to observe epigenetic changes that take place in the brains of PD subjects. However, these analyses are limited to the post-mortem brain and longitudinal studies cannot be performed to assess the dynamics of epigenetic mechanisms. The collection of more accessible peripheral tissues such as blood or CSF allows an analysis in living individuals over time. However, since the methylation profile is tissue-specific, observations in these tissues do not confirm identical modifications in neurons. Since LRRK2 expression seems to play a role in the inflammatory response, the study of the regulation of its expression in the blood immune cells of PD patients could be interesting. The use of animal models of PD for the analysis of epigenetic alterations presents some difficulties since epigenetic mechanisms are species-specific, and these models may not be representative of epigenetic alterations in humans. However, they can allow for the study of specific neuronal populations [145].

Human iPSC-derived neurons have been developed, which allows for the generation of specific neuron lineages such as dopaminergic neurons derived from PD patients [146,147]. However, the generation of these cells requires cellular reprogramming based on epigenetic modifications resulting in "epigenetic rejuvenation" [148] and may bias the study of epigenetic alterations associated with age in PD. More recent direct conversion techniques make it possible to obtain induced neurons (iNs) with the same epigenetic age as their original fibroblasts. This conversion leaves the most age-related epigenetic marks intact but nevertheless leads to a reorganization of large parts of the epigenome [149]. On the other hand, significant epigenetic changes are largely cell type specific and the value of maintaining epigenetic marks in adult fibroblasts is controversial.

Although these in vitro models have limitations, they offer the opportunity to directly study putative effects of epigenetic modifications on gene expression. Moreover, they allow the development of 3D brain organoids or spheroids (cerebral organoids) with a better reproduction of the cerebral environment. This enables for better understanding of the epigenetic modifications that take place in neurons and other cell lineages such as astrocytes and microglia, also suspected in the pathophysiology of PD [147]. In addition to providing an understanding of the pathophysiology, epigenetic modifications could also allow for the development of biomarkers for the diagnosis, prognosis and monitoring of PD [150]. The demonstration of the causality of epigenetic mechanisms in the onset or progression of the disease could allow for the emergence of new therapeutic targets.

Epigenomic identity may also be mediated by chromosome folding [151]. Recent studies reveal that the 3D organization of the genome correlates with epigenetic modifications and that these modifications predict the structure of chromatin [151,152]. Moreover, changes in the 3D architecture of chromosomes have been observed in cancer cells [153]. This new field of analysis could allow a better understanding of the involvement of epigenetic modifications in PD.

Author Contributions: Conceptualization, A.L. and S.L.; writing—original draft preparation, A.L.; writing—review and editing, L.-L.M.; P.J.M., P.R., H.C.; O.C., J.-C.C., A.B.; supervision, J.-C.C. All authors have read and agreed to the published version of the manuscript.

Funding: This research was funded by the French program "Investissements d'avenir" (ANR-10-IAIHU-06).

Institutional Review Board Statement: Not applicable.

Informed Consent Statement: Not applicable.

Data Availability Statement: Not applicable.

Conflicts of Interest: The authors declare no conflict of interest.

References

1. Shulman, J.M.; De Jager, P.L.; Feany, M.B. Parkinson's Disease: Genetics and Pathogenesis. *Annu. Rev. Pathol. Mech. Dis.* **2011**, *6*, 193–222. [CrossRef] [PubMed]
2. Bandres-Ciga, S.; Diez-Fairen, M.; Kim, J.J.; Singleton, A.B. Genetics of Parkinson's disease: An introspection of its journey towards precision medicine. *Neurobiol. Dis.* **2020**, *137*, 104782. [CrossRef] [PubMed]
3. Spillantini, M.G.; Schmidt, M.L.; Lee, V.M.; Trojanowski, J.Q.; Jakes, R.; Goedert, M. α-synuclein in Lewy bodies. *Nature* **1997**, *388*, 839–840. [CrossRef] [PubMed]
4. Nunnari, J.; Suomalainen, A. Mitochondria: In Sickness and in Health. *Cell* **2012**, *148*, 1145–1159. [CrossRef]
5. Gibney, E.R.; Nolan, C.M. Epigenetics and gene expression. *Heredity* **2010**, *105*, 4–13. [CrossRef]
6. Bannister, A.J.; Kouzarides, T. Regulation of chromatin by histone modifications. *Cell Res.* **2011**, *21*, 381–395. [CrossRef]
7. Kouzarides, T. Chromatin modifications and their function. *Cell* **2007**, *128*, 693–705. [CrossRef]
8. Goll, M.G.; Kirpekar, F.; Maggert, K.A.; Yoder, J.A.; Hsieh, C.L.; Zhang, X.; Golic, K.G.; Jacobsen, S.E.; Bestor, T.H. Methylation of tRNA^{Asp} by the DNA methyltransferase homolog Dnmt2. *Science* **2006**, *311*, 395–398. [CrossRef]
9. Laird, P.W. Principles and challenges of genome-wide DNA methylation analysis. *Nat. Rev. Genet.* **2010**, *11*, 191–203. [CrossRef]
10. Moore, L.D.; Le, T.; Fan, G. DNA Methylation and Its Basic Function. *Neuropsychopharmacology* **2013**, *38*, 23–38. [CrossRef]
11. Goll, M.G.; Bestor, T.H. Eukaryotic Cytosine Methyltransferases. *Annu. Rev. Biochem.* **2005**, *74*, 481–514. [CrossRef]
12. Xue, M.; Zhuo, Y.; Shan, B. MicroRNAs, long noncoding RNAs, and their functions in human disease. In *Bioinformatics in MicroRNA Research*; Huang, J., Borchert, G.M., Dou, D., Huan, J., Lan, W., Tan, M., Wu, B., Eds.; Methods in Molecular Biology; Springer: New York, NY, USA, 2017; Volume 1617, pp. 1–25. [CrossRef]
13. Argonaute Proteins: Functional Insights and Emerging Roles | Nature Reviews Genetics. Available online: https://www.nature.com/articles/nrg3462 (accessed on 26 October 2021).
14. miRNA Activity Inferred from Single Cell mRNA Expression | Scientific Reports. Available online: https://www.nature.com/articles/s41598-021-88480-5 (accessed on 24 November 2021).
15. Gil, N.; Ulitsky, I. Regulation of gene expression by cis-acting long non-coding RNAs. *Nat. Rev. Genet.* **2020**, *21*, 102–117. [CrossRef]
16. Gründemann, J.; Schlaudraff, F.; Haeckel, O.; Liss, B. Elevated α-synuclein mRNA levels in individual UV-laser-microdissected dopaminergic substantia nigra neurons in idiopathic Parkinson's disease. *Nucleic Acids Res.* **2008**, *36*, e38. [CrossRef]
17. Jowaed, A.; Schmitt, I.; Kaut, O.; Wüllner, U. Methylation Regulates α-Synuclein Expression and Is Decreased in Parkinson's Disease Patients' Brains. *J. Neurosci.* **2010**, *30*, 6355–6359. [CrossRef]
18. Matsumoto, L.; Takuma, H.; Tamaoka, A.; Kurisaki, H.; Date, H.; Tsuji, S.; Iwata, A. CpG Demethylation Enhances α-Synuclein Expression and Affects the Pathogenesis of Parkinson's Disease. *PLoS ONE* **2010**, *5*, e15522. [CrossRef]
19. Pihlstrøm, L.; Berge, V.; Rengmark, A.; Toft, M. Parkinson's disease correlates with promoter methylation in the α-synuclein gene. *Mov. Disord.* **2015**, *30*, 577–580. [CrossRef]
20. Ai, S.-X.; Xu, Q.; Hu, Y.-C.; Song, C.-Y.; Guo, J.-F.; Shen, L.; Wang, C.-R.; Yu, R.-L.; Yan, X.-X.; Tang, B.-S. Hypomethylation of SNCA in blood of patients with sporadic Parkinson's disease. *J. Neurol. Sci.* **2014**, *337*, 123–128. [CrossRef]
21. Tan, Y.-Y.; Wu, L.; Zhao, Z.-B.; Wang, Y.; Xiao, Q.; Liu, J.; Wang, G.; Ma, J.-F.; Chen, S.-D. Methylation of α-synuclein and leucine-rich repeat kinase 2 in leukocyte DNA of Parkinson's disease patients. *Park. Relat. Disord.* **2014**, *20*, 308–313. [CrossRef]
22. Miñones-Moyano, E.; Porta, S.; Escaramís, G.; Rabionet, R.; Iraola, S.; Kagerbauer, B.; Espinosa-Parrilla, Y.; Ferrer, I.; Estivill, X.; Martí, E. MicroRNA profiling of Parkinson's disease brains identifies early downregulation of miR-34b/c which modulate mitochondrial function. *Hum. Mol. Genet.* **2011**, *20*, 3067–3078. [CrossRef]
23. Cho, H.J.; Liu, G.; Jin, S.M.; Parisiadou, L.; Xie, C.; Yu, J.; Sun, L.; Ma, B.; Ding, J.; Vancraenenbroeck, R.; et al. MicroRNA-205 regulates the expression of Parkinson's disease-related leucine-rich repeat kinase 2 protein. *Hum. Mol. Genet.* **2013**, *22*, 608–620. [CrossRef]
24. Cook, D.A.; Kannarkat, G.T.; Cintron, A.F.; Butkovich, L.M.; Fraser, K.B.; Chang, J.; Grigoryan, N.; Factor, S.A.; West, A.B.; Boss, J.M.; et al. LRRK2 levels in immune cells are increased in Parkinson's disease. *NPJ Parkinson's Dis.* **2017**, *3*, 11. [CrossRef]

25. Beyer, K.; Domingo-Sàbat, M.; Humbert, J.; Carrato, C.; Ferrer, I.; Ariza, A. Differential expression of α-synuclein, parkin, and synphilin-1 isoforms in Lewy body disease. *Neurogenetics* **2008**, *9*, 163–172. [CrossRef]
26. Cai, M.; Tian, J.; Zhao, G.-H.; Luo, W.; Zhang, B.-R. Study of Methylation Levels of Parkin Gene Promoter in Parkinson's Disease Patients. *Int. J. Neurosci.* **2011**, *121*, 497–502. [CrossRef]
27. De Mena, L.; Cardo, L.F.; Coto, E.; Alvarez, V.; Coto, E. No differential DNA methylation of PARK2 in brain of Parkinson's disease patients and healthy controls. *Mov. Disord.* **2013**, *28*, 2032–2033. [CrossRef]
28. Eryilmaz, I.E.; Cecener, G.; Erer, S.; Egeli, U.; Tunca, B.; Zarifoglu, M.; Elibol, B.; Tokcaer, A.B.; Saka, E.; Demirkiran, M.; et al. Epigenetic approach to early-onset Parkinson's disease: Low methylation status of SNCA and PARK2 promoter regions. *Neurol. Res.* **2017**, *39*, 965–972. [CrossRef]
29. Ding, H.; Huang, Z.; Chen, M.; Wang, C.; Chen, X.; Chen, J.; Zhang, J. Identification of a panel of five serum miRNAs as a biomarker for Parkinson's disease. *Park. Relat. Disord.* **2016**, *22*, 68–73. [CrossRef]
30. Xing, R.-X.; Li, L.-G.; Liu, X.-W.; Tian, B.-X.; Cheng, Y. Down regulation of miR-218, miR-124, and miR-144 relates to Parkinson's disease via activating NF-κB signaling. *Kaohsiung J. Med. Sci.* **2020**, *36*, 786–792. [CrossRef]
31. Serafin, A.; Foco, L.; Zanigni, S.; Blankenburg, H.; Picard, A.; Zanon, A.; Giannini, G.; Pichler, I.; Facheris, M.F.; Cortelli, P.; et al. Overexpression of blood microRNAs 103a, 30b, and 29a in L-dopa-treated patients with PD. *Neurology* **2015**, *84*, 645–653. [CrossRef]
32. Muqit, M.M.K.; Abou-Sleiman, P.M.; Saurin, A.T.; Harvey, K.; Gandhi, S.; Deas, E.; Eaton, S.; Smith, M.D.P.; Venner, K.; Matilla, A.; et al. Altered cleavage and localization of PINK1 to aggresomes in the presence of proteasomal stress. *J. Neurochem.* **2006**, *98*, 156–169. [CrossRef]
33. Blackinton, J.G.; Anvret, A.; Beilina, A.; Olson, L.; Cookson, M.R.; Galter, D. Expression of PINK1 mRNA in human and rodent brain and in Parkinson's disease. *Brain Res.* **2007**, *1184*, 10–16. [CrossRef]
34. Navarro-Sánchez, L.; Águeda-Gómez, B.; Aparicio, S.; Pérez-Tur, J. Epigenetic Study in Parkinson's Disease: A Pilot Analysis of DNA Methylation in Candidate Genes in Brain. *Cells* **2018**, *7*, 150. [CrossRef] [PubMed]
35. Fazeli, S.; Motovali-Bashi, M.; Peymani, M.; Hashemi, M.-S.; Etemadifar, M.; Nasr-Esfahani, M.H.; Ghaedi, K. A compound downregulation of SRRM2 and miR-27a-3p with upregulation of miR-27b-3p in PBMCs of Parkinson's patients is associated with the early stage onset of disease. *PLoS ONE* **2020**, *15*, e0240855.
36. Dos Santos, M.C.T.; Barreto-Sanz, M.A.; Correia, B.R.S.; Bell, R.; Widnall, C.; Perez, L.T.; Berteau, C.; Schulte, C.; Scheller, D.; Berg, D.; et al. miRNA-based signatures in cerebrospinal fluid as potential diagnostic tools for early stage Parkinson's disease. *Oncotarget* **2018**, *9*, 17455–17465. [CrossRef] [PubMed]
37. Kumaran, R.; Vandrovcova, J.; Luk, C.; Sharma, S.; Renton, A.; Wood, N.; Hardy, J.A.; Lees, A.J.; Bandopadhyay, R. Differential DJ-1 gene expression in Parkinson's disease. *Neurobiol. Dis.* **2009**, *36*, 393–400. [CrossRef] [PubMed]
38. Tan, Y.; Wu, L.; Li, D.; Liu, X.; Ding, J.; Chen, S. Methylation status of DJ-1 in leukocyte DNA of Parkinson's disease patients. *Transl. Neurodegener.* **2016**, *5*, 5. [CrossRef] [PubMed]
39. Chen, Y.; Gao, C.; Sun, Q.; Pan, H.; Huang, P.; Ding, J.; Chen, S. MicroRNA-4639 Is a Regulator of DJ-1 Expression and a Potential Early Diagnostic Marker for Parkinson's Disease. *Front. Aging Neurosci.* **2017**, *9*, 232. [CrossRef] [PubMed]
40. Murphy, K.E.; Gysbers, A.M.; Abbott, S.K.; Tayebi, N.; Kim, W.S.; Sidransky, E.; Cooper, A.; Garner, B.; Halliday, G.M. Reduced glucocerebrosidase is associated with increased α-synuclein in sporadic Parkinson's disease. *Brain* **2014**, *137*, 834–848. [CrossRef]
41. Moors, T.E.; Paciotti, S.; Ingrassia, A.; Quadri, M.; Breedveld, G.; Tasegian, A.; Chiasserini, D.; Eusebi, P.; Duran-Pacheco, G.; Kremer, T.; et al. Characterization of Brain Lysosomal Activities in GBA-Related and Sporadic Parkinson's Disease and Dementia with Lewy Bodies. *Mol. Neurobiol.* **2019**, *56*, 1344–1355. [CrossRef]
42. Wong, Y.C.; Krainc, D. α-synuclein toxicity in neurodegeneration: Mechanism and therapeutic strategies. *Nat. Med.* **2017**, *23*, 1–13. [CrossRef]
43. Maroteaux, L.; Campanelli, J.T.; Scheller, R.H. Synuclein: A neuron-specific protein localized to the nucleus and presynaptic nerve terminal. *J. Neurosci.* **1988**, *8*, 2804–2815. [CrossRef]
44. Bendor, J.T.; Logan, T.P.; Edwards, R.H. The Function of α-Synuclein. *Neuron* **2013**, *79*, 1044–1066. [CrossRef]
45. Kalia, L.V.; Kalia, S.K.; McLean, P.; Lozano, A.M.; Lang, A.E. α-Synuclein oligomers and clinical implications for Parkinson disease. *Ann. Neurol.* **2012**, *73*, 155–169. [CrossRef]
46. Erskine, D.; Patterson, L.; Alexandris, A.; Hanson, P.S.; McKeith, I.G.; Attems, J.; Morris, C.M. Regional levels of physiological α-synuclein are directly associated with Lewy body pathology. *Acta Neuropathol.* **2018**, *135*, 153–154. [CrossRef]
47. Middleton, E.R.; Rhoades, E. Effects of Curvature and Composition on α-Synuclein Binding to Lipid Vesicles. *Biophys. J.* **2010**, *99*, 2279–2288. [CrossRef]
48. Davidson, W.S.; Jonas, A.; Clayton, D.F.; George, J.M. Stabilization of α-Synuclein Secondary Structure upon Binding to Synthetic Membranes. *J. Biol. Chem.* **1998**, *273*, 9443–9449. [CrossRef]
49. Fusco, G.; Chen, S.W.; Williamson, P.T.F.; Cascella, R.; Perni, M.; Jarvis, J.A.; Cecchi, C.; Vendruscolo, M.; Chiti, F.; Cremades, N.; et al. Structural basis of membrane disruption and cellular toxicity by α-synuclein oligomers. *Science* **2017**, *358*, 1440–1443. [CrossRef]
50. Kontopoulos, E.; Parvin, J.D.; Feany, M.B. α-synuclein acts in the nucleus to inhibit histone acetylation and promote neurotoxicity. *Hum. Mol. Genet.* **2006**, *15*, 3012–3023. [CrossRef]

51. Outeiro, T.F.; Kontopoulos, E.; Altmann, S.M.; Kufareva, I.; Strathearn, K.E.; Amore, A.M.; Volk, C.B.; Maxwell, M.M.; Rochet, J.-C.; McLean, P.J.; et al. Sirtuin 2 Inhibitors Rescue α-Synuclein-Mediated Toxicity in Models of Parkinson's Disease. *Science* **2007**, *317*, 516–519. [CrossRef]
52. Mutez, E.; Lepretre, F.; Le Rhun, E.; Larvor, L.; Duflot, A.; Mouroux, V.; Kerckaert, J.; Figeac, M.; Dujardin, K.; Destée, A.; et al. SNCA locus duplication carriers: From genetics to Parkinson disease phenotypes. *Hum. Mutat.* **2011**, *32*, E2079–E2090. [CrossRef]
53. Ross, O.A.; Braithwaite, A.T.; Skipper, L.M.; Kachergus, J.; Hulihan, M.M.; Middleton, F.A.; Nishioka, K.; Fuchs, J.; Gasser, T.; Maraganore, D.M.; et al. Genomic investigation of α-synuclein multiplication and parkinsonism. *Ann. Neurol.* **2008**, *63*, 743–750. [CrossRef]
54. Chiba-Falek, O.; Lopez, G.J.; Nussbaum, R.L. Levels of α-synuclein mRNA in sporadic Parkinson disease patients. *Mov. Disord.* **2006**, *21*, 1703–1708. [CrossRef]
55. McLean, J.R.; Hallett, P.J.; Cooper, O.; Stanley, M.; Isacson, O. Transcript expression levels of full-length α-synuclein and its three alternatively spliced variants in Parkinson's disease brain regions and in a transgenic mouse model of α-synuclein overexpression. *Mol. Cell. Neurosci.* **2012**, *49*, 230–239. [CrossRef]
56. Tong, J.; Wong, H.; Guttman, M.; Ang, L.C.; Forno, L.S.; Shimadzu, M.; Rajput, A.H.; Muenter, M.D.; Kish, S.J.; Hornykiewicz, O.; et al. Brain α-synuclein accumulation in multiple system atrophy, Parkinson's disease and progressive supranuclear palsy: A comparative investigation. *Brain* **2010**, *133*, 172–188. [CrossRef]
57. Nalls, M.A.; Pankratz, N.; Lill, C.M.; Do, C.B.; Hernandez, D.G.; Saad, M.; DeStefano, A.L.; Kara, E.; Bras, J.; Sharma, M.; et al. Large-scale meta-analysis of genome-wide association data identifies six new risk loci for Parkinson's disease. *Nat. Genet.* **2014**, *46*, 989–993. [CrossRef]
58. Soldner, F.; Stelzer, Y.; Shivalila, C.S.; Abraham, B.; Latourelle, J.C.; Barrasa, M.I.; Goldmann, J.; Myers, R.H.; Young, R.A.; Jaenisch, R. Parkinson-associated risk variant in distal enhancer of α-synuclein modulates target gene expression. *Nature* **2016**, *533*, 95–99. [CrossRef]
59. Desplats, P.; Spencer, B.; Coffee, E.; Patel, P.; Michael, S.; Patrick, C.; Adame, A.; Rockenstein, E.; Masliah, E. α-Synuclein Sequesters Dnmt1 from the Nucleus. *J. Biol. Chem.* **2011**, *286*, 9031–9037. [CrossRef]
60. Jiang, W.; Li, J.; Zhang, Z.; Wang, H.; Wang, Z. Epigenetic upregulation of α-synuclein in the rats exposed to methamphetamine. *Eur. J. Pharmacol.* **2014**, *745*, 243–248. [CrossRef]
61. Schmitt, I.; Kaut, O.; Khazneh, H.; Deboni, L.; Ahmad, A.; Berg, D.; Klein, C.; Fröhlich, H.; Wüllner, U. L-dopa increases α-synuclein DNA methylation in Parkinson's disease patients in vivo and in vitro. *Mov. Disord.* **2015**, *30*, 1794–1801. [CrossRef]
62. De Lau, L.M.L.; Breteler, M.M.B. Epidemiology of Parkinson's disease. *Lancet Neurol.* **2006**, *5*, 525–535. [CrossRef]
63. Guhathakurta, S.; Kim, J.; Adams, L.; Basu, S.; Song, M.K.; Adler, E.; Je, G.; Fiadeiro, M.B.; Kim, Y.-S. Targeted attenuation of elevated histone marks at SNCA alleviates α-synuclein in Parkinson's disease. *EMBO Mol. Med.* **2021**, *13*, e12188. [CrossRef]
64. Vermunt, M.W.; Reinink, P.; Korving, J.; De Bruijn, E.; Creyghton, P.M.; Basak, O.; Geeven, G.; Toonen, P.W.; Lansu, N.; Meunier, C.; et al. Large-Scale Identification of Coregulated Enhancer Networks in the Adult Human Brain. *Cell Rep.* **2014**, *9*, 767–779. [CrossRef] [PubMed]
65. Voutsinas, G.E.; Stavrou, E.F.; Karousos, G.; Dasoula, A.; Papachatzopoulou, A.; Syrrou, M.; Verkerk, A.J.; Van der Spek, P.; Patrinos, G.P.; Stöger, R.; et al. Allelic imbalance of expression and epigenetic regulation within the α-synuclein wild-type and p.Ala53Thr alleles in Parkinson disease. *Hum. Mutat.* **2010**, *31*, 685–691. [CrossRef] [PubMed]
66. Mittal, S.; Bjørnevik, K.; Soon Im, D.; Flierl, A.; Dong, X.; Locascio, J.J.; Abo, K.M.; Long, E.; Jin, M.; Xu, B.; et al. β2-Adrenoreceptor is a Regulator of the α-Synuclein Gene Driving Risk of Parkinson's Disease. *Science* **2017**, *357*, 891–898. [CrossRef] [PubMed]
67. Mudò, G.; Mäkelä, J.; Di Liberto, V.; Tselykh, T.V.; Olivieri, M.; Piepponen, P.; Eriksson, O.; Mälkiä, A.; Bonomo, A.; Kairisalo, M.; et al. Transgenic expression and activation of PGC-1α protect dopaminergic neurons in the MPTP mouse model of Parkinson's disease. *Cell. Mol. Life Sci.* **2012**, *69*, 1153–1165. [CrossRef]
68. Siddiqui, A.; Chinta, S.J.; Mallajosyula, J.K.; Rajagopolan, S.; Hanson, I.; Rane, A.; Melov, S.; Andersen, J.K. Selective binding of nuclear α-synuclein to the PGC1alpha promoter under conditions of oxidative stress may contribute to losses in mitochondrial function: Implications for Parkinson's disease. *Free Radic. Biol. Med.* **2012**, *53*, 993–1003. [CrossRef]
69. Shin, J.-H.; Ko, H.S.; Kang, H.; Lee, Y.; Lee, Y.-I.; Pletinkova, O.; Troconso, J.C.; Dawson, V.L.; Dawson, T.M. PARIS (ZNF746) Repression of PGC-1α Contributes to Neurodegeneration in Parkinson's Disease. *Cell* **2011**, *144*, 689–702. [CrossRef]
70. Junn, E.; Lee, K.-W.; Jeong, B.S.; Chan, T.W.; Im, J.-Y.; Mouradian, M.M. Repression of α-synuclein expression and toxicity by microRNA-7. *Proc. Natl. Acad. Sci. USA* **2009**, *106*, 13052–13057. [CrossRef]
71. Doxakis, E. Post-transcriptional Regulation of α-Synuclein Expression by mir-7 and mir-153. *J. Biol. Chem.* **2010**, *285*, 12726–12734. [CrossRef]
72. Fragkouli, A.; Doxakis, E. miR-7 and miR-153 protect neurons against MPP+-induced cell death via upregulation of mTOR pathway. *Front. Cell. Neurosci.* **2014**, *8*, 182. [CrossRef]
73. Choi, D.C.; Yoo, M.; Kabaria, S.; Junn, E. MicroRNA-7 facilitates the degradation of α-synuclein and its aggregates by promoting autophagy. *Neurosci. Lett.* **2018**, *678*, 118–123. [CrossRef]
74. Villar-Menéndez, I.; Porta, S.; Buira, S.P.; Pereira-Veiga, T.; Díaz-Sánchez, S.; Albasanz, J.L.; Ferrer, I.; Martín, M.; Barrachina, M. Increased striatal adenosine A2A receptor levels is an early event in Parkinson's disease-related pathology and it is potentially regulated by miR-34b. *Neurobiol. Dis.* **2014**, *69*, 206–214. [CrossRef]

75. Kabaria, S.; Choi, D.C.; Chaudhuri, A.D.; Mouradian, M.M.; Junn, E. Inhibition of miR-34b and miR-34c enhances α-synuclein expression in Parkinson's disease. *FEBS Lett.* **2014**, *589*, 319–325. [CrossRef]
76. Rhinn, H.; Qiang, L.; Yamashita, T.; Rhee, D.; Zolin, A.; Vanti, W.; Abeliovich, A. α-Synuclein transcript alternative 3′UTR usage as a convergent mechanism in Parkinson's disease pathology. *Nat. Commun.* **2012**, *3*, 1084. [CrossRef]
77. Consales, C.; Cirotti, C.; Filomeni, G.; Panatta, M.; Butera, A.; Merla, C.; Lopresto, V.; Pinto, R.; Marino, C.; Benassi, B. Fifty-Hertz Magnetic Field Affects the Epigenetic Modulation of the miR-34b/c in Neuronal Cells. *Mol. Neurobiol.* **2018**, *55*, 5698–5714. [CrossRef]
78. Mizuta, I.; Takafuji, K.; Ando, Y.; Satake, W.; Kanagawa, M.; Kobayashi, K.; Nagamori, S.; Shinohara, T.; Ito, C.; Yamamoto, M.; et al. YY1 binds to α-synuclein 3′-flanking region SNP and stimulates antisense noncoding RNA expression. *J. Hum. Genet.* **2013**, *58*, 711–719. [CrossRef]
79. Segal, T.; Salmon-Divon, M.; Gerlitz, G. The Heterochromatin Landscape in Migrating Cells and the Importance of H3K27me3 for Associated Transcriptome Alterations. *Cells* **2018**, *7*, 205. [CrossRef]
80. Cedar, H.; Bergman, Y. Linking DNA methylation and histone modification: Patterns and paradigms. *Nat. Rev. Genet.* **2009**, *10*, 295–304. [CrossRef]
81. Hughes, A.L.; Kelley, J.R.; Klose, R.J. Understanding the interplay between CpG island-associated gene promoters and H3K4 methylation. *Biochim. Biophys. Acta Gene Regul. Mech.* **2020**, *1863*, 194567. [CrossRef]
82. Zimprich, A.; Biskup, S.; Leitner, P.; Lichtner, P.; Farrer, M.; Lincoln, S.; Kachergus, J.; Hulihan, M.; Uitti, R.J.; Calne, D.B.; et al. Mutations in LRRK2 Cause Autosomal-Dominant Parkinsonism with Pleomorphic Pathology. *Neuron* **2004**, *44*, 601–607. [CrossRef]
83. Marchand, A.; Drouyer, M.; Sarchione, A.; Chartier-Harlin, M.-C.; Taymans, J.-M. LRRK2 Phosphorylation, More Than an Epiphenomenon. *Front. Neurosci.* **2020**, *14*, 527. [CrossRef]
84. Deyaert, E.; Wauters, L.; Guaitoli, G.; Konijnenberg, A.; Leemans, M.; Terheyden, S.; Petrovic, A.; Gallardo, R.; Nederveen-Schippers, L.M.; Athanasopoulos, P.; et al. A homologue of the Parkinson's disease-associated protein LRRK2 undergoes a monomer-dimer transition during GTP turnover. *Nat. Commun.* **2017**, *8*, 1008. [CrossRef]
85. Esteves, A.R.; Swerdlow, R.H.; Cardoso, S.M. LRRK2, a puzzling protein: Insights into Parkinson's disease pathogenesis. *Exp. Neurol.* **2014**, *261*, 206–216. [CrossRef]
86. Minakaki, G.; Krainc, D.; Burbulla, L.F. The Convergence of α-Synuclein, Mitochondrial, and Lysosomal Pathways in Vulnerability of Midbrain Dopaminergic Neurons in Parkinson's Disease. *Front. Cell Dev. Biol.* **2020**, *8*, 1465. [CrossRef]
87. Berwick, D.C.; Heaton, G.R.; Azeggagh, S.; Harvey, K. LRRK2 Biology from structure to dysfunction: Research progresses, but the themes remain the same. *Mol. Neurodegener.* **2019**, *14*, 49. [CrossRef] [PubMed]
88. O'Hara, D.M.; Pawar, G.; Kalia, S.K.; Kalia, L.V. LRRK2 and α-Synuclein: Distinct or Synergistic Players in Parkinson's Disease? *Front. Neurosci.* **2020**, *14*, 577. [CrossRef] [PubMed]
89. Gehrke, S.; Imai, Y.; Sokol, N.; Lu, B. Pathogenic LRRK2 negatively regulates microRNA-mediated translational repression. *Nature* **2010**, *466*, 637–641. [CrossRef] [PubMed]
90. Han, K.A.; Shin, W.H.; Jung, S.; Seol, W.; Seo, H.; Ko, C.; Chung, K.C. Leucine-rich repeat kinase 2 exacerbates neuronal cytotoxicity through phosphorylation of histone deacetylase 3 and histone deacetylation. *Hum. Mol. Genet.* **2017**, *26*, 1–18. [CrossRef] [PubMed]
91. Chen, M.-L.; Wu, R.-M. LRRK 2 gene mutations in the pathophysiology of the ROCO domain and therapeutic targets for Parkinson's disease: A review. *J. Biomed. Sci.* **2018**, *25*, 52. [CrossRef]
92. Trabzuni, D.; Ryten, M.; Emmett, W.; Ramasamy, A.; Lackner, K.J.; Zeller, T.; Walker, R.; Smith, C.; Lewis, P.; Mamais, A.; et al. Fine-Mapping, Gene Expression and Splicing Analysis of the Disease Associated LRRK2 Locus. *PLoS ONE* **2013**, *8*, e70724. [CrossRef]
93. Galter, D.; Westerlund, M.; Carmine, A.; Lindqvist, E.; Sydow, O.; Olson, L. LRRK2 expression linked to dopamine-innervated areas. *Ann. Neurol.* **2006**, *59*, 714–719. [CrossRef]
94. Sharma, S.; Bandopadhyay, R.; Lashley, T.; Renton, A.E.M.; Kingsbury, A.E.; Kumaran, R.; Kallis, C.; Vilariño-Güell, C.; O'Sullivan, S.S.; Lees, A.J.; et al. LRRK2 expression in idiopathic and G2019S positive Parkinson's disease subjects: A morphological and quantitative study: LRRK2 expression in Parkinson's disease. *Neuropathol. Appl. Neurobiol.* **2011**, *37*, 777–790. [CrossRef]
95. Ferrari, E.; Gandellini, P. Unveiling the ups and downs of miR-205 in physiology and cancer: Transcriptional and post-transcriptional mechanisms. *Cell Death Dis.* **2020**, *11*, 980. [CrossRef]
96. Camps, C.; Saini, H.K.; Mole, D.R.; Choudhry, H.; Reczko, M.; Guerra-Assunção, J.A.; Tian, Y.-M.; Buffa, F.M.; Harris, A.L.; Hatzigeorgiou, A.G.; et al. Integrated analysis of microRNA and mRNA expression and association with HIF binding reveals the complexity of microRNA expression regulation under hypoxia. *Mol. Cancer* **2014**, *13*, 28. [CrossRef]
97. Wang, X.; Yu, M.; Zhao, K.; He, M.; Ge, W.; Sun, Y.; Wang, Y.; Sun, H.; Yihua, W. Upregulation of MiR-205 under hypoxia promotes epithelial–mesenchymal transition by targeting ASPP2. *Cell Death Dis.* **2016**, *7*, e2517. [CrossRef]
98. Hoover, A.R.; Dozmorov, I.; MacLeod, J.; Du, Q.; De la Morena, M.T.; Forbess, J.; Guleserian, K.; Cleaver, O.B.; Van Oers, N.S.C. MicroRNA-205 Maintains T Cell Development following Stress by Regulating Forkhead Box N1 and Selected Chemokines. *J. Biol. Chem.* **2016**, *291*, 23237–23247. [CrossRef]
99. Wang, J.; Song, W. Regulation of LRRK2 promoter activity and gene expression by Sp1. *Mol. Brain* **2016**, *9*, 33. [CrossRef]

100. Wang, S.; Zhang, X.; Guo, Y.; Rong, H.; Liu, T. The long noncoding RNA HOTAIR promotes Parkinson's disease by upregulating LRRK2 expression. *Oncotarget* **2017**, *8*, 24449–24456. [CrossRef]
101. Lee, H.; James, W.; Cowley, S. LRRK2 in peripheral and central nervous system innate immunity: Its link to Parkinson's disease. *Biochem. Soc. Trans.* **2017**, *45*, 131–139. [CrossRef]
102. Gillardon, F.; Schmid, R.; Draheim, H. Parkinson's disease-linked leucine-rich repeat kinase 2(R1441G) mutation increases proinflammatory cytokine release from activated primary microglial cells and resultant neurotoxicity. *Neuroscience* **2012**, *208*, 41–48. [CrossRef]
103. Badanjak, K.; Mulica, P.; Smajic, S.; Delcambre, S.; Tranchevent, L.-C.; Diederich, N.; Rauen, T.; Schwamborn, J.C.; Glaab, E.; Cowley, S.A.; et al. iPSC-Derived Microglia as a Model to Study Inflammation in Idiopathic Parkinson's Disease. *Front. Cell Dev. Biol.* **2021**, *9*, 740758. [CrossRef]
104. Ge, P.; Dawson, V.L.; Dawson, T.M. PINK1 and Parkin mitochondrial quality control: A source of regional vulnerability in Parkinson's disease. *Mol. Neurodegener.* **2020**, *15*, 20. [CrossRef]
105. Kamienieva, I.; Duszyński, J.; Szczepanowska, J. Multitasking guardian of mitochondrial quality: Parkin function and Parkinson's disease. *Transl. Neurodegener.* **2021**, *10*, 5. [CrossRef]
106. Murillo-González, F.E.; García-Aguilar, R.; Vega, L.; Elizondo, G. Regulation of Parkin expression as the key balance between neural survival and cancer cell death. *Biochem. Pharmacol.* **2021**, *190*, 114650. [CrossRef]
107. Corti, O.; Lesage, S.; Brice, A. What Genetics Tells us About the Causes and Mechanisms of Parkinson's Disease. *Physiol. Rev.* **2011**, *91*, 1161–1218. [CrossRef]
108. Lesage, S.; Magali, P.; Lohmann, E.; Lacomblez, L.; Teive, H.; Janin, S.; Cousin, P.-Y.; Dürr, A.; Brice, A. Deletion of the parkin and PACRG gene promoter in early-onset parkinsonism. *Hum. Mutat.* **2007**, *28*, 27–32. [CrossRef]
109. Pankratz, N.; Kissell, D.K.; Pauciulo, M.W.; Halter, C.A.; Rudolph, A.; Pfeiffer, R.F.; Marder, K.S.; Foroud, T.; Nichols, W.C.; For the Parkinson Study Group-PROGENI Investigators. Parkin dosage mutations have greater pathogenicity in familial PD than simple sequence mutations. *Neurology* **2009**, *73*, 279–286. [CrossRef]
110. Itier, J.-M.; Ibáñez, P.; Mena, M.A.; Abbas, N.; Cohen-Salmon, C.; Bohme, G.A.; Laville, M.; Pratt, J.; Corti, O.; Pradier, L.; et al. Parkin gene inactivation alters behaviour and dopamine neurotransmission in the mouse. *Hum. Mol. Genet.* **2003**, *12*, 2277–2291. [CrossRef]
111. Petrucelli, L.; O'Farrell, C.; Lockhart, P.; Baptista, M.; Kehoe, K.; Vink, L.; Choi, P.; Wolozin, B.; Farrer, M.; Hardy, J.; et al. Parkin Protects against the Toxicity Associated with Mutant α-Synuclein: Proteasome Dysfunction Selectively Affects Catecholaminergic Neurons. *Neuron* **2002**, *36*, 1007–1019. [CrossRef]
112. Liu, B.; Traini, R.; Killinger, B.; Schneider, B.; Moszczynska, A. Overexpression of parkin in the rat nigrostriatal dopamine system protects against methamphetamine neurotoxicity. *Exp. Neurol.* **2013**, *247*, 359–372. [CrossRef]
113. Staropoli, J.F.; McDermott, C.; Martinat, C.; Schulman, B.; Demireva, E.; Abeliovich, A. Parkin Is a Component of an SCF-like Ubiquitin Ligase Complex and Protects Postmitotic Neurons from Kainate Excitotoxicity. *Neuron* **2003**, *37*, 735–749. [CrossRef]
114. Chung, K.K.K.; Thomas, B.; Li, X.; Pletnikova, O.; Troncoso, J.C.; Marsh, L.; Dawson, V.L.; Dawson, T.M. S-Nitrosylation of Parkin Regulates Ubiquitination and Compromises Parkin's Protective Function. *Science* **2004**, *304*, 1328–1331. [CrossRef] [PubMed]
115. Meng, F.; Yao, D.; Shi, Y.; Kabakoff, J.; Wu, W.; Reicher, J.; Ma, Y.; Moosmann, B.; Masliah, E.; Lipton, S.A.; et al. Oxidation of the cysteine-rich regions of parkin perturbs its E3 ligase activity and contributes to protein aggregation. *Mol. Neurodegener.* **2011**, *6*, 34. [CrossRef] [PubMed]
116. Wang, C.; Ko, H.S.; Thomas, B.; Tsang, F.; Chew, K.C.; Tay, S.-P.; Ho, M.W.; Lim, T.-M.; Soong, T.-W.; Pletnikova, O.; et al. Stress-induced alterations in parkin solubility promote parkin aggregation and compromise parkin's protective function. *Hum. Mol. Genet.* **2005**, *14*, 3885–3897. [CrossRef] [PubMed]
117. La Cognata, V.; Iemmolo, R.; D'Agata, V.; Scuderi, S.; Drago, F.; Zappia, M.; Cavallaro, S. Increasing the Coding Potential of Genomes Through Alternative Splicing: The Case of PARK2 Gene. *Curr. Genom.* **2014**, *15*, 203–216. [CrossRef] [PubMed]
118. Scuderi, S.; La Cognata, V.; Drago, F.; Cavallaro, S.; D'Agata, V. Alternative Splicing Generates Different Parkin Protein Isoforms: Evidences in Human, Rat, and Mouse Brain. *BioMed Res. Int.* **2014**, *2014*, e690796. [CrossRef] [PubMed]
119. Mockenhaupt, S.; Makeyev, E.V. Non-coding functions of alternative pre-mRNA splicing in development. *Semin. Cell Dev. Biol.* **2015**, *47–48*, 32–39. [CrossRef]
120. Pisignano, G.; Ladomery, M. Epigenetic Regulation of Alternative Splicing: How LncRNAs Tailor the Message. *Non-Coding RNA* **2021**, *7*, 21. [CrossRef]
121. D'Amico, A.G.; Maugeri, G.; Reitano, R.; Cavallaro, S.; D'Agata, V. Proteomic Analysis of Parkin Isoforms Expression in Different Rat Brain Areas. *J. Protein Chem.* **2016**, *35*, 354–362. [CrossRef]
122. Kasap, M.; Akpinar, G.; Sazci, A.; Idrisoglu, H.A.; Vahaboğlu, H. Evidence for the presence of full-length PARK2 mRNA and Parkin protein in human blood. *Neurosci. Lett.* **2009**, *460*, 196–200. [CrossRef]
123. Shires, S.E.; Quiles, J.M.; Najor, R.H.; Leon, L.J.; Cortez, M.Q.; Lampert, M.A.; Mark, A.; Gustafsson, B. Nuclear Parkin Activates the ERRα Transcriptional Program and Drives Widespread Changes in Gene Expression Following Hypoxia. *Sci. Rep.* **2020**, *10*, 8499. [CrossRef]
124. Brudek, T.; Winge, K.; Rasmussen, N.B.; Bahl, J.M.C.; Tanassi, J.T.; Agander, T.K.; Hyde, T.M.; Pakkenberg, B. Altered α-synuclein, parkin, and synphilin isoform levels in multiple system atrophy brains. *J. Neurochem.* **2016**, *136*, 172–185. [CrossRef]

Table 1. Genotype-phenotype summary for monogenic forms of Parkinson's disease.

Gene	Mode of Inheritance	Frequency	Ethnic Population Distribution	Types of Mutations	Clinical Phenotype	Response to PD Medication	Response to DBS	Pathological Findings
SNCA	AD	Rare, with a frequency from 0.045% to 1.1% in recent studies [9]	Majority European, then Asians and Hispanics [5]	Missense, duplications, and triplications	Range of age at onset, prominent motor fluctuations, range of complications including cognitive impairment and psychiatric manifestations	Initial good response	Few examples, appears to have a good response for duplications, poor response for missense mutations	α-synuclein-positive and LB pathology [10]
LRRK2	AD	1% of PD but higher in North African Berber Arab and Ashkenazi Jewish populations	The p.G2019S mutation found in Europeans with high prevalence in North African Berbers and Ashkenazi Jews	7 missense variants described	Resembles idiopathic PD	Vast majority show a good response to levodopa [11]	DBS is effective [12]	Most patients with the p.G2019S mutation show LB pathology, whereas this finding is rare for other mutations
VPS35	AD	Rare (overall prevalence of 0.115%)	European, Asian, Ashkenazi Jewish [5]	1 missense mutation described, p.D620N	Resembles idiopathic PD	Good response [11]	Small numbers reported, at least 2 had a good outcome [11]	Not available
PRKN	AR	Most common cause of EOPD, 12.5% of recessive PD [13]	Majority Asian, followed by Caucasians and Hispanics [14]	Missense mutations, frameshift mutations, structural variants	EOPD, lower limb dystonia, absence of cognitive impairment	Good response to levodopa therapy, frequent motor fluctuations, and dyskinesias	Good outcome in all patients [11]	Substantia nigra pars compacta loss with the notable absence of LB pathology

Table 1. Cont.

Gene	Mode of Inheritance	Frequency	Ethnic Population Distribution	Types of Mutations	Clinical Phenotype	Response to PD Medication	Response to DBS	Pathological Findings
PINK1	AR	Second most common cause of EOPD, 1.9% of recessive PD [13]	European, Asian, may be frequent in Arab Berber and Polynesian populations [9,14,15]	Missense mutations, nonsense mutations, structural variants	EOPD, typical PD, dyskinesias, dystonia, and motor fluctuations can occur	Vast majority show a good outcome [11]	Good or moderate [11]	LB pathology may or may not be present in the handful of autopsy cases reported
PARK7 (DJ1)	AR	0.16% of recessive PD [13]	Most patients are from Italy, Iran, and Turkey [14]	Missense, splice site, frameshift, and structural variants	EOPD	50% show a good response, others moderate or minimal [11]	No reports identified [11]	LB pathology [16]
TAF1	X-linked	0.34 per 100,000 in the Philippines, Island of Panay 5.24 per 100,000	Philippines, high prevalence on the Island of Panay	Insertion of a SINE-VNTR-Alu type retrotransposon in intron 32 of the TAF1 gene	Parkinsonism, dystonia	May be responsive to levodopa, particularly for those with pure parkinsonism [17]	DBS results in an improvement in dystonia and to a lesser extent parkinsonism [18]	Accumulation of lipofuscin in the neurons and glia, but absence of LB pathology
ATP13A2	AR	Rare	Spread across the globe [19]	Frameshift, missense, and splice site mutations [19]	KRS, clinical triad of spasticity, dementia, and supranuclear gaze palsy [20], facial-faucial-finger mini-myoclonus [21], other phenotypes include HSP	Variable response to levodopa [19]	May respond well, variable [22]	Accumulation of lipofuscin, absence of LB pathology [23]

Table 1. Cont.

Gene	Mode of Inheritance	Frequency	Ethnic Population Distribution	Types of Mutations	Clinical Phenotype	Response to PD Medication	Response to DBS	Pathological Findings
DCTN1	AD	Rare	Spread across the globe	10 different heterozygous missense mutations [19]	Perry syndrome—rapidly progressive parkinsonism, depression and mood changes, weight loss, and progressive respiratory changes	May be levodopa-responsive	No reports identified	Selective loss of putative respiratory neurons in the ventrolateral medulla and in the raphe nucleus, no or few LBs, TDP43-positive inclusions [24,25]
DNAJC6	AR	Rare	Mainly found in Middle Eastern populations, although families of European origin have also been found to harbor DNAJC6 mutations	5 different homozygous mutations, largest family carries a nonsense mutation [19]	Juvenile PD with complicating features, EOPD	Poor	Good outcome [26]	No reports identified
FBXO7	AR	Rare	Reported in the Iranian, Turkish, Italian, Dutch, Pakistani, and Chinese populations	Biallelic missense, splice site, and nonsense mutations	Juvenile PD, EOPD, parkinsonian-pyramidal syndrome, can overlap with NBIA [27]	Variable	No reports identified	No reports identified
PLA2G6	AR	Rare	Various ethnic groups, including Indian, Pakistani, European, Japanese, Chinese, and Korean populations	54 mutations associated with parkinsonism [28]	Adult-onset dystonia-parkinsonism with cognitive and psychiatric symptoms [28], other phenotypes include NBIA	Variable	May benefit from DBS [28]	Mixed Lewy and Tau pathology [28]

Table 1. *Cont.*

Gene	Mode of Inheritance	Frequency	Ethnic Population Distribution	Types of Mutations	Clinical Phenotype	Response to PD Medication	Response to DBS	Pathological Findings
SYNJ1	AR	Rare	Reported in Iranian, Italian, German, Algerian, Senegalese, and Chinese populations	Missense, frameshift	Parkinsonism in the third decade of life, complicating features such as dystonia, seizures, or cognitive impairment	Poor	No reports identified	No reports identified
CHCHD2	AD	Rare	Japanese and Chinese patients	Missense, splice site	Typical PD	Good	No reports identified	A brain autopsy revealed widespread α-synuclein pathology with Lewy bodies present in the brainstem, neocortex, and limbic regions [29]
LRP10	AD	Rare	Italy, Taiwan	Loss of function and missense variants	Late onset PD, PD dementia, dementia with Lewy Bodies [30–32]	Good [32]	Excellent response for a patient with a *LRP10* and *GBA* variant in trans [33]	Severe LB pathology
TMEM230	AD	Rare	Identified in a Canadian Mennonite family	Missense variant	Typical PD	Responds to levodopa in most cases	No report identified	Typical LB pathology [34]
UQCRC1	AD	Rare	Taiwan, may not be in European populations	Missense variants	Parkinsonism with polyneuropathy	Good	No report identified	No report identified
VPS13C	AR	Rare	Turkish, French	Truncating mutations	EOPD, rapid progression, complicating features including dysphagia, cognitive impairment, hyperreflexia [19,35]	Initial good response	Poor	Resembles diffuse LB disease [35]

AD: autosomal dominant, AR: autosomal recessive, DBS: deep brain stimulation, EOPD: early-onset Parkinson's disease, HSP: hereditary spastic paraplegia, KRS: Kufor Rakeb syndrome, LB: Lewy body, NBIA: neurodegeneration with brain iron accumulation, PD: Parkinson's disease, XDP: X-linked dystonia parkinsonism.

2.1.2. Pathophysiology

The discovery of dominant mutations in *SNCA* as a cause of PD is consistent with the critical role the α-synuclein protein plays in PD pathogenesis. The molecular effects may vary according to the type of *SNCA* mutation [36]. The p.A30P, p.A53T, and p.E46K mutations all affect the N-terminal domain of the α-synuclein protein [36]. The p.A30P and p.A53T mutations stimulate protofibril formation and smaller to larger aggregates [36]. The p.E46K mutation increases the N-terminal positive charge and enhances N-terminal and C-terminal contacts, whereas the opposite is seen for the p.A30P and p.A53T mutations [36]. A recent study showed impaired mitochondrial respiration, energy deficits, vulnerability to rotenone, and altered lipid metabolism in dopaminergic neurons derived from a patient with the p.A30P mutation in *SNCA*, with a comparison to gene-corrected clones, highlighting the numerous effects of these mutations [37].

Table 2. Levodopa-responsiveness stratified according to Parkinson's disease monogenic forms.

Good Response to Levodopa	Poor, Variable, or Uncertain Response to Levodopa
SNCA	TAF1
LRRK2	ATP13A2
VPS35	DCTN1
PRKN	DNAJC6
PINK1	FBXO7
DJ1	PLA2G6
CHCHD2	SYNJ1
LRP10	
TMEM230	
UQCRC1	
VPS13C	

2.2. LRRK2

2.2.1. Genotype-Phenotype

At least seven missense variants in *LRRK2* have been described as causing PD (p.N1437H, p.R1441C/G/H, p.Y1699C, p.G2019S, and p.I2020T) [3]. On an individual level, *LRRK2*-PD is clinically indistinguishable from idiopathic PD. However, as a group, it may be considered as having a milder phenotype [38,39]. For example, *LRRK2* mutation carriers are less likely to have non-motor symptoms such as olfactory impairment, cognitive features, and REM-behavior sleep disorder [39]. Furthermore, patients with *LRRK2*-PD may be susceptible to certain cancers [40–42]. A very recent study provides evidence that *LRRK2*-PD is associated with a significantly higher risk of stroke [43]. Additionally, recent evidence suggests that regular use of non-steroidal anti-inflammatory drugs may be associated with reduced penetrance of PD in both pathogenic and risk variant carriers [44].

The most common and well-characterized *LRRK2* mutation is the p.G2019S mutation. It has a prevalence of 1% in the PD population with a high prevalence in North African Berber Arab (39%) and Ashkenazi Jewish (approximately 18%) populations [45–47]. The penetrance of this mutation is incomplete and variable and influenced by age, environment, and genetic background [48].

Other mutations in *LRRK2* may be relevant to different ethnic and regional populations. For example, the p.R1441C variant has a founder effect in Basque populations and may be higher in Southern Italy and Belgium [38]. The p.G2019S mutation is very rare in Chinese populations, whereas the p.G2385R and p.R1628P variants are common (5–10% in patients, 2–5% in controls) [49–51].

Recent reports suggest that loss of function variants in *LRRK2* are not associated with PD, arguing that haploinsufficiency is neither causative nor protective of PD [52].

2.2.2. Pathophysiology

All the definite *LRRK2* mutations are in the catalytic domains and may result in hyperactivation of the kinase domain [3,53]. LRRK2 is involved in a large array of cell biological processes, and the disease mechanism may reflect important roles in microtubule function and Rab proteins as phosphorylation substrates [2,54].

2.3. VPS35

2.3.1. Genotype-Phenotype

VPS35 is implicated in autosomal dominant PD [55,56], with the missense variant p.D620N being the only mutation confirmed to date. This variant appears to be a mutational hotspot identified in different ethnic populations [57]. The mutation has an overall prevalence of 0.115% from the reported studies but may be as high as 1% in autosomal dominant PD [57–59]. The phenotype resembles idiopathic PD with a median age at onset of 49 years, levodopa responsiveness, and predominant tremor [5,58]. A recent study suggests that disease progression may be slow, with minimal cognitive impairment even after more than 10 years of disease onset [60].

2.3.2. Pathophysiology

VPS35 plays a critical role in endosomal trafficking, but there is emerging evidence for a role in mitochondrial function [61]. The p.D620N mutation impairs the sorting function of the retromer complex, resulting in a disturbance of maturation of endolysosomes and autophagy, membrane receptor recycling, and mitochondrial-derived vesicle formation [2,59,62]. There may also be a role in neurotransmission and an interaction with other genes causing monogenic PD (such as *SNCA*, *LRRK2*, and *PRKN*) [62].

3. Autosomal Recessive Forms

3.1. PRKN

3.1.1. Genotype-Phenotype

Mutations in *PRKN* are the most common cause of early-onset PD (EOPD), particularly in European populations. A recent study by Lesage and colleagues demonstrated that *PRKN* mutations account for 27.6% of autosomal recessive families [13]. They found that the proportion of probands with *PRKN* mutations is higher the younger the age at onset (AAO), as follows: 42.2% for AAO less than or equal to 20 years, 29% for 21 to 30 years, 13% for 31 to 40 years, but only 4.4% for 41 to 60 years [13].

A variety of different mutation types are described, including structural variants (43.2%, including exonic deletions, duplications, and triplications), missense mutations (22.3%), and frameshift mutations (16.5%) [14,63]. Deletions in exon 3 are the most common mutation [14]. Furthermore, a deletion of the *PRKN* and *PACRG* gene promoter has also been described in autosomal recessive PD [63].

PD-*PRKN* is characterized phenotypically by an early age at disease onset, lower limb dystonia at presentation, absence of cognitive impairment, a good and sustained response to levodopa, and frequent motor fluctuations and dyskinesias [64].

3.1.2. Pathophysiology

Mutations in *PRKN* and *PINK1* likely disturb PINK1/parkin-mediated mitophagy, which is the selective degradation of mitochondria, a function essential for mitochondrial homeostasis [65]. In brief, parkin is a E3 ubiquitin ligase that ubiquinates outer mitochondrial membrane proteins such as mitofusin 1 and 2 [66]. PINK1 phosphorylates parkin and maintains its mitochondrial stabilization and translocation, mediating parkin activation [2,66].

3.2. PINK1

3.2.1. Genotype-Phenotype

PINK1 is the second most common cause of autosomal recessive PD and is characterized by typical Parkinson's features such as tremor, bradykinesia, and rigidity, with a median age of onset of 32 [14,67]. Additional phenotypic features include dyskinesias in 39%, dystonia in 21%, and motor fluctuations in 34%, with cognitive impairment and psychosis occurring rarely (14% and 9%, respectively) [14]. The disease is slowly progressive, with a sustained response to levodopa therapy, although with an increased tendency for levodopa-induced dyskinesias.

The main mutation type was missense mutations (47.6%), then structural variants (19.1%), followed by nonsense mutations (14.3%) [14]. The most common specific mutation was a missense mutation, c.1040T>C (p.Leu347Pro) [14].

A recent paper suggests that the c.1040T>C mutation is frequently found in patients from the Pacific Islands [15]. The allele frequency was particularly high in West Polynesians (2.8%), which would translate to a homozygosity of 1 in 5000 people, suggesting that this could have a major contribution to EOPD in the region [15].

3.2.2. Pathophysiology

See *PRKN* above.

3.3. PARK7

3.3.1. Genotype-Phenotype

Mutations in *PARK7* can cause early-onset autosomal recessive parkinsonism, with at least 20 mutations in the *PARK7* gene identified. The majority of *PARK7* mutation carriers have EOPD (83%), whereas 13% have juvenile onset and 4% have late onset [14]. Recently, a Turkish family with juvenile PD was found to have a novel deletion of the neighboring genes of *PARK7* and *TNFRSF9*, raising the possibility of TNFRSF9 as a disease modifier [68].

3.3.2. Pathophysiology

DJ-1 is ubiquitously expressed and is highly expressed in cells with high energy demands. DJ-1 exerts an antioxidative stress function through scavenging reactive oxygen species, regulation of transcription and signal transduction pathways, and acting as a molecular chaperone and enzyme [69]. Mutations within the *PARK7* gene substantially affect the survival of cells in oxidative environments, potentially leading to PD [70,71].

4. X-Linked Dystonia-Parkinsonism

4.1. Genotype-Phenotype

X-linked dystonia-parkinsonism (XDP), also referred to as Lubag, is a movement disorder initially described in Filipino males, caused by the insertion of a SINE-VNTR-Alu (SVA)-type retrotransposon in intron 32 of the *TAF1* gene [72,73]. The prevalence is 0.34 per 100,000 in the Philippines, with a high prevalence on the Island of Panay of 5.24 per 100,000 [74]. It initially presents with dystonia, and predominantly involves the craniocervical region that can become generalized at a later stage [72,75]. It may also present with parkinsonism, or this can develop later in the disease course [75]. Therefore, it can show longitudinal evolution from a hyperkinetic to a hypokinetic movement disorder. Although it primarily affects males, manifesting female carriers have been reported. The median age at onset is 40 years from a recent MDSGene review [72].

4.2. Pathophysiology

Recent evidence suggests that probands with XDP have reduced expression of the canonical TAF1 transcript [73]. De novo assembly of multiple neuronal lineages derived from pluripotent stem cells showed reduced expression due to alternative splicing and intron retention close to the SVA [73]. CRISPR/Cas 9 excision of the SVA was able to rescue TAF1 expression, providing evidence of abnormal transcription mediated by the

SVA in the pathophysiology of XDP [73]. Further evidence suggests that a hexanucleotide repeat within the SVA modifies disease expressivity, with the number of repeats showing an inverse correlation with the age at onset [76].

5. Complex or Atypical Forms

5.1. ATP13A2

5.1.1. Genotype-Phenotype

Biallelic mutations in *ATP13A2* have been found to cause a complex form of parkinsonism known as Kufor-Rakeb syndrome (KRS), characterized by juvenile onset parkinsonism, cognitive impairment, and a supranuclear gaze palsy. *ATP13A2* mutations can also cause a range of phenotypes, including neuronal ceroid lipofuscinosis, hereditary spastic paraplegia, and juvenile amyotrophic lateral sclerosis.

Recently, perhaps the first postmortem study of KRS was reported [77]. This showed accumulation of lipofuscin in the neurons and glia, but an absence of Lewy body pathology as well as alpha-synuclein, TDP43, tau, and beta amyloid pathology. This provides evidence for a pathological link with neuronal lipofuscinosis rather than the typical findings in PD [77].

5.1.2. Pathophysiology

ATP13A2 mutations impair lysosomal and mitochondrial function. The mechanism may involve impaired lysosomal polyamine transport resulting in lysosome-dependent cell death [78].

5.2. DCTN1

5.2.1. Genotype-Phenotype

DCTN1-associated Parkinson-plus disorder, also called Perry syndrome, is a rare autosomal dominant disorder characterized by rapidly progressive parkinsonism, depression and mood changes, weight loss, and progressive respiratory changes, chiefly tachypnoea and nocturnal hypoventilation [79].

The disease is linked to mutations in exon 2 of the *DCTN1* gene. The mean age at onset of disease is 48 years (range: 35–61) and the mean duration to death is 5 years since diagnosis, from either respiratory failure, sudden unexplained death, or suicide [80]. *DCTN1* mutations have been associated with additional phenotypes, including distal spinal and bulbar muscular atrophy and amyotrophic lateral sclerosis.

5.2.2. Pathophysiology

DCTN1 encodes p150glued, the major subunit of the dynactin complex which binds to the motor protein dynein which binds directly to microtubules and different dynactin subunits [80]. Mutations in *DCTN1* diminish microtubule binding and lead to intracytoplasmic inclusions [81].

5.3. DNAJC6

5.3.1. Genotype-Phenotype

Biallelic mutations in *DNAJC6* cause juvenile-onset, atypical parkinsonism with onset during childhood and a very rapid disease progression with loss of ambulation within 10 years from onset [82,83]. Patients are poorly responsive to levodopa therapy and have additional manifestations such as developmental delay, intellectual disability, seizures, and other movement disorders (e.g., dystonia, spasticity, myoclonus). A minority of patients have early-onset parkinsonism, with symptom onset in the third to fourth decade of life and an absence of additional features [84]. These patients generally have a slower rate of disease progression and a favorable response to levodopa therapy.

5.3.2. Pathophysiology

DNAJC6 encodes for auxilin 1, a brain-specific form of auxilin and a co-chaperone protein involved in the clathrin-mediated synaptic vesicle endocytosis. Auxilin deficiency has been found in animal models to result in impaired synaptic vesicle endocytosis, and thus negatively impacts synaptic neurotransmission, homeostasis, and signaling [85]. However, the exact mechanism by which auxilin deficiency leads to dopaminergic neurodegeneration and atypical neurological symptoms remains unclear.

5.4. FBXO7
5.4.1. Genotype-Phenotype

Mutations in *FBXO7* cause autosomal recessive, juvenile/early-onset parkinsonian-pyramidal syndrome (also called PARK15). Missense, splice site, and nonsense mutations have been reported. The median age at onset was 17 years, with a range of 10 to 52 years. The typical presenting symptoms were bradykinesia and tremor, and patients affected by this atypical parkinsonism frequently show pyramidal signs, dysarthria, and dyskinesia. Psychiatric manifestations, such as visual hallucination, agitation, aggression, disinhibition, and impulsive control disorder, are prominent in these patients as a complication of dopaminergic therapy [86–89].

5.4.2. Pathophysiology

FBXO7 is expressed in various tissues, including the gray and white matters of the brain. It directly interacts with PINK1 and parkin to engage in mitophagy [90]. The loss of *FBXO7* expression has been shown to lead to a significant inhibition of parkin recruitment to depolarized mitochondria [90].

5.5. PLA2G6
5.5.1. Genotype-Phenotype

Mutations in *PLA2G6* have been linked to a variety of neurological disorders, including infantile neuroaxonal dystrophy, neurodegeneration with brain iron accumulation 2B, and Karak syndrome. *PLA2G6* mutations may also result in another phenotype—autosomal recessive, adult-onset dystonia-parkinsonism (also called PARK14) [91].

Patients with *PLA2G6*-related parkinsonism first show symptoms in their childhood or early adulthood, with an age at onset ranging from 8 to 36. In addition to parkinsonism, the majority have dystonia [92,93]. Neuropsychiatric presentations such as depression, psychosis, and cognitive decline are common. There is a good response to levodopa therapy. Magnetic resonance imaging of the brain in most patients showed an absence of iron deposition, and if iron was present, it was found in the substantia nigra or globus pallidus, or both [94].

5.5.2. Pathophysiology

PLA2G6, a phospholipase 2, catalyzes the hydrolysis of the sn-2 acyl-ester bonds in phospholipids to form arachidonic acid and other fatty acids. This is involved in the phospholipid remodeling, apoptosis, and prostaglandin and leukotriene synthesis. The exact mechanism of *PLA2G6* in neurodegenerative diseases remains obscure, however defective phospholipases have been implicated in the pathogenesis of neurodegenerative conditions with iron dyshomeostasis.

5.6. SYNJ1
5.6.1. Genotype-Phenotype

Mutations in *SYNJ1* are linked to autosomal recessive, early-onset Parkinson disease-20 (PARK20). Individuals affected by *SYNJ1*-associated parkinsonism generally show symptoms in the third decade of life, and manifest parkinsonism (tremor, bradykinesia) with a poor response to levodopa treatment, as well as additional atypical signs such as dystonia, seizures, cognitive impairment, and developmental delay [95].

5.6.2. Pathophysiology

Synaptojanin-1 plays a crucial role in synaptic vesicle dynamics, including endocytosis and recycling. SJ1-knockout mice display endocytic defects and a remarkable accumulation of clathrin-coated intermediates [96]. Fasano et al. further showed that SYNJ1 is critically involved in early endosome function, and that a loss of *SYNJ1* leads to impaired recycling of the transferrin receptor to the plasma membrane, highlighting the important role that the autophagy-lysosome pathway plays in PD pathogenesis [92].

6. Recently Described Parkinson's Disease Genes

6.1. CHCHD2

6.1.1. Genotype-Phenotype

Mutations in the *CHCHD2* gene were linked to an autosomal dominant, late-onset form of PD (PARK22) in the Japanese population in 2015 by Funayama et al., who reported two missense mutations (p.T61I, p.R145Q) and a splice-site mutation (c.300 + 5G > A) in the *CHCHD2* gene [93]. Both missense mutations were also reported in the Chinese population [97,98], although were not found in a study on a large cohort of PD patients of western European ancestry [99]. Instead, three rare variants (p.A32T, p.P34L, and p.I80V) in the *CHCHD2* gene were found in the western European cohort, occurring in highly conserved residues [99]. A homozygous missense mutation (p.A71P) has also been reported in a 26-year-old Caucasian woman with recessive early-onset PD [100]. Patients affected by *CHCHD2*-associated PD typically present with typical parkinsonian features, with a significant response to levodopa.

6.1.2. Pathophysiology

CHCHD2 contains a mitochondrial-targeting sequence at the N-terminus and localizes to the mitochondrial intermembrane space. Its close homologue CHCHD10 is enriched at crista junctions of the mitochondria and is believed to be involved in oxidative phosphorylation or in maintenance of crista morphology [101]. The loss of CHCHD2 in flies leads to mitochondrial and neural phenotypes associated with PD pathology and causes chronic oxidative stress and thus age-dependent neurodegeneration in the dopaminergic neurons [102].

6.2. LRP10

6.2.1. Genotype-Phenotype

Through genome-wide linkage analysis of an Italian family with autosomal dominant PD, Quadri and colleagues implicated the *LRP10* gene on chromosome 14 as a possible causative disease gene [31]. This was verified through analysis of a larger cohort of patients, where rare, potential mutations in *LRP10* were found to be associated with PD and dementia with Lewy bodies [31]. These findings were unable to be replicated in a study by Tesson et al., whose co-segregation analysis did not support a causal role for *LRP10* in PD [103]. Since then, several additional variants in the *LRP10* have been identified in patients with PD, progressive supranuclear palsy, frontotemporal dementia, and amyotrophic lateral sclerosis, although the correlation of *LRP10* variants with the development of α-synucleinopathies and other neurodegenerative diseases has been debated [104–106].

6.2.2. Pathophysiology

LRP10 is a single-pass transmembrane protein and a member of a subfamily of LDL receptors. Grochowska et al. discovered that *LRP10* expression was high in non-neuronal cells but undetectable in neurons, and that it was present in the trans-Golgi network, plasma membrane, retromer, and early endosomes in astrocytes [107]. They suggested that LRP10-mediated pathogenicity involves the interaction of LRP10 and SORL1 in vesicle tracking pathways, as they were shown to co-localize and interact, and that disturbed vesicle trafficking and loss of LRP10 function were crucial in the pathogenesis of neurodegenerative diseases [107].

6.3. TMEM230

6.3.1. Genotype-Phenotype

The link with PD was first proposed in 2016 by Deng et al., who investigated a large Canadian Mennonite pedigree with autosomal dominant, typical PD, and discovered a p.R141L mutation in *TMEM230* which reportedly fully co-segregated with disease [34]. The same pedigree was investigated by Vilarino-Guell and colleagues, who identified a heterozygous missense variant in *DNAJC13* (p.N855S) which did not fully co-segregate with disease [108]. Whilst *TMEM230* variants have been identified in further studies on PD patient groups, other follow-up genetic studies have failed to detect PD-associated *TMEM230* variants, and whether evidence exists for 'proof of pathogenecity' has been debated [109,110].

6.3.2. Pathophysiology

TMEM230 is a transmembrane protein with ubiquitous expression. It is a trafficking protein of secretory and recycling vesicles, including neuronal synaptic vesicles. Expression of mutant *TMEM230* was found to lead to increased α-synuclein levels [34]. Loss of function of TMEM230 impairs secretory autophagy, Golgi-derived vesicle secretion, and retromer trafficking [111].

6.4. UQCRC1

6.4.1. Genotype-Phenotype

An association between *UQCRC1* mutations and familial PD was first reported by Lin et al. in 2020, who identified a novel heterozygous mutation (p.Y314S) in the *UQCRC1* gene which co-segregated with disease in a Taiwanese family with autosomal dominant parkinsonism with polyneuropathy [112]. An additional variant in *UQCRC1* (p.I311L) also co-segregated with disease [112]. In a subsequent study, no common variant was found to be significantly associated with PD in the European population [113].

6.4.2. Pathophysiology

UQCRC1 is a core component of complex III in the respiratory chain. In Drosophila and mouse models, URCRC1 p.Y314S knock-in organisms showed dopaminergic neuronal loss, age-dependent locomotor deficits, and peripheral neuropathy [112]. Disruption of the *Uqcrc1* gene in mice causes embryonic lethality [114], and deficiency of Uqcrc1 in Drosophila increases the cytochrome c in the cytoplasmic fraction and activates the caspase cascade, thus causing a reduction of dopaminergic neurons and neurodegeneration [115].

6.5. VPS13C

6.5.1. Genotype-Phenotype

Lesage et al. first reported five truncating mutations in *VPS13C* in three unrelated PD patients [35]. These probands were either homozygous or compound heterozygous and had a distinct phenotype of EOPD which progressed rapidly and showed a good but transient initial response to levodopa treatment. Additional variants in *VPS13C* have been identified in further reports on autosomal recessive, early-onset forms of parkinsonism, although not in late-onset PD [116].

6.5.2. Pathophysiology

VPS13C is part of the family of conserved VPS13 proteins and behaves similarly to VPS35 (see above). VPS13C is a phospholipid transporter and localizes to the contact sites between the endoplasmic reticulum (ER) and late endosome [117]. VPS13 proteins are thought to mediate endoplasmic reticulum-phagy at late endosomes [117].

7. Rare, Atypical, and Unconfirmed Forms

There are many genes that can cause parkinsonian phenotypes, and comprehensive lists can be found elsewhere, with over 70 different genes causing early-onset parkinsonism or parkinsonism as part of a complex neurological disorder [118]. Clinicians should be especially vigilant for treatable causes such as Wilson's disease [118]. Mutations in *GCH1* can cause dopa-responsive dystonia and PD and should also be considered. *POLG* mutations can cause movement disorders including parkinsonism and dystonia. Mutations in *PTRHD1* can cause autosomal recessive PD with intellectual impairment but are rare [119]. *RAB39B* mutations can cause X-linked intellectual impairment and parkinsonism with classic Lewy body pathology on autopsy studies [120]. Several additional reported genes have not been independently replicated and perhaps require further validation before being considered PD genes, such as *DNAJC13*, *EIF4G1*, *GIGYF2*, *HTRA2*, and *UCHL1* [121].

8. Risk Variants versus Monogenic Forms

When discussing genetic risk in PD, one should differentiate risk variants from causative monogenic ones. Risk variants are relatively common, each with an individual small effect size, yet collectively they significantly increase disease risk. A recent large meta-analysis of genome-wide association studies (GWAS) identified 90 such genome-wide risk alleles that collectively account for 16–36% of PD heritability [122]. A causative monogenic variant, on the other hand, is a rare variant with a large effect size, that is considered the causative culprit of the disease. Complicating this oversimplified dichotomic differentiation is the fact that autosomal dominant forms of monogenic PD have incomplete age-dependent penetrance to a variable extent, which may be affected by the causative gene and the specific pathogenic variant as well as the patient's ethnicity. Moreover, a complex interplay between monogenic causative variants and risk variants may affect disease penetrance, as exemplified by a recent study which showed that disease penetrance of the *LRRK2* variant p.G2019S is modified by a polygenic risk score [45].

9. *GBA* Variants

A notable issue is the one related to pathogenic *GBA* (or *GBA1*) variants, which constitute the most common genetic risk factor for PD. These variants are found in approximately 8.5% of PD patients [123]. However, this number varies significantly across different ethnic groups, ranging between 2.3% and 12% in populations of non-Ashkenazi Jewish origin to 10–31% in Ashkenazi Jews [124]. *GBA* variants were more common in patients with early-onset disease (<50 years), more rapid development of dementia, and a more aggressive motor course [125,126]. Pathogenic variants in this gene have a low, age-dependent penetrance in PD, which is highly variable across different reports, ranging between 8% and 30% by age 80 years [127–130]. In a recent study, the authors used a kin-cohort design to evaluate the penetrance of pathogenic *GBA* variants in a cohort of unselected PD patients, showing that the risk to develop disease by age 60, 70, and 80 years was 10%, 16%, and 19%, respectively [131]. This study also found a trend towards a greater PD penetrance for severe pathogenic variants compared to mild pathogenic variants in the *GBA* gene, although this difference did not reach statistical significance [131].

Adding to the complexity of *GBA*-associated PD, a recent study demonstrated an association between PD polygenic risk score and both penetrance and age at onset in individuals carrying a disease-associated *GBA* variant [132]. Another study examined PD clustering in eight families of non-Parkinsonian *GBA*-p.N370S homozygote Gaucher patients, showing that all PD cases in these families stemmed from only one of the proband's parents, further highlighting the potential role of genetic modifiers in PD risk among carriers of *GBA* variants [133].

Furthermore, a recent study showed that both pathogenic (i.e., associated with Gaucher disease) and non-pathogenic (i.e., not associated with Gaucher) variants in *GBA* are common in PD, with a more aggressive course in terms of dementia and motor progression [126].

In summary, *GBA* variants are a common risk factor for PD. They should be clearly differentiated as such from monogenic causes for PD, to avoid ambiguity and terminological and conceptual perplexity when discussing PD risk with patients and clinicians.

10. Genetic Testing in Parkinson's Disease

Genetic workup is not routinely performed as part of PD evaluation, and movement disorder specialists only very occasionally suggest genetic testing to PD patients. This is due to a combination of factors related to cost, lack of physician's perceived impact on patient's management, and physician's discomfort regarding test selection and its results or their impact on the patients and their family members [134]. The field of genetic testing in PD is rapidly evolving during recent years, due to the better availability of next-generation sequencing (NGS)-based molecular tests and the initiation of genetic diagnosis-based interventional clinical trials.

10.1. Who Should Be Offered Genetic Testing in Parkinson's Disease?

Traditionally, a monogenic cause would most probably be suspected, and therefore a genetic test considered, in patients with early-onset PD before age 50 years, and particularly before age 40 years. Furthermore, although polygenic risk and multifactorial inheritance would probably explain most cases with familial clustering of PD, a striking familial history, either of autosomal dominant or autosomal recessive pattern, is yet another clue for a possible monogenic cause that may suggest that a genetic test should be considered. Ethnic origin may also affect the decision to perform genetic testing, for example in patients of Ashkenazi Jewish or African Berber origin. As opposed to this traditional case-by-case approach, as molecular testing is becoming more available, a recently suggested permissive approach supports a more widespread use of genetic testing in PD to improve patient care, to allow inclusion of patients in molecular diagnosis-based clinical trials, and to benefit therapeutic insights and strategies for the larger PD population, including patients with idiopathic disease [135]. This notion can tremendously benefit PD patients both individually and collectively. However, it should be backed up by thorough knowledge of the different evolving aspects of genetic testing in PD, and by an individually tailored explanation to patients and potential carriers in their family prior to testing as well as when returning them the test results, regarding the test and the potential implications of its results for them and for their family members.

10.2. The Implications of a Genetic Diagnosis in Parkinson's Disease

A genetic diagnosis may have significant implications for PD patients, both for expected disease course and response to therapeutic interventions. As mentioned, several monogenic forms are expected to respond well to levodopa medication (e.g., *PRKN*), whereas others are poorly responsive (e.g., *DNAJC6*) (Table 2). Additionally, a recent study found that the rate of cognitive decline for *GBA* mutation carriers after bilateral subthalamic nucleus deep brain stimulation (STN-DBS) is higher than that of carriers of *PRKN* and *LRRK2* mutations and those without identified disease-associated pathogenic variants [136] (Figure 1A). These findings were further corroborated by a new study which suggests that STN-DBS is associated with a greater rate of cognitive decline in *GBA* mutation carriers [137]. A recommendation that arose from this study is that PD patients should be screened for *GBA* pathogenic variants prior to DBS surgery, and that carriers of such variants should be counseled on the greater risk of cognitive decline [137].

For *SNCA*-PD, the response to DBS may also differ according to the type of mutation (Figure 1B). A recent report of four patients with *SNCA* mutations showed a good response in the three patients with duplications and a poor response in the patient with a missense mutation (p.A53E) [138] (Table 1).

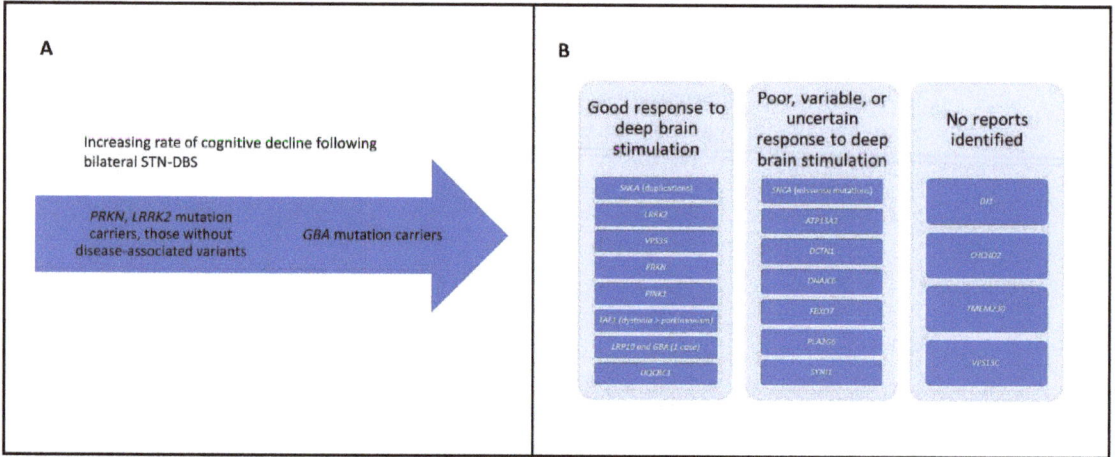

Figure 1. (**A**) Increasing cognitive decline in *GBA* carriers versus *PRKN*, *LRRK2*, and those without disease-associated variants. (**B**) Outcome of deep brain stimulation stratified according to Parkinson's disease monogenic forms.

In addition to implications for DBS, the emerging importance of a genetic diagnosis in PD is also related to new gene-based targeted approaches that are being developed in recent years [3], since a specific molecular genetic diagnosis may allow inclusion in interventional clinical trials that target a genetically determined subgroup of PD patients. Moreover, a genetic diagnosis for additional family members at risk of developing PD allows for a more accurate estimation of recurrence risk and informs genetic counseling and family planning. Moreover, some patients are greatly distressed just by the uncertainty regarding the cause for their condition and a genetic diagnosis may bring them great relief.

10.3. Challenges in Genetic Testing

The challenges in genetic testing in PD are related to the patient, the choice of genetic test, and the test results. Patients may be reluctant to perform genetic testing due to different reasons, including a lack of perceived benefit, concern regarding the implications of the test results for them or their family members, or cost. Genetic counseling prior to performing a genetic test is non-directive, meaning that patients or their relatives cannot be directed to have a genetic test, however it should include a thorough, individually tailored explanation regarding the reason why a genetic test is offered, the test itself, its advantages and limitations, and the potential implications of the test results for the patient and their family members. This type of pre-test discussion with the patient is necessary to address the patient's concerns and to ensure that they are given all the required information to make a knowledge-based decision on whether to proceed with genetic testing or not.

Many types of genetic tests are available in clinical and research settings, ranging from focused testing for a single gene or a specific variant, through variant panels and gene panels, to exome or genome sequencing. Due to the increase in availability and decrease in cost of NGS-based tests, the traditional approach of testing one gene at a time was largely replaced in recent years with broader tests, such as gene panels and exome or genome sequencing, except when a known pathogenic variant has been previously found in the patient's family, or in uncommon cases where a very high suspicion is raised for a specific gene. When choosing to use a gene panel, one should consider the considerable variability in gene content of different panels. A recent study evaluated the types of clinical genetic tests that are used in PD, revealing notable differences in gene panel size, ranging from 5 to 62 genes. That study showed that five genes were included in all panels (*SNCA*, *PRKN*, *PINK1*, *PARK7* (*DJ1*), and *LRRK2*), while *VPS35* and *GBA* were only variably included, and that the differences between panels were mainly the result of the variable inclusion of genes associated with atypical parkinsonism and dystonia disorders, or genes with an uncertain association with PD [139]. The selected gene panel should ideally include all established genes for PD with both sequence and deletion/duplication analysis. In cases where the patient presents a combined or an atypical phenotype, a broader approach should be considered, either by using a more comprehensive gene panel or by a genomic analysis with exome or genome sequencing, depending on the specific clinical indicators. Notable limitations that should be taken into consideration are the ones associated with the *GBA* gene, for which a related pseudogene and structural variations may complicate the detection of pathogenic variants. A novel approach is to use long-read sequencing to assess this gene, with the GridION nanopore sequencing platform recently used in a New Zealand cohort of patients [139]. Another factor to consider is the cost of genetic tests, which might not be covered by the patient's insurance and therefore may inevitably affect decisions in the molecular workup in some cases. In summary, the decision regarding which genetic test should be used depends on case-specific factors and requires to consider the different types of tests available, their advantages and limitations, and their suitability for each individual patient.

11. Role for Heterozygosity in Autosomal Recessive Parkinson Genes

The possibility that monoallelic pathogenic variants in autosomal recessive PD genes constitute a risk factor for PD is controversial, and conflicting evidence regarding this issue has been reported.

11.1. PRKN Heterozygotes

A recent population-based study analyzed data of 164 confirmed heterozygous *PRKN* mutation carriers and 2582 controls from South Tyrol in Northern Italy. This study showed a significantly higher number of carriers than controls with a reported akinesia-related phenotype based on a validated PD screening questionnaire [140]. Another study evaluated *PRKN* as a risk factor for PD in three large independent case-control cohorts and revealed a 1.55-fold risk increase in heterozygous carriers, who also had a younger age of disease onset [141]. However, ~70% of potentially monoallelic cases were not assessed for a second *PRKN* mutation. To further address this, the authors conducted a meta-analysis of available cohorts and studies of individuals from European ancestry, demonstrating a significant 1.65-fold increase in PD risk in monoallelic *PRKN* mutation carriers. Nevertheless, when excluding from the analysis studies which did not search for biallelic carriers and those that focused on early-onset PD, no association between monoallelic *PRKN* mutation and disease risk was found, highlighting the importance of confounding factors that might bias this association [141]. In a recent study, full sequencing and CNV analysis of *PRKN* in 2809 PD patients and 3629 controls revealed no association between all types of heterozygous *PRKN* variants and PD risk [142].

11.2. PINK1 Heterozygotes

Several studies have previously suggested that heterozygous *PINK1* variants may act as a risk factor for late-onset PD. Of note, one study in a large German family suggested that heterozygous *PINK1* mutations may increase the risk for the development of at least subtle motor and non-motor signs of PD [143]. Puschmann et al. investigated the functional effects of the heterozygous *PINK1* p.G411S variant and concluded that it acts as a risk factor for PD, which confers its effect by a partial dominant-negative mechanism [144]. A recent comprehensive analysis contradicted these studies. By harnessing combined data from several large datasets totaling 13,708 cases and 362,850 control individuals, this investigation found no evidence of association between heterozygous *PINK1* mutations and PD risk [145], further highlighting the complexity and controversy in this field.

11.3. Conclusion on Heterozygous Carriers

The evidence for the role of heterozygous carriers is conflicting—some studies which were based largely on findings in specific cases or families suggested a possible association, while newer studies that utilized large datasets mostly refuted this possibility.

A hidden trans-acting pathogenic variant on the other allele of the gene may at least partly explain these contradictory findings. This may occur in cases where the chosen methodology could not identify these variants, for example when a deletion/duplication analysis was not performed or when the second allele harbored a disease-associated non-coding, structural, or mosaic variant which the molecular testing strategy that was used could not reveal. In these cases, an apparent association between a monoallelic variant and disease may be erroneously concluded. This scenario, however, would not explain cases of families with a clear autosomal dominant inheritance pattern across several generations. Another possible explanation for the conflicting evidence may be that monoallelic deleterious variants in autosomal recessive Parkinson-related genes confer an increased disease risk to some extent as part of a multifactorial inheritance, where each individual, family, or ethnic group are affected by a certain genetic background and/or environmental factors. In this potential scenario, while a monoallelic pathogenic variant may indeed increase the risk for PD, the threshold for disease expression may vary substantially between different individuals, families, or ethnic groups, depending on other genetic variants and environmental factors. This might be missed when analyzing very large, grouped datasets or data that are limited to specific ethnic groups. Other potential factors that might contribute to those contradictory findings may stem from data collection-related biases, such as a recall bias or cases of subtle signs of parkinsonism in reportedly healthy individuals which are considered in the analysis as unaffected controls.

12. Dual LRRK2 and GBA Mutation Carriers

It would be anticipated that having a mutation in both *LRRK2* and *GBA* would have an added deleterious effect, as suggested by laboratory studies [146,147]. However, a recent longitudinal study of a large PD sample measuring progression using the Montreal Cognitive Assessment and Movement Disorders Society—Unified Parkinson Disease Rating Scale–Part III, showed that patients with both the p.G2019S mutation and *GBA*-PD had a slower rate of decline than those with *GBA*-PD alone, which was no different from *LRRK2*-G2019S alone [148]. Similarly, a retrospective observational study of Ashkenazi Jewish patients revealed that patients with mutations in *LRRK2* and *GBA* (described by the authors as "*GBA-LRRK2*-PD") were less frequently affected by dementia, probable REM-behavior sleep disorder, and psychosis, compared to other groups (*GBA*-PD, *LRRK2*-PD, mutation-negative PD) [149]. This raises the possibility of a protective effect of having the *LRRK2* p.G2019S mutation in *GBA* mutation carriers [149].

13. Conclusions

There have been major advances in research into monogenic PD in recent years. There have been multiple PD gene discoveries, although we highlight the importance of independent validation of these findings. There have been greater insights into genotype–phenotype relationships, and laboratory studies have translated the genetic discoveries into an improved understanding of the pathophysiological mechanism underlying PD.

It has become apparent that there are major ethnic and regional differences in the distribution of mutations in PD genes. There has been further evidence on the role of heterozygous carriers in autosomal recessive PD genes, and the effect of having mutations in both *LRRK2* and *GBA* in the same individual. Additionally, there is a suggestion that the underlying monogenic cause may influence the disease course as well as the response to levodopa and DBS.

Advances in genomic technology provide individuals with PD with greater access to genetic testing through both clinical and research pathways. Global efforts will play a key role in exploiting this genomic data. Worldwide studies can pool many patients to identify rare genetic causes of PD and can also be used to attempt to replicate important genetic discoveries. Furthermore, they offer greater representation of underrepresented populations from different ethnic groups and geographic regions. There are several major global projects to identify new disease genes in PD, including established initiatives such as the International Parkinson Disease Genomics Consortium [150] and newer initiatives such as the Global Parkinson's disease genetics program (GP2) [151].

PD currently remains an incurable disorder but advances in our understanding of the genetics of PD may inform our understanding of the pathophysiology and thus help with efforts to develop targeted therapies.

Author Contributions: All authors made substantial contributions to the conception and design, drafting of the work and revision, approved the submitted version, and agree to be personally accountable for the authors' own contributions and for ensuring that questions related to the accuracy or integrity of any part of the work, even ones in which the author was not personally involved, are appropriately investigated, resolved, and documented in the literature (F.J., A.F. and K.R.K.). All authors have read and agreed to the published version of the manuscript.

Funding: This research received no external funding.

Institutional Review Board Statement: Not applicable.

Informed Consent Statement: Not applicable.

Acknowledgments: K.R.K. receives funding to study dystonia from the Paul Ainsworth Family Foundation and receives a Working Group Co-Lead Award from the Michael J. Fox Foundation, Aligning Science Across Parkinson's (ASAP) initiative, which is unrelated to the current paper.

Conflicts of Interest: The authors declare no conflict of interest.

References

1. Kumar, K.R.; Lohmann, K.; Klein, C. Genetics of Parkinson disease and other movement disorders. *Curr. Opin. Neurol.* **2012**, *25*, 466–474. [CrossRef] [PubMed]
2. Panicker, N.; Ge, P.; Dawson, V.L.; Dawson, T.M. The cell biology of Parkinson's disease. *J. Cell Biol.* **2021**, *220*, e202012095. [CrossRef] [PubMed]
3. Senkevich, K.; Rudakou, U.; Gan-Or, Z. New therapeutic approaches to Parkinson's disease targeting GBA, LRRK2 and Parkin. *Neuropharmacology* **2022**, *202*, 108822. [CrossRef] [PubMed]
4. Prasuhn, J.; Davis, R.L.; Kumar, K.R. Targeting Mitochondrial Impairment in Parkinson's Disease: Challenges and Opportunities. *Front. Cell Dev. Biol.* **2020**, *8*, 615361. [CrossRef]
5. Trinh, J.; Zeldenrust, F.M.J.; Huang, J.; Kasten, M.; Schaake, S.; Petkovic, S.; Madoev, H.; Grunewald, A.; Almuammar, S.; Konig, I.R.; et al. Genotype-phenotype relations for the Parkinson's disease genes SNCA, LRRK2, VPS35: MDS Gene systematic review. *Mov. Disord.* **2018**, *33*, 1857–1870. [CrossRef]
6. Blauwendraat, C.; Kia, D.A.; Pihlstrom, L.; Gan-Or, Z.; Lesage, S.; Gibbs, J.R.; Ding, J.; Alcalay, R.N.; Hassin-Baer, S.; Pittman, A.M.; et al. Insufficient evidence for pathogenicity of SNCA His50Gln (H50Q) in Parkinson's disease. *Neurobiol. Aging* **2018**, *64*, 159.e5–159.e8. [CrossRef]
7. Liu, H.; Koros, C.; Strohaker, T.; Schulte, C.; Bozi, M.; Varvaresos, S.; Ibanez de Opakua, A.; Simitsi, A.M.; Bougea, A.; Voumvourakis, K.; et al. A Novel SNCA A30G Mutation Causes Familial Parkinson's Disease. *Mov. Disord.* **2021**, *36*, 1624–1633. [CrossRef]
8. Book, A.; Guella, I.; Candido, T.; Brice, A.; Hattori, N.; Jeon, B.; Farrer, M.J.; SNCA Multiplication Investigators of the GEoPD Consortium. A Meta-Analysis of alpha-Synuclein Multiplication in Familial Parkinsonism. *Front. Neurol.* **2018**, *9*, 1021. [CrossRef]
9. Lesage, S.; Houot, M.; Mangone, G.; Tesson, C.; Bertrand, H.; Forlani, S.; Anheim, M.; Brefel-Courbon, C.; Broussolle, E.; Thobois, S.; et al. Genetic and Phenotypic Basis of Autosomal Dominant Parkinson's Disease in a Large Multi-Center Cohort. *Front. Neurol.* **2020**, *11*, 682. [CrossRef]
10. Schneider, S.A.; Alcalay, R.N. Neuropathology of genetic synucleinopathies with parkinsonism: Review of the literature. *Mov. Disord.* **2017**, *32*, 1504–1523. [CrossRef]
11. Over, L.; Bruggemann, N.; Lohmann, K. Therapies for Genetic Forms of Parkinson's Disease: Systematic Literature Review. *J. Neuromuscul. Dis.* **2021**, *8*, 341–356. [CrossRef] [PubMed]
12. Leaver, K.; Viser, A.; Kopell, B.H.; Ortega, R.A.; Miravite, J.; Okun, M.S.; Elango, S.; Raymond, D.; Bressman, S.B.; Saunders-Pullman, R.; et al. Clinical profiles and outcomes of deep brain stimulation in G2019S LRRK2 Parkinson disease. *J. Neurosurg.* **2021**, 1–8. [CrossRef] [PubMed]

13. Lesage, S.; Lunati, A.; Houot, M.; Romdhan, S.B.; Clot, F.; Tesson, C.; Mangone, G.; Toullec, B.L.; Courtin, T.; Larcher, K.; et al. Characterization of recessive Parkinson's disease in a large multicenter study. *Ann. Neurol.* **2020**, *88*, 843–850. [CrossRef] [PubMed]
14. Kasten, M.; Hartmann, C.; Hampf, J.; Schaake, S.; Westenberger, A.; Vollstedt, E.J.; Balck, A.; Domingo, A.; Vulinovic, F.; Dulovic, M.; et al. Genotype-Phenotype Relations for the Parkinson's Disease Genes Parkin, PINK1, DJ1: MDSGene Systematic Review. *Mov. Disord.* **2018**, *33*, 730–741. [CrossRef] [PubMed]
15. Patel, S.G.; Buchanan, C.M.; Mulroy, E.; Simpson, M.; Reid, H.A.; Drake, K.M.; Merriman, M.E.; Phipps-Green, A.; Cadzow, M.; Merriman, T.R.; et al. Potential PINK1 Founder Effect in Polynesia Causing Early-Onset Parkinson's Disease. *Mov. Disord.* **2021**, *36*, 2199–2200. [CrossRef]
16. Taipa, R.; Pereira, C.; Reis, I.; Alonso, I.; Bastos-Lima, A.; Melo-Pires, M.; Magalhaes, M. DJ-1 linked parkinsonism (PARK7) is associated with Lewy body pathology. *Brain* **2016**, *139*, 1680–1687. [CrossRef]
17. Evidente, V.G.H. X-Linked Dystonia-Parkinsonism. In *GeneReviews(R)*; Adam, M.P., Ardinger, H.H., Pagon, R.A., Wallace, S.E., Bean, L.J.H., Gripp, K.W., Mirzaa, G.M., Amemiya, A., Eds.; University of Washington: Seattle, WA, USA, 1993.
18. Bruggemann, N.; Domingo, A.; Rasche, D.; Moll, C.K.E.; Rosales, R.L.; Jamora, R.D.G.; Hanssen, H.; Munchau, A.; Graf, J.; Weissbach, A.; et al. Association of Pallidal Neurostimulation and Outcome Predictors With X-linked Dystonia Parkinsonism. *JAMA Neurol.* **2019**, *76*, 211–216. [CrossRef]
19. Wittke, C.; Petkovic, S.; Dobricic, V.; Schaake, S.; Group, M.D.-e.P.S.; Respondek, G.; Weissbach, A.; Madoev, H.; Trinh, J.; Vollstedt, E.J.; et al. Genotype-Phenotype Relations for the Atypical Parkinsonism Genes: MDS Gene Systematic Review. *Mov. Disord.* **2021**, *36*, 1499–1510. [CrossRef]
20. Park, J.S.; Blair, N.F.; Sue, C.M. The role of ATP13A2 in Parkinson's disease: Clinical phenotypes and molecular mechanisms. *Mov. Disord.* **2015**, *30*, 770–779. [CrossRef]
21. Williams, D.R.; Hadeed, A.; al-Din, A.S.; Wreikat, A.L.; Lees, A.J. Kufor Rakeb disease: Autosomal recessive, levodopa-responsive parkinsonism with pyramidal degeneration, supranuclear gaze palsy, and dementia. *Mov. Disord.* **2005**, *20*, 1264–1271. [CrossRef]
22. Wang, D.; Gao, H.; Li, Y.; Jiang, S.; Yang, X. ATP13A2 Gene Variants in Patients with Parkinson's Disease in Xinjiang. *BioMed Res. Int.* **2020**, *2020*, 6954820. [CrossRef] [PubMed]
23. Chien, H.F.; Rodriguez, R.D.; Bonifati, V.; Nitrini, R.; Pasqualucci, C.A.; Gelpi, E.; Barbosa, E.R. Neuropathologic Findings in a Patient with Juvenile-Onset Levodopa-Responsive Parkinsonism Due to ATP13A2 Mutation. *Neurology* **2021**, *97*, 763–766. [CrossRef] [PubMed]
24. Tsuboi, Y.; Dickson, D.W.; Nabeshima, K.; Schmeichel, A.M.; Wszolek, Z.K.; Yamada, T.; Benarroch, E.E. Neurodegeneration involving putative respiratory neurons in Perry syndrome. *Acta Neuropathol.* **2008**, *115*, 263–268. [CrossRef]
25. Tsuboi, Y.; Mishima, T.; Fujioka, S. Perry Disease: Concept of a New Disease and Clinical Diagnostic Criteria. *J. Mov. Disord.* **2021**, *14*, 1–9. [CrossRef]
26. Kurian, M.A.; Abela, L. DNAJC6 Parkinson Disease. In *GeneReviews(R)*; Adam, M.P., Ardinger, H.H., Pagon, R.A., Wallace, S.E., Bean, L.J.H., Gripp, K.W., Mirzaa, G.M., Amemiya, A., Eds.; University of Washington: Seattle, WA, USA, 1993.
27. Correa-Vela, M.; Lupo, V.; Montpeyo, M.; Sancho, P.; Marce-Grau, A.; Hernandez-Vara, J.; Darling, A.; Jenkins, A.; Fernandez-Rodriguez, S.; Tello, C.; et al. Impaired proteasome activity and neurodegeneration with brain iron accumulation in FBXO7 defect. *Ann. Clin. Transl. Neurol.* **2020**, *7*, 1436–1442. [CrossRef]
28. Magrinelli, F.; Mehta, S.; Di Lazzaro, G.; Latorre, A.; Edwards, M.J.; Balint, B.; Basu, P.; Kobylecki, C.; Groppa, S.; Hegde, A.; et al. Dissecting the Phenotype and Genotype of PLA2G6-Related Parkinsonism. *Mov. Disord.* **2022**, *37*, 148–161. [CrossRef]
29. Ikeda, A.; Nishioka, K.; Meng, H.; Takanashi, M.; Hasegawa, I.; Inoshita, T.; Shiba-Fukushima, K.; Li, Y.; Yoshino, H.; Mori, A.; et al. Mutations in CHCHD2 cause alpha-synuclein aggregation. *Hum. Mol. Genet.* **2019**, *28*, 3895–3911. [CrossRef]
30. Liao, T.W.; Wang, C.C.; Chung, W.H.; Su, S.C.; Chin, S.H.; Fung, H.C.; Wu, Y.R. Role of LRP10 in Parkinson's disease in a Taiwanese cohort. *Parkinsonism Relat. Disord.* **2021**, *89*, 79–83. [CrossRef]
31. Quadri, M.; Mandemakers, W.; Grochowska, M.M.; Masius, R.; Geut, H.; Fabrizio, E.; Breedveld, G.J.; Kuipers, D.; Minneboo, M.; Vergouw, L.J.M.; et al. LRP10 genetic variants in familial Parkinson's disease and dementia with Lewy bodies: A genome-wide linkage and sequencing study. *Lancet Neurol.* **2018**, *17*, 597–608. [CrossRef]
32. Manini, A.; Straniero, L.; Monfrini, E.; Percetti, M.; Vizziello, M.; Franco, G.; Rimoldi, V.; Zecchinelli, A.; Pezzoli, G.; Corti, S.; et al. Screening of LRP10 mutations in Parkinson's disease patients from Italy. *Parkinsonism Relat. Disord.* **2021**, *89*, 17–21. [CrossRef]
33. Neri, M.; Braccia, A.; Panteghini, C.; Garavaglia, B.; Gualandi, F.; Cavallo, M.A.; Scerrati, A.; Ferlini, A.; Sensi, M. Parkinson's disease-dementia in trans LRP10 and GBA variants: Response to deep brain stimulation. *Parkinsonism Relat. Disord.* **2021**, *92*, 72–75. [CrossRef] [PubMed]
34. Deng, H.X.; Shi, Y.; Yang, Y.; Ahmeti, K.B.; Miller, N.; Huang, C.; Cheng, L.; Zhai, H.; Deng, S.; Nuytemans, K.; et al. Identification of TMEM230 mutations in familial Parkinson's disease. *Nat. Genet.* **2016**, *48*, 733–739. [CrossRef] [PubMed]
35. Lesage, S.; Drouet, V.; Majounie, E.; Deramecourt, V.; Jacoupy, M.; Nicolas, A.; Cormier-Dequaire, F.; Hassoun, S.M.; Pujol, C.; Ciura, S.; et al. Loss of VPS13C Function in Autosomal-Recessive Parkinsonism Causes Mitochondrial Dysfunction and Increases PINK1/Parkin-Dependent Mitophagy. *Am. J. Hum. Genet.* **2016**, *98*, 500–513. [CrossRef] [PubMed]
36. Serratos, I.N.; Hernandez-Perez, E.; Campos, C.; Aschner, M.; Santamaria, A. An Update on the Critical Role of alpha-Synuclein in Parkinson's Disease and Other Synucleinopathies: From Tissue to Cellular and Molecular Levels. *Mol. Neurobiol.* **2022**, *59*, 620–642. [CrossRef] [PubMed]

37. Barbuti, P.A.; Ohnmacht, J.; Santos, B.F.R.; Antony, P.M.; Massart, F.; Cruciani, G.; Dording, C.M.; Pavelka, L.; Casadei, N.; Kwon, Y.J.; et al. Gene-corrected p.A30P SNCA patient-derived isogenic neurons rescue neuronal branching and function. *Sci. Rep.* **2021**, *11*, 21946. [CrossRef] [PubMed]
38. Saunders-Pullman, R.; Raymond, D.; Elango, S. LRRK2 Parkinson Disease. In *GeneReviews(R)*; Adam, M.P., Ardinger, H.H., Pagon, R.A., Wallace, S.E., Bean, L.J.H., Gripp, K.W., Mirzaa, G.M., Amemiya, A., Eds.; University of Washington: Seattle, WA, USA, 1993.
39. Kestenbaum, M.; Alcalay, R.N. Clinical Features of LRRK2 Carriers with Parkinson's Disease. *Adv. Neurobiol.* **2017**, *14*, 31–48. [CrossRef]
40. Waro, B.J.; Aasly, J.O. Exploring cancer in LRRK2 mutation carriers and idiopathic Parkinson's disease. *Brain Behav.* **2018**, *8*, e00858. [CrossRef]
41. Agalliu, I.; San Luciano, M.; Mirelman, A.; Giladi, N.; Waro, B.; Aasly, J.; Inzelberg, R.; Hassin-Baer, S.; Friedman, E.; Ruiz-Martinez, J.; et al. Higher frequency of certain cancers in LRRK2 G2019S mutation carriers with Parkinson disease: A pooled analysis. *JAMA Neurol.* **2015**, *72*, 58–65. [CrossRef]
42. Agalliu, I.; Ortega, R.A.; Luciano, M.S.; Mirelman, A.; Pont-Sunyer, C.; Brockmann, K.; Vilas, D.; Tolosa, E.; Berg, D.; Waro, B.; et al. Cancer outcomes among Parkinson's disease patients with leucine rich repeat kinase 2 mutations, idiopathic Parkinson's disease patients, and nonaffected controls. *Mov. Disord.* **2019**, *34*, 1392–1398. [CrossRef]
43. Macias-Garcia, D.; Perinan, M.T.; Munoz-Delgado, L.; Jesus, S.; Jimenez-Jaraba, M.V.; Buiza-Rueda, D.; Bonilla-Toribio, M.; Adarmes-Gomez, A.; Carrillo, F.; Gomez-Garre, P.; et al. Increased Stroke Risk in Patients with Parkinson's Disease with LRRK2 Mutations. *Mov. Disord.* **2022**, *37*, 225–227. [CrossRef]
44. San Luciano, M.; Tanner, C.M.; Meng, C.; Marras, C.; Goldman, S.M.; Lang, A.E.; Tolosa, E.; Schule, B.; Langston, J.W.; Brice, A.; et al. Nonsteroidal Anti-inflammatory Use and LRRK2 Parkinson's Disease Penetrance. *Mov. Disord.* **2020**, *35*, 1755–1764. [CrossRef] [PubMed]
45. Iwaki, H.; Blauwendraat, C.; Makarious, M.B.; Bandres-Ciga, S.; Leonard, H.L.; Gibbs, J.R.; Hernandez, D.G.; Scholz, S.W.; Faghri, F.; International Parkinson's Disease Genomics Consortium; et al. Penetrance of Parkinson's Disease in LRRK2 p.G2019S Carriers Is Modified by a Polygenic Risk Score. *Mov. Disord.* **2020**, *35*, 774–780. [CrossRef] [PubMed]
46. Ozelius, L.J.; Senthil, G.; Saunders-Pullman, R.; Ohmann, E.; Deligtisch, A.; Tagliati, M.; Hunt, A.L.; Klein, C.; Henick, B.; Hailpern, S.M.; et al. LRRK2 G2019S as a cause of Parkinson's disease in Ashkenazi Jews. *N. Engl. J. Med.* **2006**, *354*, 424–425. [CrossRef] [PubMed]
47. Lesage, S.; Durr, A.; Tazir, M.; Lohmann, E.; Leutenegger, A.L.; Janin, S.; Pollak, P.; Brice, A.; French Parkinson's Disease Genetics Study Group. LRRK2 G2019S as a cause of Parkinson's disease in North African Arabs. *N. Engl. J. Med.* **2006**, *354*, 422–423. [CrossRef] [PubMed]
48. Healy, D.G.; Falchi, M.; O'Sullivan, S.S.; Bonifati, V.; Durr, A.; Bressman, S.; Brice, A.; Aasly, J.; Zabetian, C.P.; Goldwurm, S.; et al. Phenotype, genotype, and worldwide genetic penetrance of LRRK2-associated Parkinson's disease: A case-control study. *Lancet Neurol.* **2008**, *7*, 583–590. [CrossRef]
49. Lim, S.Y.; Tan, A.H.; Ahmad-Annuar, A.; Klein, C.; Tan, L.C.S.; Rosales, R.L.; Bhidayasiri, R.; Wu, Y.R.; Shang, H.F.; Evans, A.H.; et al. Parkinson's disease in the Western Pacific Region. *Lancet Neurol.* **2019**, *18*, 865–879. [CrossRef]
50. Xie, C.L.; Pan, J.L.; Wang, W.W.; Zhang, Y.; Zhang, S.F.; Gan, J.; Liu, Z.G. The association between the LRRK2 G2385R variant and the risk of Parkinson's disease: A meta-analysis based on 23 case-control studies. *Neurol. Sci.* **2014**, *35*, 1495–1504. [CrossRef]
51. Zhang, Y.; Sun, Q.; Yi, M.; Zhou, X.; Guo, J.; Xu, Q.; Tang, B.; Yan, X. Genetic Analysis of LRRK2 R1628P in Parkinson's Disease in Asian Populations. *Parkinsons Dis.* **2017**, *2017*, 8093124. [CrossRef]
52. Blauwendraat, C.; Reed, X.; Kia, D.A.; Gan-Or, Z.; Lesage, S.; Pihlstrom, L.; Guerreiro, R.; Gibbs, J.R.; Sabir, M.; Ahmed, S.; et al. Frequency of Loss of Function Variants in LRRK2 in Parkinson Disease. *JAMA Neurol.* **2018**, *75*, 1416–1422. [CrossRef]
53. Bryant, N.; Malpeli, N.; Ziaee, J.; Blauwendraat, C.; Liu, Z.; Consortium, A.P.; West, A.B. Identification of LRRK2 missense variants in the accelerating medicines partnership Parkinson's disease cohort. *Hum. Mol. Genet.* **2021**, *30*, 454–466. [CrossRef]
54. Berwick, D.C.; Heaton, G.R.; Azeggagh, S.; Harvey, K. LRRK2 Biology from structure to dysfunction: Research progresses, but the themes remain the same. *Mol. Neurodegener.* **2019**, *14*, 49. [CrossRef] [PubMed]
55. Zimprich, A.; Benet-Pages, A.; Struhal, W.; Graf, E.; Eck, S.H.; Offman, M.N.; Haubenberger, D.; Spielberger, S.; Schulte, E.C.; Lichtner, P.; et al. A mutation in VPS35, encoding a subunit of the retromer complex, causes late-onset Parkinson disease. *Am. J. Hum. Genet.* **2011**, *89*, 168–175. [CrossRef] [PubMed]
56. Vilarino-Guell, C.; Wider, C.; Ross, O.A.; Dachsel, J.C.; Kachergus, J.M.; Lincoln, S.J.; Soto-Ortolaza, A.I.; Cobb, S.A.; Wilhoite, G.J.; Bacon, J.A.; et al. VPS35 mutations in Parkinson disease. *Am. J. Hum. Genet.* **2011**, *89*, 162–167. [CrossRef] [PubMed]
57. Ando, M.; Funayama, M.; Li, Y.; Kashihara, K.; Murakami, Y.; Ishizu, N.; Toyoda, C.; Noguchi, K.; Hashimoto, T.; Nakano, N.; et al. VPS35 mutation in Japanese patients with typical Parkinson's disease. *Mov. Disord.* **2012**, *27*, 1413–1417. [CrossRef]
58. Kumar, K.R.; Weissbach, A.; Heldmann, M.; Kasten, M.; Tunc, S.; Sue, C.M.; Svetel, M.; Kostic, V.S.; Segura-Aguilar, J.; Ramirez, A.; et al. Frequency of the D620N mutation in VPS35 in Parkinson disease. *Arch. Neurol.* **2012**, *69*, 1360–1364. [CrossRef]
59. Rahman, A.A.; Morrison, B.E. Contributions of VPS35 Mutations to Parkinson's Disease. *Neuroscience* **2019**, *401*, 1–10. [CrossRef]
60. Ishiguro, M.; Li, Y.; Yoshino, H.; Daida, K.; Ishiguro, Y.; Oyama, G.; Saiki, S.; Funayama, M.; Hattori, N.; Nishioka, K. Clinical manifestations of Parkinson's disease harboring VPS35 retromer complex component p.D620N with long-term follow-up. *Parkinsonism Relat. Disord.* **2021**, *84*, 139–143. [CrossRef]

61. Cutillo, G.; Simon, D.K.; Eleuteri, S. VPS35 and the mitochondria: Connecting the dots in Parkinson's disease pathophysiology. *Neurobiol. Dis.* **2020**, *145*, 105056. [CrossRef]
62. Sassone, J.; Reale, C.; Dati, G.; Regoni, M.; Pellecchia, M.T.; Garavaglia, B. The Role of VPS35 in the Pathobiology of Parkinson's Disease. *Cell. Mol. Neurobiol.* **2021**, *41*, 199–227. [CrossRef]
63. Lesage, S.; Magali, P.; Lohmann, E.; Lacomblez, L.; Teive, H.; Janin, S.; Cousin, P.Y.; Durr, A.; Brice, A.; French Parkinson Disease Genetics Study Group. Deletion of the parkin and PACRG gene promoter in early-onset parkinsonism. *Hum. Mutat.* **2007**, *28*, 27–32. [CrossRef]
64. Bruggemann, N.; Klein, C. Parkin Type of Early-Onset Parkinson Disease. In *GeneReviews(R)*; Adam, M.P., Ardinger, H.H., Pagon, R.A., Wallace, S.E., Bean, L.J.H., Stephens, K., Amemiya, A., Eds.; University of Washington: Seattle, WA, USA, 1993.
65. Grunewald, A.; Kumar, K.R.; Sue, C.M. New insights into the complex role of mitochondria in Parkinson's disease. *Prog. Neurobiol.* **2019**, *177*, 73–93. [CrossRef] [PubMed]
66. Bradshaw, A.V.; Campbell, P.; Schapira, A.H.V.; Morris, H.R.; Taanman, J.W. The PINK1-Parkin mitophagy signalling pathway is not functional in peripheral blood mononuclear cells. *PLoS ONE* **2021**, *16*, e0259903. [CrossRef] [PubMed]
67. Schneider, S.A.; Klein, C. PINK1 Type of Young-Onset Parkinson Disease. In *GeneReviews(R)*; Adam, M.P., Ardinger, H.H., Pagon, R.A., Wallace, S.E., Bean, L.J.H., Gripp, K.W., Mirzaa, G.M., Amemiya, A., Eds.; University of Washington: Seattle, WA, USA, 1993.
68. Guler, S.; Gul, T.; Guler, S.; Haerle, M.C.; Basak, A.N. Early-Onset Parkinson's Disease: A Novel Deletion Comprising the DJ-1 and TNFRSF9 Genes. *Mov. Disord.* **2021**, *36*, 2973–2976. [CrossRef] [PubMed]
69. Zhang, L.; Wang, J.; Wang, J.; Yang, B.; He, Q.; Weng, Q. Role of DJ-1 in Immune and Inflammatory Diseases. *Front. Immunol.* **2020**, *11*, 994. [CrossRef]
70. Kim, R.H.; Smith, P.D.; Aleyasin, H.; Hayley, S.; Mount, M.P.; Pownall, S.; Wakeham, A.; You-Ten, A.J.; Kalia, S.K.; Horne, P.; et al. Hypersensitivity of DJ-1-deficient mice to 1-methyl-4-phenyl-1,2,3,6-tetrahydropyridine (MPTP) and oxidative stress. *Proc. Natl. Acad. Sci. USA* **2005**, *102*, 5215–5220. [CrossRef]
71. Repici, M.; Giorgini, F. DJ-1 in Parkinson's Disease: Clinical Insights and Therapeutic Perspectives. *J. Clin. Med.* **2019**, *8*, 1377. [CrossRef]
72. Pauly, M.G.; Ruiz Lopez, M.; Westenberger, A.; Saranza, G.; Bruggemann, N.; Weissbach, A.; Rosales, R.L.; Diesta, C.C.; Jamora, R.D.G.; Reyes, C.J.; et al. Expanding Data Collection for the MDSGene Database: X-linked Dystonia-Parkinsonism as Use Case Example. *Mov. Disord.* **2020**, *35*, 1933–1938. [CrossRef]
73. Aneichyk, T.; Hendriks, W.T.; Yadav, R.; Shin, D.; Gao, D.; Vaine, C.A.; Collins, R.L.; Domingo, A.; Currall, B.; Stortchevoi, A.; et al. Dissecting the Causal Mechanism of X-Linked Dystonia-Parkinsonism by Integrating Genome and Transcriptome Assembly. *Cell* **2018**, *172*, 897–909.e21. [CrossRef]
74. Lee, L.V.; Maranon, E.; Demaisip, C.; Peralta, O.; Borres-Icasiano, R.; Arancillo, J.; Rivera, C.; Munoz, E.; Tan, K.; Reyes, M.T. The natural history of sex-linked recessive dystonia parkinsonism of Panay, Philippines (XDP). *Parkinsonism Relat. Disord.* **2002**, *9*, 29–38. [CrossRef]
75. Santiano, R.A.S.; Rosales, R.L. A Cross-Cultural Validation of the Filipino and Hiligaynon Versions of the Parts IIIB (Non-Motor Features) and IV (Activities of Daily Living) of the X-Linked Dystonia-Parkinsonism- MDSP Rating Scale. *Clin. Parkinsonism Relat. Disord.* **2021**, *5*, 100100. [CrossRef]
76. Westenberger, A.; Reyes, C.J.; Saranza, G.; Dobricic, V.; Hanssen, H.; Domingo, A.; Laabs, B.H.; Schaake, S.; Pozojevic, J.; Rakovic, A.; et al. A hexanucleotide repeat modifies expressivity of X-linked dystonia parkinsonism. *Ann. Neurol.* **2019**, *85*, 812–822. [CrossRef] [PubMed]
77. Nybo, C.J.; Gustavsson, E.K.; Farrer, M.J.; Aasly, J.O. Neuropathological findings in PINK1-associated Parkinson's disease. *Parkinsonism Relat. Disord.* **2020**, *78*, 105–108. [CrossRef] [PubMed]
78. van Veen, S.; Martin, S.; Van den Haute, C.; Benoy, V.; Lyons, J.; Vanhoutte, R.; Kahler, J.P.; Decuypere, J.P.; Gelders, G.; Lambie, E.; et al. ATP13A2 deficiency disrupts lysosomal polyamine export. *Nature* **2020**, *578*, 419–424. [CrossRef] [PubMed]
79. Richardson, D.; McEntagart, M.M.; Isaacs, J.D. DCTN1-related Parkinson-plus disorder (Perry syndrome). *Pract. Neurol.* **2020**, *20*, 317–319. [CrossRef] [PubMed]
80. Wider, C.; Dachsel, J.C.; Farrer, M.J.; Dickson, D.W.; Tsuboi, Y.; Wszolek, Z.K. Elucidating the genetics and pathology of Perry syndrome. *J. Neurol. Sci.* **2010**, *289*, 149–154. [CrossRef] [PubMed]
81. Farrer, M.J.; Hulihan, M.M.; Kachergus, J.M.; Dachsel, J.C.; Stoessl, A.J.; Grantier, L.L.; Calne, S.; Calne, D.B.; Lechevalier, B.; Chapon, F.; et al. DCTN1 mutations in Perry syndrome. *Nat. Genet.* **2009**, *41*, 163–165. [CrossRef] [PubMed]
82. Edvardson, S.; Cinnamon, Y.; Ta-Shma, A.; Shaag, A.; Yim, Y.I.; Zenvirt, S.; Jalas, C.; Lesage, S.; Brice, A.; Taraboulos, A.; et al. A deleterious mutation in DNAJC6 encoding the neuronal-specific clathrin-uncoating co-chaperone auxilin, is associated with juvenile parkinsonism. *PLoS ONE* **2012**, *7*, e36458. [CrossRef]
83. Koroglu, C.; Baysal, L.; Cetinkaya, M.; Karasoy, H.; Tolun, A. DNAJC6 is responsible for juvenile parkinsonism with phenotypic variability. *Parkinsonism Relat. Disord.* **2013**, *19*, 320–324. [CrossRef]
84. Olgiati, S.; Quadri, M.; Fang, M.; Rood, J.P.; Saute, J.A.; Chien, H.F.; Bouwkamp, C.G.; Graafland, J.; Minneboo, M.; Breedveld, G.J.; et al. DNAJC6 Mutations Associated with Early-Onset Parkinson's Disease. *Ann. Neurol.* **2016**, *79*, 244–256. [CrossRef]
85. Yim, Y.I.; Sun, T.; Wu, L.G.; Raimondi, A.; De Camilli, P.; Eisenberg, E.; Greene, L.E. Endocytosis and clathrin-uncoating defects at synapses of auxilin knockout mice. *Proc. Natl. Acad. Sci. USA* **2010**, *107*, 4412–4417. [CrossRef]

86. Di Fonzo, A.; Dekker, M.C.; Montagna, P.; Baruzzi, A.; Yonova, E.H.; Correia Guedes, L.; Szczerbinska, A.; Zhao, T.; Dubbel-Hulsman, L.O.; Wouters, C.H.; et al. FBXO7 mutations cause autosomal recessive, early-onset parkinsonian-pyramidal syndrome. *Neurology* **2009**, *72*, 240–245. [CrossRef] [PubMed]
87. Wei, L.; Ding, L.; Li, H.; Lin, Y.; Dai, Y.; Xu, X.; Dong, Q.; Lin, Y.; Long, L. Juvenile-onset parkinsonism with pyramidal signs due to compound heterozygous mutations in the F-Box only protein 7 gene. *Parkinsonism Relat. Disord.* **2018**, *47*, 76–79. [CrossRef] [PubMed]
88. Yalcin-Cakmakli, G.; Olgiati, S.; Quadri, M.; Breedveld, G.J.; Cortelli, P.; Bonifati, V.; Elibol, B. A new Turkish family with homozygous FBXO7 truncating mutation and juvenile atypical parkinsonism. *Parkinsonism Relat. Disord.* **2014**, *20*, 1248–1252. [CrossRef] [PubMed]
89. Lohmann, E.; Coquel, A.S.; Honore, A.; Gurvit, H.; Hanagasi, H.; Emre, M.; Leutenegger, A.L.; Drouet, V.; Sahbatou, M.; Guven, G.; et al. A new F-box protein 7 gene mutation causing typical Parkinson's disease. *Mov. Disord.* **2015**, *30*, 1130–1133. [CrossRef] [PubMed]
90. Burchell, V.S.; Nelson, D.E.; Sanchez-Martinez, A.; Delgado-Camprubi, M.; Ivatt, R.M.; Pogson, J.H.; Randle, S.J.; Wray, S.; Lewis, P.A.; Houlden, H.; et al. The Parkinson's disease-linked proteins Fbxo7 and Parkin interact to mediate mitophagy. *Nat. Neurosci.* **2013**, *16*, 1257–1265. [CrossRef] [PubMed]
91. Paisan-Ruiz, C.; Bhatia, K.P.; Li, A.; Hernandez, D.; Davis, M.; Wood, N.W.; Hardy, J.; Houlden, H.; Singleton, A.; Schneider, S.A. Characterization of PLA2G6 as a locus for dystonia-parkinsonism. *Ann. Neurol.* **2009**, *65*, 19–23. [CrossRef]
92. Fasano, D.; Parisi, S.; Pierantoni, G.M.; De Rosa, A.; Picillo, M.; Amodio, G.; Pellecchia, M.T.; Barone, P.; Moltedo, O.; Bonifati, V.; et al. Alteration of endosomal trafficking is associated with early-onset parkinsonism caused by SYNJ1 mutations. *Cell Death Dis.* **2018**, *9*, 385. [CrossRef]
93. Funayama, M.; Ohe, K.; Amo, T.; Furuya, N.; Yamaguchi, J.; Saiki, S.; Li, Y.; Ogaki, K.; Ando, M.; Yoshino, H.; et al. CHCHD2 mutations in autosomal dominant late-onset Parkinson's disease: A genome-wide linkage and sequencing study. *Lancet Neurol.* **2015**, *14*, 274–282. [CrossRef]
94. Karkheiran, S.; Shahidi, G.A.; Walker, R.H.; Paisan-Ruiz, C. PLA2G6-associated Dystonia-Parkinsonism: Case Report and Literature Review. *Tremor Other Hyperkinet. Mov.* **2015**, *5*, 317. [CrossRef]
95. Lesage, S.; Mangone, G.; Tesson, C.; Bertrand, H.; Benmahdjoub, M.; Kesraoui, S.; Arezki, M.; Singleton, A.; Corvol, J.C.; Brice, A. Clinical Variability of SYNJ1-Associated Early-Onset Parkinsonism. *Front. Neurol.* **2021**, *12*, 648457. [CrossRef]
96. Cao, M.; Wu, Y.; Ashrafi, G.; McCartney, A.J.; Wheeler, H.; Bushong, E.A.; Boassa, D.; Ellisman, M.H.; Ryan, T.A.; De Camilli, P. Parkinson Sac Domain Mutation in Synaptojanin 1 Impairs Clathrin Uncoating at Synapses and Triggers Dystrophic Changes in Dopaminergic Axons. *Neuron* **2017**, *93*, 882–896.e5. [CrossRef] [PubMed]
97. Shi, C.H.; Mao, C.Y.; Zhang, S.Y.; Yang, J.; Song, B.; Wu, P.; Zuo, C.T.; Liu, Y.T.; Ji, Y.; Yang, Z.H.; et al. CHCHD2 gene mutations in familial and sporadic Parkinson's disease. *Neurobiol. Aging* **2016**, *38*, 217.e9–217.e13. [CrossRef] [PubMed]
98. Yang, X.; Zhao, Q.; An, R.; Zheng, J.; Tian, S.; Chen, Y.; Xu, Y. Mutational scanning of the CHCHD2 gene in Han Chinese patients with Parkinson's disease and meta-analysis of the literature. *Parkinsonism Relat. Disord.* **2016**, *29*, 42–46. [CrossRef] [PubMed]
99. Jansen, I.E.; Bras, J.M.; Lesage, S.; Schulte, C.; Gibbs, J.R.; Nalls, M.A.; Brice, A.; Wood, N.W.; Morris, H.; Hardy, J.A.; et al. CHCHD2 and Parkinson's disease. *Lancet Neurol.* **2015**, *14*, 678–679. [CrossRef]
100. Lee, R.G.; Sedghi, M.; Salari, M.; Shearwood, A.J.; Stentenbach, M.; Kariminejad, A.; Goullee, H.; Rackham, O.; Laing, N.G.; Tajsharghi, H.; et al. Early-onset Parkinson disease caused by a mutation in CHCHD2 and mitochondrial dysfunction. *Neurol. Genet.* **2018**, *4*, e276. [CrossRef] [PubMed]
101. Bannwarth, S.; Ait-El-Mkadem, S.; Chaussenot, A.; Genin, E.C.; Lacas-Gervais, S.; Fragaki, K.; Berg-Alonso, L.; Kageyama, Y.; Serre, V.; Moore, D.G.; et al. A mitochondrial origin for frontotemporal dementia and amyotrophic lateral sclerosis through CHCHD10 involvement. *Brain* **2014**, *137*, 2329–2345. [CrossRef]
102. Meng, H.; Yamashita, C.; Shiba-Fukushima, K.; Inoshita, T.; Funayama, M.; Sato, S.; Hatta, T.; Natsume, T.; Umitsu, M.; Takagi, J.; et al. Loss of Parkinson's disease-associated protein CHCHD2 affects mitochondrial crista structure and destabilizes cytochrome c. *Nat. Commun.* **2017**, *8*, 15500. [CrossRef]
103. Tesson, C.; Brefel-Courbon, C.; Corvol, J.C.; Lesage, S.; Brice, A.; French Parkinson's Disease Genetics Study Group. LRP10 in alpha-synucleinopathies. *Lancet Neurol.* **2018**, *17*, 1034. [CrossRef]
104. Chen, Y.; Cen, Z.; Zheng, X.; Pan, Q.; Chen, X.; Zhu, L.; Chen, S.; Wu, H.; Xie, F.; Wang, H.; et al. LRP10 in autosomal-dominant Parkinson's disease. *Mov. Disord.* **2019**, *34*, 912–916. [CrossRef]
105. Daida, K.; Nishioka, K.; Li, Y.; Yoshino, H.; Kikuchi, A.; Hasegawa, T.; Funayama, M.; Hattori, N. Mutation analysis of LRP10 in Japanese patients with familial Parkinson's disease, progressive supranuclear palsy, and frontotemporal dementia. *Neurobiol. Aging* **2019**, *84*, 235.e11–235.e16. [CrossRef]
106. Li, C.; Chen, Y.; Ou, R.; Gu, X.; Wei, Q.; Cao, B.; Zhang, L.; Hou, Y.; Liu, K.; Chen, X.; et al. Mutation analysis of LRP10 in a large Chinese familial Parkinson disease cohort. *Neurobiol. Aging* **2021**, *99*, 99.e1–99.e6. [CrossRef] [PubMed]
107. Grochowska, M.M.; Carreras Mascaro, A.; Boumeester, V.; Natale, D.; Breedveld, G.J.; Geut, H.; van Cappellen, W.A.; Boon, A.J.W.; Kievit, A.J.A.; Sammler, E.; et al. LRP10 interacts with SORL1 in the intracellular vesicle trafficking pathway in non-neuronal brain cells and localises to Lewy bodies in Parkinson's disease and dementia with Lewy bodies. *Acta Neuropathol.* **2021**, *142*, 117–137. [CrossRef] [PubMed]

108. Vilarino-Guell, C.; Rajput, A.; Milnerwood, A.J.; Shah, B.; Szu-Tu, C.; Trinh, J.; Yu, I.; Encarnacion, M.; Munsie, L.N.; Tapia, L.; et al. DNAJC13 mutations in Parkinson disease. *Hum. Mol. Genet.* **2014**, *23*, 1794–1801. [CrossRef] [PubMed]
109. Farrer, M.J. Doubts about TMEM230 as a gene for parkinsonism. *Nat. Genet.* **2019**, *51*, 367–368. [CrossRef]
110. Deng, H.X.; Pericak-Vance, M.A.; Siddique, T. Reply to 'TMEM230 variants in Parkinson's disease' and 'Doubts about TMEM230 as a gene for parkinsonism'. *Nat. Genet.* **2019**, *51*, 369–371. [CrossRef]
111. Kim, M.J.; Deng, H.X.; Wong, Y.C.; Siddique, T.; Krainc, D. The Parkinson's disease-linked protein TMEM230 is required for Rab8a-mediated secretory vesicle trafficking and retromer trafficking. *Hum. Mol. Genet.* **2017**, *26*, 729–741. [CrossRef]
112. Lin, C.H.; Tsai, P.I.; Lin, H.Y.; Hattori, N.; Funayama, M.; Jeon, B.; Sato, K.; Abe, K.; Mukai, Y.; Takahashi, Y.; et al. Mitochondrial UQCRC1 mutations cause autosomal dominant parkinsonism with polyneuropathy. *Brain* **2020**, *143*, 3352–3373. [CrossRef]
113. Senkevich, K.; Bandres-Ciga, S.; Gan-Or, Z.; Krohn, L.; International Parkinson's Disease Genomics Consortium. Lack of evidence for association of UQCRC1 with Parkinson's disease in Europeans. *Neurobiol. Aging* **2021**, *101*, 297.e1–297.e4. [CrossRef]
114. Shan, W.; Li, J.; Xu, W.; Li, H.; Zuo, Z. Critical role of UQCRC1 in embryo survival, brain ischemic tolerance and normal cognition in mice. *Cell. Mol. Life Sci.* **2019**, *76*, 1381–1396. [CrossRef]
115. Hung, Y.C.; Huang, K.L.; Chen, P.L.; Li, J.L.; Lu, S.H.; Chang, J.C.; Lin, H.Y.; Lo, W.C.; Huang, S.Y.; Lee, T.T.; et al. UQCRC1 engages cytochrome c for neuronal apoptotic cell death. *Cell Rep.* **2021**, *36*, 109729. [CrossRef]
116. Rudakou, U.; Ruskey, J.A.; Krohn, L.; Laurent, S.B.; Spiegelman, D.; Greenbaum, L.; Yahalom, G.; Desautels, A.; Montplaisir, J.Y.; Fahn, S.; et al. Analysis of common and rare VPS13C variants in late-onset Parkinson disease. *Neurol. Genet.* **2020**, *6*, 385. [CrossRef] [PubMed]
117. Chen, S.; Mari, M.; Parashar, S.; Liu, D.; Cui, Y.; Reggiori, F.; Novick, P.J.; Ferro-Novick, S. Vps13 is required for the packaging of the ER into autophagosomes during ER-phagy. *Proc. Natl. Acad. Sci. USA* **2020**, *117*, 18530–18539. [CrossRef] [PubMed]
118. Morales-Briceno, H.; Mohammad, S.S.; Post, B.; Fois, A.F.; Dale, R.C.; Tchan, M.; Fung, V.S.C. Clinical and neuroimaging phenotypes of genetic parkinsonism from infancy to adolescence. *Brain* **2020**, *143*, 751–770. [CrossRef] [PubMed]
119. Khodadadi, H.; Azcona, L.J.; Aghamollaii, V.; Omrani, M.D.; Garshasbi, M.; Taghavi, S.; Tafakhori, A.; Shahidi, G.A.; Jamshidi, J.; Darvish, H.; et al. PTRHD1 (C2orf79) mutations lead to autosomal-recessive intellectual disability and parkinsonism. *Mov. Disord.* **2017**, *32*, 287–291. [CrossRef] [PubMed]
120. Wilson, G.R.; Sim, J.C.; McLean, C.; Giannandrea, M.; Galea, C.A.; Riseley, J.R.; Stephenson, S.E.; Fitzpatrick, E.; Haas, S.A.; Pope, K.; et al. Mutations in RAB39B cause X-linked intellectual disability and early-onset Parkinson disease with alpha-synuclein pathology. *Am. J. Hum. Genet.* **2014**, *95*, 729–735. [CrossRef] [PubMed]
121. Saini, P.; Rudakou, U.; Yu, E.; Ruskey, J.A.; Asayesh, F.; Laurent, S.B.; Spiegelman, D.; Fahn, S.; Waters, C.; Monchi, O.; et al. Association study of DNAJC13, UCHL1, HTRA2, GIGYF2, and EIF4G1 with Parkinson's disease. *Neurobiol. Aging* **2021**, *100*, 119.e7–119.e13. [CrossRef]
122. Nalls, M.A.; Blauwendraat, C.; Vallerga, C.L.; Heilbron, K.; Bandres-Ciga, S.; Chang, D.; Tan, M.; Kia, D.A.; Noyce, A.J.; Xue, A.; et al. Identification of novel risk loci, causal insights, and heritable risk for Parkinson's disease: A meta-analysis of genome-wide association studies. *Lancet Neurol.* **2019**, *18*, 1091–1102. [CrossRef]
123. Skrahina, V.; Gaber, H.; Vollstedt, E.J.; Forster, T.M.; Usnich, T.; Curado, F.; Bruggemann, N.; Paul, J.; Bogdanovic, X.; Zulbahar, S.; et al. The Rostock International Parkinson's Disease (ROPAD) Study: Protocol and Initial Findings. *Mov. Disord.* **2021**, *36*, 1005–1010. [CrossRef]
124. Neumann, J.; Bras, J.; Deas, E.; O'Sullivan, S.S.; Parkkinen, L.; Lachmann, R.H.; Li, A.; Holton, J.; Guerreiro, R.; Paudel, R.; et al. Glucocerebrosidase mutations in clinical and pathologically proven Parkinson's disease. *Brain* **2009**, *132*, 1783–1794. [CrossRef]
125. Lim, J.L.; Lohmann, K.; Tan, A.H.; Tay, Y.W.; Ibrahim, K.A.; Abdul Aziz, Z.; Mawardi, A.S.; Puvanarajah, S.D.; Lim, T.T.; Looi, I.; et al. Glucocerebrosidase (GBA) gene variants in a multi-ethnic Asian cohort with Parkinson's disease: Mutational spectrum and clinical features. *J. Neural Transm.* **2022**, *129*, 37–48. [CrossRef]
126. Stoker, T.B.; Camacho, M.; Winder-Rhodes, S.; Liu, G.; Scherzer, C.R.; Foltynie, T.; Evans, J.; Breen, D.P.; Barker, R.A.; Williams-Gray, C.H. Impact of GBA1 variants on long-term clinical progression and mortality in incident Parkinson's disease. *J. Neurol. Neurosurg. Psychiatry* **2020**, *91*, 695–702. [CrossRef] [PubMed]
127. McNeill, A.; Duran, R.; Hughes, D.A.; Mehta, A.; Schapira, A.H. A clinical and family history study of Parkinson's disease in heterozygous glucocerebrosidase mutation carriers. *J. Neurol. Neurosurg. Psychiatry* **2012**, *83*, 853–854. [CrossRef] [PubMed]
128. Anheim, M.; Elbaz, A.; Lesage, S.; Durr, A.; Condroyer, C.; Viallet, F.; Pollak, P.; Bonaiti, B.; Bonaiti-Pellie, C.; Brice, A.; et al. Penetrance of Parkinson disease in glucocerebrosidase gene mutation carriers. *Neurology* **2012**, *78*, 417–420. [CrossRef] [PubMed]
129. Rana, H.Q.; Balwani, M.; Bier, L.; Alcalay, R.N. Age-specific Parkinson disease risk in GBA mutation carriers: Information for genetic counseling. *Genet. Med.* **2013**, *15*, 146–149. [CrossRef]
130. Alcalay, R.N.; Dinur, T.; Quinn, T.; Sakanaka, K.; Levy, O.; Waters, C.; Fahn, S.; Dorovski, T.; Chung, W.K.; Pauciulo, M.; et al. Comparison of Parkinson risk in Ashkenazi Jewish patients with Gaucher disease and GBA heterozygotes. *JAMA Neurol.* **2014**, *71*, 752–757. [CrossRef]
131. Balestrino, R.; Tunesi, S.; Tesei, S.; Lopiano, L.; Zecchinelli, A.L.; Goldwurm, S. Penetrance of Glucocerebrosidase (GBA) Mutations in Parkinson's Disease: A Kin Cohort Study. *Mov. Disord.* **2020**, *35*, 2111–2114. [CrossRef]
132. Blauwendraat, C.; Reed, X.; Krohn, L.; Heilbron, K.; Bandres-Ciga, S.; Tan, M.; Gibbs, J.R.; Hernandez, D.G.; Kumaran, R.; Langston, R.; et al. Genetic modifiers of risk and age at onset in GBA associated Parkinson's disease and Lewy body dementia. *Brain* **2020**, *143*, 234–248. [CrossRef]

133. Dinur, T.; Becker-Cohen, M.; Revel-Vilk, S.; Zimran, A.; Arkadir, D. Parkinson's Clustering in Families of Non-Neuronopathic N370S GBA Mutation Carriers Indicates the Presence of Genetic Modifiers. *J. Parkinsons Dis.* 2021, *11*, 615–618. [CrossRef]
134. Alcalay, R.N.; Kehoe, C.; Shorr, E.; Battista, R.; Hall, A.; Simuni, T.; Marder, K.; Wills, A.M.; Naito, A.; Beck, J.C.; et al. Genetic testing for Parkinson disease: Current practice, knowledge, and attitudes among US and Canadian movement disorders specialists. *Genet. Med.* 2020, *22*, 574–580. [CrossRef]
135. Cook, L.; Schulze, J.; Kopil, C.; Hastings, T.; Naito, A.; Wojcieszek, J.; Payne, K.; Alcalay, R.N.; Klein, C.; Saunders-Pullman, R.; et al. Genetic Testing for Parkinson Disease: Are We Ready. *Neurol. Clin. Pract.* 2021, *11*, 69–77. [CrossRef]
136. Mangone, G.; Bekadar, S.; Cormier-Dequaire, F.; Tahiri, K.; Welaratne, A.; Czernecki, V.; Pineau, F.; Karachi, C.; Castrioto, A.; Durif, F.; et al. Early cognitive decline after bilateral subthalamic deep brain stimulation in Parkinson's disease patients with GBA mutations. *Parkinsonism Relat. Disord.* 2020, *76*, 56–62. [CrossRef] [PubMed]
137. Pal, G.; Mangone, G.; Hill, E.J.; Ouyang, B.; Liu, Y.; Lythe, V.; Ehrlich, D.; Saunders-Pullman, R.; Shanker, V.; Bressman, S.; et al. Parkinson Disease and Subthalamic Nucleus Deep Brain Stimulation: Cognitive Effects in GBA Mutation Carriers. *Ann. Neurol.* 2022, *91*, 424–435. [CrossRef] [PubMed]
138. Youn, J.; Oyama, G.; Hattori, N.; Shimo, Y.; Kuusimaki, T.; Kaasinen, V.; Antonini, A.; Kim, D.; Lee, J.I.; Cho, K.R.; et al. Subthalamic deep brain stimulation in Parkinson's disease with SNCA mutations: Based on the follow-up to 10 years. *Brain Behav.* 2022, *12*, e2503. [CrossRef] [PubMed]
139. Cook, L.; Schulze, J.; Verbrugge, J.; Beck, J.C.; Marder, K.S.; Saunders-Pullman, R.; Klein, C.; Naito, A.; Alcalay, R.N.; ClinGen Parkinson's Disease Gene Curation Expert Panel; et al. The commercial genetic testing landscape for Parkinson's disease. *Parkinsonism Relat. Disord.* 2021, *92*, 107–111. [CrossRef] [PubMed]
140. Castelo Rueda, M.P.; Raftopoulou, A.; Gogele, M.; Borsche, M.; Emmert, D.; Fuchsberger, C.; Hantikainen, E.M.; Vukovic, V.; Klein, C.; Pramstaller, P.P.; et al. Frequency of Heterozygous Parkin (PRKN) Variants and Penetrance of Parkinson's Disease Risk Markers in the Population-Based CHRIS Cohort. *Front. Neurol.* 2021, *12*, 706145. [CrossRef]
141. Lubbe, S.J.; Bustos, B.I.; Hu, J.; Krainc, D.; Joseph, T.; Hehir, J.; Tan, M.; Zhang, W.; Escott-Price, V.; Williams, N.M.; et al. Assessing the relationship between monoallelic PRKN mutations and Parkinson's risk. *Hum. Mol. Genet.* 2021, *30*, 78–86. [CrossRef]
142. Yu, E.; Rudakou, U.; Krohn, L.; Mufti, K.; Ruskey, J.A.; Asayesh, F.; Estiar, M.A.; Spiegelman, D.; Surface, M.; Fahn, S.; et al. Analysis of Heterozygous PRKN Variants and Copy-Number Variations in Parkinson's Disease. *Mov. Disord.* 2021, *36*, 178–187. [CrossRef]
143. Eggers, C.; Schmidt, A.; Hagenah, J.; Bruggemann, N.; Klein, J.C.; Tadic, V.; Kertelge, L.; Kasten, M.; Binkofski, F.; Siebner, H.; et al. Progression of subtle motor signs in PINK1 mutation carriers with mild dopaminergic deficit. *Neurology* 2010, *74*, 1798–1805. [CrossRef]
144. Puschmann, A.; Fiesel, F.C.; Caulfield, T.R.; Hudec, R.; Ando, M.; Truban, D.; Hou, X.; Ogaki, K.; Heckman, M.G.; James, E.D.; et al. Heterozygous PINK1 p.G411S increases risk of Parkinson's disease via a dominant-negative mechanism. *Brain* 2017, *140*, 98–117. [CrossRef]
145. Krohn, L.; Grenn, F.P.; Makarious, M.B.; Kim, J.J.; Bandres-Ciga, S.; Roosen, D.A.; Gan-Or, Z.; Nalls, M.A.; Singleton, A.B.; Blauwendraat, C.; et al. Comprehensive assessment of PINK1 variants in Parkinson's disease. *Neurobiol. Aging* 2020, *91*, 168.e1–168.e5. [CrossRef]
146. Ysselstein, D.; Nguyen, M.; Young, T.J.; Severino, A.; Schwake, M.; Merchant, K.; Krainc, D. LRRK2 kinase activity regulates lysosomal glucocerebrosidase in neurons derived from Parkinson's disease patients. *Nat. Commun.* 2019, *10*, 5570. [CrossRef]
147. Sanyal, A.; DeAndrade, M.P.; Novis, H.S.; Lin, S.; Chang, J.; Lengacher, N.; Tomlinson, J.J.; Tansey, M.G.; LaVoie, M.J. Lysosome and Inflammatory Defects in GBA1-Mutant Astrocytes Are Normalized by LRRK2 Inhibition. *Mov. Disord.* 2020, *35*, 760–773. [CrossRef] [PubMed]
148. Ortega, R.A.; Wang, C.; Raymond, D.; Bryant, N.; Scherzer, C.R.; Thaler, A.; Alcalay, R.N.; West, A.B.; Mirelman, A.; Kuras, Y.; et al. Association of Dual LRRK2 G2019S and GBA Variations with Parkinson Disease Progression. *JAMA Netw. Open* 2021, *4*, e215845. [CrossRef] [PubMed]
149. Yahalom, G.; Greenbaum, L.; Israeli-Korn, S.; Fay-Karmon, T.; Livneh, V.; Ruskey, J.A.; Ronciere, L.; Alam, A.; Gan-Or, Z.; Hassin-Baer, S. Carriers of both GBA and LRRK2 mutations, compared to carriers of either, in Parkinson's disease: Risk estimates and genotype-phenotype correlations. *Parkinsonism Relat. Disord.* 2019, *62*, 179–184. [CrossRef] [PubMed]
150. International Parkinson Disease Genomics Consortium. Ten Years of the International Parkinson Disease Genomics Consortium: Progress and Next Steps. *J. Parkinsons Dis.* 2020, *10*, 19–30. [CrossRef]
151. Global Parkinson's Genetics Program. GP2: The Global Parkinson's Genetics Program. *Mov. Disord.* 2021, *36*, 842–851. [CrossRef]

Article

Elucidating Hexanucleotide Repeat Number and Methylation within the X-Linked Dystonia-Parkinsonism (XDP)-Related SVA Retrotransposon in *TAF1* with Nanopore Sequencing

Theresa Lüth [1,†], Joshua Laß [1,†], Susen Schaake [1], Inken Wohlers [2,3], Jelena Pozojevic [1], Roland Dominic G. Jamora [4], Raymond L. Rosales [5], Norbert Brüggemann [1,6], Gerard Saranza [7], Cid Czarina E. Diesta [8], Kathleen Schlüter [1], Ronnie Tse [1], Charles Jourdan Reyes [1], Max Brand [1], Hauke Busch [2,3], Christine Klein [1], Ana Westenberger [1] and Joanne Trinh [1,*]

[1] Institute of Neurogenetics, University of Luebeck, 23538 Luebeck, Germany; theresa.lueth@neuro.uni-luebeck.de (T.L.); joshua.lass@student.uni-luebeck.de (J.L.); susen.schaake@neuro.uni-luebeck.de (S.S.); jelena.pozojevic@neuro.uni-luebeck.de (J.P.); norbert.brueggemann@neuro.uni-luebeck.de (N.B.); kathleen.schlueter@student.uni-luebeck.de (K.S.); ronnie.tse@neuro.uni-luebeck.de (R.T.); charles.reyes@neuro.uni-luebeck.de (C.J.R.); max.brand@student.uni-luebeck.de (M.B.); christine.klein@neuro.uni-luebeck.de (C.K.); ana.westenberger@neuro.uni-luebeck.de (A.W.)

[2] Medical Systems Biology Division, Luebeck Institute of Experimental Dermatology, University of Luebeck, 23538 Luebeck, Germany; Inken.Wohlers@uni-luebeck.de (I.W.); Hauke.Busch@uni-luebeck.de (H.B.)

[3] Institute for Cardiogenetics, University of Luebeck, 23538 Luebeck, Germany

[4] Department of Neurosciences, College of Medicine, Philippine General Hospital, University of the Philippines Manila, Manila 1000, Philippines; rgjamora@up.edu.ph

[5] Department of Neurology and Psychiatry, The Hospital Neuroscience Institute, University of Santo Tomas, Manila 1008, Philippines; rlrosales@ust.edu.ph

[6] Department of Neurology, University of Luebeck, 23538 Luebeck, Germany

[7] Section of Neurology, Department of Internal Medicine, Chong Hua Hospital, Cebu City 6000, Philippines; gerardsaranza@gmail.com

[8] Department of Neurosciences, Movement Disorders Clinic, Makati Medical Center, Makati 1229, Philippines; ciddiesta@gmail.com

* Correspondence: joanne.trinh@neuro.uni-luebeck.de

† These authors contributed equally to this work.

Abstract: Background: X-linked dystonia-parkinsonism (XDP) is an adult-onset neurodegenerative disorder characterized by progressive dystonia and parkinsonism. It is caused by a SINE-VNTR-Alu (SVA) retrotransposon insertion in the *TAF1* gene with a polymorphic $(CCCTCT)_n$ domain that acts as a genetic modifier of disease onset and expressivity. Methods: Herein, we used Nanopore sequencing to investigate SVA genetic variability and methylation. We used blood-derived DNA from 96 XDP patients for amplicon-based deep Nanopore sequencing and validated it with fragment analysis which was performed using fluorescence-based PCR. To detect methylation from blood- and brain-derived DNA, we used a Cas9-targeted approach. Results: High concordance was observed for hexanucleotide repeat numbers detected with Nanopore sequencing and fragment analysis. Within the SVA locus, there was no difference in genetic variability other than variations of the repeat motif between patients. We detected high CpG methylation frequency (MF) of the SVA and flanking regions (mean MF = 0.94, SD = ± 0.12). Our preliminary results suggest only subtle differences between the XDP patient and the control in predicted enhancer sites directly flanking the SVA locus. Conclusions: Nanopore sequencing can reliably detect SVA hexanucleotide repeat numbers, methylation and, lastly, variation in the repeat motif.

Keywords: X-linked dystonia-parkinsonism; nanopore sequencing; repeat motif; CpG methylation

1. Introduction

X-linked dystonia-parkinsonism (XDP) is a neurodegenerative movement disorder and its phenomenology was first described in the literature in 1976 [1]. Patients originate mainly from the Philippines or are of Filipino descent and mainly aggregate on the island of Panay. A known family history of the disease is present for ~94% of the patients. XDP originated through a founder mutation approximately 1000 years ago [2]. The disease is characterized by dystonic movements and postures as well as parkinsonism due to an insertion of the retrotransposon SINE-VNTR-Alu (SVA) in intron 32 of the *TAF1* (TATA-binding protein-associated factor 1) gene [3,4].

The *TAF1* SVA insertion has five domains. At the 5' end, there is a hexanucleotide repeat domain, which consists of the repeat sequence $(CCCTCT)_n$ [5]. This hexanucleotide repeat $(CCCTCT)_n$ domain varies in repeat numbers among patients, ranging from 30 to 55. The repeat number is inversely correlated with age at onset and disease severity [5,6]. In addition, somatic mosaicism has been observed, with a higher median number of repeats detected in the cerebellum and basal ganglia compared to blood [7]. In XDP patients, seven variants have been found on the X chromosome: five single-nucleotide variants (SNVs), a 48-bp deletion and the SVA insertion [8,9]. Within the SVA, no variants have been reported besides the $(CCCTCT)_n$ repeat polymorphism [5].

The *TAF1* SVA insertion is also associated with decreased *TAF1* expression [6]. The reduced *TAF1* expression that has been observed in blood and patient-derived induced pluripotent stem cells can be rescued by excision of the retrotransposon insertion [10,11]. Thus, *TAF1* reduction is a consequence of the presence of the SVA insertion. However, how the *TAF1* SVA insertion influences gene expression levels still remains an enigma. Of note, two enhancers are predicted to be located upstream and ten enhancers downstream of the *TAF1* SVA insertion. The SVA itself is highly methylated due to the high "GC" content (~60%) within the variable number tandem repeat (VNTR) region, also known as "mobile CpG-island" [9,12]. Therefore, the SVA retrotransposon insertion may affect *TAF1* expression by changing the methylation status (causing hypo- or hypermethylation) of the surrounding genomic region across several enhancer sites. There are approximately 2700 SVA elements within the human reference genome (hg19) [13], and specific characterization of the *TAF1* SVA insertion in XDP patients has been hard to achieve with short-read sequencing technologies. *TAF1* SVA is a non-reference mobile element. Recently, mobile element insertions have been investigated in the context of the Simons Genome Diversity Project, and on average, 47 non-reference mobile element insertions are present per individual [14]. Similar to XDP, the insertion of SVAs have been implicated in many diseases such as neurofibromatosis type 1 and haemophilia B [15]. To our knowledge, the full-length *TAF1* SVA and flanking regions (>22 kb) have not been sequenced and investigated.

In this study, we establish a straightforward Nanopore sequencing workflow to investigate the genetic architecture of *TAF1* SVA by characterizing: (1) genomic variants within the SVA, (2) variations of the hexanucleotide repeat number and (3) CpG methylation by utilizing Nanopore long-read sequencing.

2. Materials and Methods

2.1. Patient Demographics

The study was approved by the Ethics Committees of the University of Luebeck, Germany and the Metropolitan Medical Center, Manila, Philippines (REF: IRB-MMC #: 10-073). For the analysis of genomic variants within the SVA and the detection of variations of the hexanucleotide repeat domain, $n = 96$ patients with XDP were investigated. As XDP follows an X-linked recessive inheritance pattern, only male patients were included. The mean age at onset (AAO) was 40.66 (SD = ±8.75), and the mean age at examination (AAE) and sample collection was 45.4 (SD = ±10.24) (Supplementary Table S1).

The CpG methylation was investigated in blood-derived DNA from one deceased XDP patient (L-7995) and one control (L-14529). The control was matched according to age, gender and ethnicity. For the patient (L-7995), brain tissue samples derived from the basal ganglia (BG) and cerebellum (CRB) were also available. The patient had an AAO of 31 years. The AAE was 36 years in the patient and the control.

2.2. Single-Nucleotide Variants and Repeat Detection

DNA was extracted with the Blood and Cell Culture DNA Midi kit (Qiagen). Long-range PCR was performed to amplify the *TAF1* SVA (amplicon Size: 3.2 kb) in XDP patients, as previously described [7], using the PrimeSTAR GXL DNA Polymerase® (Takara Bio). The primer sequences are documented in Supplementary Table S2. Subsequently, 1 µg of each patient-derived PCR product was barcoded with the Native 96 Barcoding Kit (EXP-NBD196) and multiplexed. Two libraries with the Ligation Sequencing Kit (LSK109) were generated for Nanopore sequencing on two R9.4.1 flow cells on a GridION. The input for library preparation was 200 fmol of DNA per sample.

Validation by fragment analysis to determine the repeat length of the hexanucleotide $(CCCTCT)_n$ was performed with a fluorescein amidites (FAM) labeled primer, as previously described [5,6].

2.3. Methylation Detection

To obtain the epigenetic information and to enrich the target region, Cas9-targeted sequencing from Oxford Nanopore Technology (ONT) was performed. For the specific ligation of the sequencing adapter, the blunt ends with $5'$ phosphates resulting from the Cas9 ribonucleoprotein complex, cleaving out the region of interest. The CRISPR RNAs (crRNAs) were designed with the ChopChop tool (https://chopchop.cbu.uib.no, accessed on 10 December 2021) [16]. Four crRNAs were used upstream of the *TAF1* SVA insertion, and four crRNAs were used downstream (Supplementary Table S3 and Figure S1A). Two different library preparations were used for the Cas9-targeted enrichment. The first library consisted of the full ~22 kb region of interest (crRNA 1, 2, 7 and 8). The second library targeted a 5.5 kb product specifically around the SVA (crRNA 3, 4, 5 and 6). For the control without an SVA insertion, the second target was 2.8 kb in size. We prepared multiple libraries for the DNA derived from one patient (L-7995) or one control (L-14529). To prepare the individual libraries, two crRNAs were used to cut upstream of the target region and two downstream to enhance the efficiency of Cas9 DNA cleavage. For the blood-derived DNA of the patient with XDP, we have used five flow cells (R9.4.1) loaded with six libraries (5×5 µg and 1×1 µg). For the BG-derived DNA of the patient with XDP, we have used five flow cells (R9.4.1) loaded with seven libraries (2×5 µg, 3×3 µg, 1×2 µg and 1×1 µg). For the CRB-derived DNA of the patient with XDP, we have used four flow cells (R9.4.1) loaded with eight libraries (7×5 µg and 1×1 µg). For the blood-derived DNA of the healthy control, we have used five flow cells (R9.4.1) loaded with six libraries (5×5 µg and 1×1 µg).

The enriched DNA was prepared with the Nanopore Ligation Sequencing Kit (SQK-LSK109), loaded on an R9.4.1 flow cell and sequenced with the MinION or GridION. For methylation analysis, all sequencing data obtained were combined to maximize coverage depth.

2.4. Data Analysis

Base-calling was performed with Guppy version 5.0.11, and the base-calling software is available for Nanopore community members (https://community.nanoporetech.com, accessed on 10 December 2021). For the detection of the repeat length, the super accuracy model (DNA_r9.4.1_450bps_sup.cfg) and the fast model (DNA_r9.4.1_450bps_fast.cfg) were used. The corresponding configuration file names were provided as a parameter to the Guppy software. The expected base-calling accuracy for the super accuracy model is 98.3% and 95.8% for the fast model. (https://community.nanoporetech.com/posts/guppy-v5-0-7-release-note, accessed on 10 December 2021). Base-calling for the methylation detection was performed with the fast model. All reads were mapped to the reference sequence with the software Minimap2 (v2.17). The coverage was determined with the software Samtools (v1.9).

Variants were identified with the software Bcftools (v1.9) (https://github.com/samtools/bcftools, accessed on 10 December 2021). To prevent false-positive results, all reported positions by Bcftools were controlled in the VCF file. We filtered for hemizygous allelic frequency (>90%) and good quality (Phred score Q > 20). Lastly, variants were evaluated in the Integrative Genomics Viewer (IGV) to exclude erroneously called variants within homopolymeric stretches.

Detection of the repeat length was conducted using the NCRF software (Noise-canceling repeat finder) (v1.01.02) [17]. To determine the repeat length for one patient, the median of all reads was calculated as previously described [7]. The NCRF alignment was used additionally to explore the frequency of deletions, insertions and mismatches within the repeat motif.

For the Cas9-targeted sequencing data, methylation was called using the software Nanopolish (v0.13.2), which can detect 5′-methylcytosine (5 mC) in a CpG context. Nanopolish requires, besides the FASTQ and FAST5 files, the alignment in a BAM format as an input. To counteract potential off-target effects of the CRISPR-Cas9 enrichment, the BAM file was filtered for reads with an alignment length >3 kb in the patient- or >1.5 kb in control-derived samples. Only CpG sites covered by >10 reads were included in the analysis.

2.5. Statistical Analysis

Spearman correlation was performed to assess the concordance of the detected hexanucleotide repeat number between Nanopore sequencing and fragment analysis. The median repeat number and the interquartile range detected by NCRF from the Nanopore data, and the number of repeats detected with fragment analysis were used for the correlation plot.

In addition, we used the NCRF report for each sample to assess repeat motif interruptions. To determine matches and mismatches (i.e., deletions, insertions and substitutions) between the Nanopore reads and the hexanucleotide repeat motif, NCRF uses a Smith–Waterman aligner approach and affine gap penalties [17]. The software reports the total number of deletions, insertions and substitutions per read. Subsequently, the mean number of repeat motif interruptions per read across all samples were calculated, as reported by NCRF, to explore accumulations of deletions, insertions and substitutions within the SVA hexanucleotide repeat domain.

DNA methylation was compared across different tissues of a patient with XDP and a control. These differences were assessed by a non-parametric Mann–Whitney U-test, as previously described in Ewing et al. 2020 [12].

3. Results

We first analyzed the sequencing data generated by PCR amplification and subsequent multiplexing on the Nanopore of the *TAF1* SVA insertion (n = 96 XDP patients). Across all individuals, we obtained a mean coverage of 17,645X (SD = ±12,392X) per barcode. The mean coverage of the samples ranged between 1690X (SD = ±190X) and 47,919X (SD = ±5074X) due to the variable sequencing efficiency of the barcodes. However, the coverage of the amplified region was even within the samples (Supplementary Figure S1B and Table S4). The mean sequence quality (Phred score) was 15.88 (SD = ±0.44), and the mean N50 was 3.38 kb (SD = ±66.42 bp) per barcoded sample.

3.1. Single-Nucleotide Polymorphisms within the SVA TAF1 Insertion

SNVs located within the *TAF1* SVA insertion were called from the amplicon sequencing data of all 96 patients. After quality filtering and the final evaluation with IGV, no SNVs were detected.

3.2. Assessment of the Hexanucleotide Repeat Length

The hexanucleotide repeat number detection with long-read sequencing amplicon data resulted in a mean of 45.17 (SD = ±4.24) repeats, ranging from 35 to 57, using super accuracy base-calling (Figure 1A). Fragment analysis as an independent validation showed a mean number of 42.21 (SD = ±4.23) repeats, ranging from 33 to 54. The detected repeat numbers were highly concordant between the two methods (Spearman's r = 0.9765, Spearman's exploratory p-value $< 1 \times 10^{-15}$, Figure 1A). However, the repeat number detected from long-read sequencing was consistently 1–4 repeat numbers higher compared to fragment analysis. Using Guppy fast base-calling for Nanopore sequencing resulted in a mean of 42.77 (SD = ±4.05) repeats, ranging from 33 to 54. Thus, we observed a higher concordance with fast base-calling between both methods (Spearman's r = 0.9883, Spearman's exploratory p-value $< 1 \times 10^{-15}$, Figure 1B). There was an identical repeat number in n = 47 patients and a difference of ~1–2 repeats in n = 48 patients.

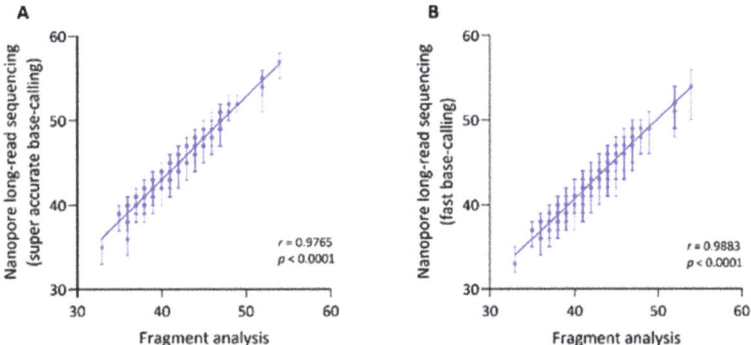

Figure 1. Repeat number detection using Nanopore long-read sequencing is highly concordant with the results from fragment analysis. Correlation between the median repeat numbers per individual of the $(CCCTCT)_n$ SVA domain, detected with fragment analysis and Nanopore sequencing using super accurate (**A**) or fast base-calling (**B**). Bars indicate the interquartile range of the detected repeat number with Nanopore sequencing. R = Spearman's rank correlation coefficient, p = Spearman's exploratory p-value.

To further validate our workflow, we analyzed the previously shown negative association between the AAO and the repeat number [5,6]. The repeat number detected with Nanopore sequencing negatively correlated with AAO in patients with XDP (Spearman's $r < -0.80$, Spearman's exploratory p-value $< 1 \times 10^{-15}$).

Next, we explored the continuity of the repeat motif. As reported by NCRF, the mean number of deletions per read within the repetitive sequence was 6.05 (SD = ±1.00), the mean number of insertions was 2.89 (SD = ±0.36) and of substitutions 0.73 (SD = ±0.09). Thus, deletions were the most common type of interruptions detected in the hexanucleotide repeat sequence of XDP patients (Figure 2).

3.3. Methylation within the SVA and in the Flanking Regions

To assess the DNA CpG methylation of the SVA, we enriched the *TAF1* SVA insertion and flanking regions with a Cas9-targeted approach. We included blood- and brain-derived DNA from one XDP patient and blood-derived DNA from one age-matched control participant. We used two Cas9 enrichment strategies: (1) the *TAF1* SVA insertion and a short flanking region (~5.5 kb) and (2) the *TAF1* SVA insertion and a longer flanking region (~22 kb), including 12 predicted enhancer sites.

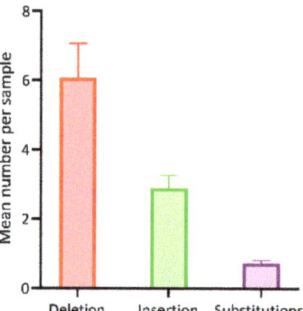

Figure 2. Occurrence of repeat motif interruptions. The bar chart shows the mean number of repeat motif interruptions per patient sample, stratified by type (i.e., deletion, insertion, substitution). The bars and whiskers represent the mean and upper limit of the standard deviation.

The enrichment of the shorter fragment resulted in an N50 of 4.5 kb for the patient-derived samples and an N50 on 2.0 kb for the control-derived sample. The mean Phred score of the reads ranged from 10.0 to 10.9 and the mean coverage from 126.9X (SD = ±79.8X) to 1226.0X (SD = ±554.9X) (Supplementary Figure S1C).

The enrichment of the longer fragment resulted in an N50 of 4.7 kb for the patient-derived samples and an N50 on 8.8 kb for the control-derived sample. The mean Phred score of the reads ranged from 12.3 to 13.5 and the mean coverage from 22.1X (SD = ±11.5X) to 591.0X (SD = ±1202.0X). The sequencing quality statistics were summarized in Supplementary Table S4.

Overall, the methylation levels within the SVA as well as in the up- and downstream flanking regions were high in the patient-derived samples (Figure 3A). However, the mean MF was lower in the brain-derived samples (BG: mean MF ± SD = 0.88 ± 0.15, CRB: mean MF ± SD = 0.90 ± 0.14) compared to the blood-derived sample (mean MF ± SD = 0.94 ± 0.12). There were $n = 153$ CpG sites within the SVA *TAF1* insertion (Figure 3B). Consistent with the overall methylation level across the 22 kb region, the mean CpG MF within the SVA specifically was still lower in the brain-derived samples (BG: mean MF ± SD = 0.87 ± 0.14, CRB: mean MF ± SD = 0.93 ± 0.08) compared to the blood-derived sample (mean MF ± SD = 0.96 ± 0.07) (exploratory Mann–Whitney U-test $p < 1.2 \times 10^{-6}$) (Supplementary Figure S2A). In addition to patient-derived DNA, we analyzed blood-derived DNA from one healthy control (Figure 3C). The overall MF across the SVA flanking region in the control sample was at 0.83 ± 0.17, which was lower than the patient-derived sample (MF ± SD = 0.93 ± 0.15) (exploratory Mann–Whitney U-test $p < 1 \times 10^{-15}$, Supplementary Figure S2B). Despite a significant difference, the effect size was small.

There were 12 predicted enhancer sites located in the targeted region, 2 upstream and 10 downstream of the *TAF1* SVA insertion, according to the ENCODE project (reference number: wgEncodeEH000790). There was no CpG site located within enhancer eight, and this predicted enhancer was excluded from the analysis. The mean MF of the enhancer sites ranged from 0.65 to 0.99 in the blood-derived sample, from 0.46 to 0.99 in the BG-derived sample and from 0.37 to 0.99 in the CRB-derived sample (Figure 4A, Supplementary Table S5). In comparison, the mean MF of these enhancers ranged from 0.69 to 0.95 in the healthy control (Figure 4B, Supplementary Table S5). We detected significantly lower methylation of the enhancer sites within the *TAF1* SVA flanking region in the control compared to the patient-derived blood sample (exploratory Mann–Whitney U-test $p = 0.0033$, Supplementary Figure S2D). With the exception of enhancer two, all enhancers showed lower methylation levels in the control subject. The most pronounced difference was observed at enhancer site six (mean MF patient: 0.91, mean MF control: 0.71) and nine (mean MF patient: 0.98, mean MF control: 0.70). However, the sample size is small, the effect sizes are small, and differences remain difficult to interpret (see details in Section 4).

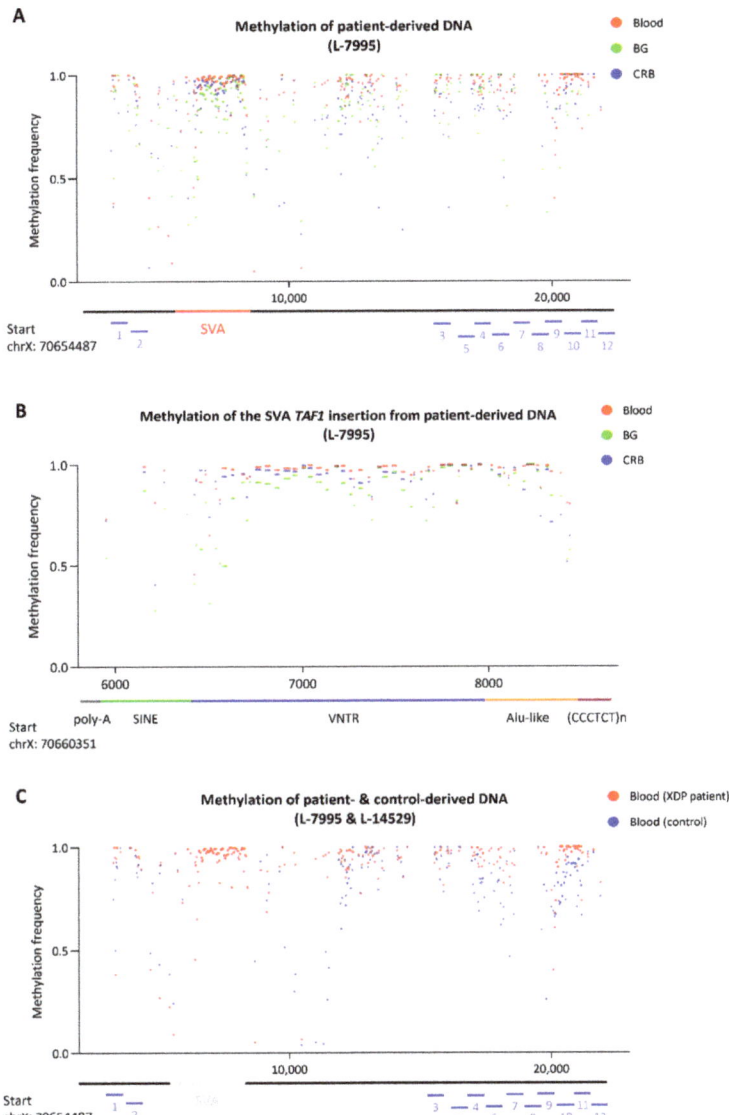

Figure 3. Methylation frequency of the *TAF1* SVA insertion and flanking regions. (**A**) Methylation frequency of two different brain tissues and blood-derived from a patient with XDP. Red indicates methylation from blood-derived DNA, green from basal ganglia-derived (BG) DNA and blue from cerebellum-derived (CRB) DNA. (**B**) Methylation frequency of *TAF1* SVA insertion with indicated SVA domains of the same patient-derived DNA samples. Red indicates methylation from blood-derived DNA, green from BG-derived DNA and blue from CRB-derived DNA. (**C**) Methylation frequency of blood-derived DNA from a patient with XDP (red) and a control participant (blue). The *x*-axis indicates the position in the reference sequence. The bars indicate the location of predicted enhancers, the *TAF1* SVA insertion and the insertion's subunits.

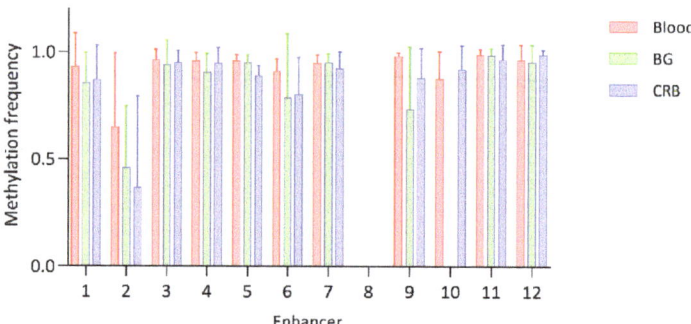

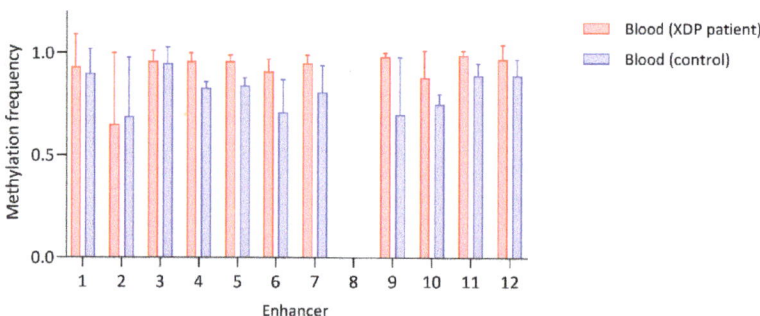

Figure 4. Methylation levels of the predicted enhancers flanking the *TAF1* SVA. (**A**) The bar plot shows the CpG methylation frequency of predicted enhancer sites within the targeted region. DNA was derived from the blood, basal ganglia and cerebellum of a XDP patient (**A**) or derived from the blood of a patient and a healthy control participant (**B**). The bars and whiskers represent the mean and upper limit of the standard deviation of the methylation frequency from the CpG sites within a predicted enhancer.

4. Discussion

In this study, we performed Nanopore single-molecule sequencing to examine the genetics and epigenetics of the full-length *TAF1* SVA insertion in patients with XDP. The novelty of our study lies in: (1) a new multiplexed workflow to quantify the *TAF1* SVA repeat number in patients that shows high concordance with fragment analysis which can be used as a cost-effective diagnostic tool in the future; (2) the exploration of novel variants within the SVA besides the repeat motif across 96 patients which has not been possible with older technologies and lastly; (3) the detection of direct CpG methylation across the full-length SVA and flanking regions up to 22 kb which incorporates 12 putative enhancer sites.

4.1. Examination of the Hexanucleotide Repeat Domain

Deep Nanopore sequencing (>5000X) of the GC-rich *TAF1* SVA insertion allowed better alignment accuracy for the low-complexity repetitive regions of the SVA than short-read sequencing approaches [18]. From our sequencing data analysis, there was no evidence of any other genetic variability besides the repeat domain within the investigated locus.

Thus, variability of AAO and disease severity largely result from the hexanucleotide repeat number within the SVA [5,6]. As the sequence of the SVA is identical between patients except for the polymorphic repeat, this further validates the notion that the insertion of this repeat sequence in intron 32 of *TAF1* causes XDP [19].

The software tool, NCRF, has been specifically designed to explore repetitive sequences in noisy long-read sequencing data [17]. More specifically, the noise in the Nanopore signal trace is due to indel and homopolymer errors, reducing sequencing accuracy [17]. To decrease the noise even further, we performed base-calling with the novel "super-high accuracy" model with improved read accuracy provided by the Nanopore software Guppy (v.5.0.11). Indeed, the repeat number resulting from the NCRF analysis was highly concordant with the results from the independent fragment analysis method. In concordance with the literature [5,6], the repeat number detected by Nanopore sequencing was negatively associated with the AAO of patients with XDP as well, which further validates our workflow. Interestingly, the detected repeat number by NCRF was consistently 1–4 repeats longer than the number obtained from the fragment analysis. The slightly larger repeat number, detected with Nanopore sequencing, could be due to deviations in the repeat pattern. We did not detect a consistently higher repeat number when we compared the results from the fast base-calling to the fragment analysis. The higher concordance with the repeat number detected with fast base-calling could result from general repeat number detection of fragment analysis without the resolution of mismatches in the repeat motif. In fact, there was a noticeable increase in the frequency of deletions in the long-read data that require further exploration as interruption of the repeat motif has also been reported for other repeat expansion diseases such as Friedreich's Ataxia or Huntington disease [20,21]. Therefore, further investigation of this issue, including single nucleotide resolution of the repeat interruption, is required.

4.2. Methylation Status of the TAF1 SVA Insertion and Adjacent Enhancer Sites

Nanopore sequencing has been used to investigate DNA methylation in the context of other repeat expansions before [12,22]. Recently, in the context of cancer, Nanopore sequencing has been used to assess the genetics and epigenetics of transposable elements simultaneously, including the CG-rich VNTR domain of SVAs, also known as "mobile CpG-island" [12]. To maintain DNA methylation, an amplification-free Cas9-guided approach for Nanopore sequencing was introduced [23]. This targeted approach has been shown to efficiently enrich repeat expansions causing frontotemporal dementia and amyotrophic lateral sclerosis or fragile X syndrome [22].

In this study, the target region was enriched against the genomic background DNA, and methylation was maintained, using a Cas9-guided approach. Coverage of the SVA insertion specifically was high; however, lower sequencing depth was obtained in the flanking regions. The variability in coverage could be due to the limitation of different targeting efficiencies of the crRNAs used in the Cas9-approach. To counteract potential off-target effects, we included only reads with sufficient alignment length to the reference sequence.

We observed hypermethylation of the *TAF1* SVA insertion in concordance with the literature [9]. Although the overall MF in all patient-derived samples was high, it was mildly reduced in the brain-derived samples. There have been speculations that neuron-specific expression of *TAF1* is reduced in patients with XDP [9]. However, recently published studies could not confirm a significant decrease of neuron-specific *TAF1* in XDP patients [24,25]. In addition, it is unclear if the slight change of the methylation level in the brain-derived samples that we observed could affect the expression level of *TAF1* and whether it would be relevant for the disease to develop. Furthermore, we did not specifically analyze the methylation level of neurons, which could be a perspective for future experiments.

There is the possibility that retrotransposon insertions introduce methylation changes into the flanking regions [26]. There have been speculations that the hypermethylation of the SVA could also affect the methylation of adjacent *cis*-regulatory elements [9]. The 12 predicted enhancer sites in the target region showed mostly comparable methylation levels in the blood and brain-derived samples, and only 2 enhancer sites showed a slight methylation decrease in the brain. Interestingly, lower methylation levels of the BG-derived sample, in particular, were present in the SVA insertion, as well (Figure 3B). This difference is most pronounced in the VNTR domain of the retrotransposon. The lower methylation observed in the BG can be explained by tissue-specific differences and is not necessarily a disease-related phenomenon. Indeed, tissue-specific methylation patterns of transposable elements have been investigated with Nanopore sequencing [12,27]. Of particular interest is the reduced methylation of intergenic subfamily SVA$_f$ insertion in the X-chromosome in tumor tissues compared to non-tumor tissues [12]. In general, overall CpG methylation of brain tissue and peripheral tissues is highly correlated across participants. On the other hand, differences between the methylation level of the brain and peripheral tissues within a particular individual are possible, which can contribute to tissue-specific gene expression [28–30].

The goal of this study was to establish a new workflow for direct CpG methylation detection using Nanopore sequencing. The study shows that it is indeed possible to detect methylation across a large region, including the *TAF1* SVA of 22 kb. There has been evidence that DNA methylation can be a molecular mechanism in XDP pathogenesis. Due to the abolishment or introduction of CpG dinucleotides by disease-specific single-nucleotide changes (DSC), significant differences in methylation between XDP patients and controls at these positions have been reported, suggesting a potential effect on *TAF1* expression [31]. These three reported DSCs are located ~700 kb away from the SVA. Our study focused on regions within and adjacent to the SVA insertion. As the GC-content of the VNTR is high and the SVA is known to be hypermethylated, the change in the direct genetic environment could lead to altered methylation patterns within the SVA and flanking regions. As a pilot study, we observed overall lower methylation of the control sample compared to the XDP patient in the SVA flanking region as well as in 10 out of the 12 predicted enhancer sites. The observed methylation differences in this study are only preliminary and should be interpreted with caution. Another limitation of this study was the lack of brain-derived DNA from controls. Thus, more individuals and tissue types should be investigated using the Cas9-targeted approach across this 22 kb region. Still, our results show the utility of long reads in detecting the full-length SVA in *TAF1* in the context of XDP.

5. Conclusions

In this study, we present a straightforward and scalable long-read deep sequencing workflow to quantify the hexanucleotide repeat number of the *TAF1* SVA in patients with XDP. The high concordance of the results obtained from Nanopore sequencing to independent fragment analysis highlights the accuracy of our workflow. The long reads were also utilized to investigate variations within the SVA locus other than the repeat motif. As the sequence of the SVA locus was identical between patients besides the hexanucleotide repeat domain, our results further underline that the insertion of this repeat sequence is associated with the variability in AAO and expressivity in XDP. Lastly, an amplification-free Cas9 targeted enrichment of the SVA locus and the flanking regions allowed us to comprehensively assess the (epi-) genetics of the *TAF1* SVA locus.

Supplementary Materials: The following are available online at https://www.mdpi.com/article/10.3390/genes13010126/s1, Table S1: Overview of patients with XDP for variant and repeat detection; Table S2: Primers for the long-range PCR; Table S3: Sequences of the crRNAs for the Cas9-targeted sequencing; Table S4: Overview of the Nanopore sequencing quality parameters; Table S5: Overview of CpG methylation of enhancer sites located in the target region; Figure S1: Location of XDP-16153 F and XDP-19345R primers and Cas9 guide RNAs and coverage after PCR or Cas9-targeted enrichment; Figure S2: Comparison of the methylation levels of the *TAF1* SVA and flanking regions.

Author Contributions: Conceptualization, C.K., A.W. and J.T.; methodology, S.S., J.P., R.T. and C.J.R.; software, T.L. and J.L.; formal analysis, T.L., J.L., J.P., K.S., M.B., A.W. and J.T.; investigation, T.L., J.L., S.S., I.W., J.P., R.D.G.J., R.L.R., N.B., G.S., C.C.E.D., R.T., K.S., M.B., C.J.R., H.B., C.K. and A.W.; resources, R.D.G.J., R.L.R., N.B., G.S., C.C.E.D., H.B., C.K., A.W. and J.T.; data curation, S.S., J.P., R.T. and C.J.R.; writing—original draft preparation, T.L., J.L. and J.T.; writing—review and editing, S.S., I.W., J.P., R.D.G.J., R.L.R., N.B., G.S., C.C.E.D., R.T., K.S., M.B., C.J.R., H.B., C.K. and A.W.; visualization, T.L. and J.L.; supervision, J.T.; project administration, C.K., A.W. and J.T.; funding acquisition, C.K., A.W. and J.T. All authors have read and agreed to the published version of the manuscript.

Funding: Funding has been obtained from the German Research Foundation ("ProtectMove"; FOR 2488, J.T., A.W., N.B. and C.K.; Germany's Excellence Strategy—EXC 22167-390884018, I.W. and H.B.), the Hermann and Lilly Schilling Foundation, the European Community (SysMedPD), the Canadian Institutes of Health Research (CIHR) (J.T.), Peter and Traudl Engelhorn Foundation and the Center for XDP (CCXDP) at Massachusetts General Hospital (N.B.).

Institutional Review Board Statement: The study was conducted according to the guidelines of the Declaration of Helsinki, and approved by the Ethics Committees of the University of Luebeck, Germany and the Metropolitan Medical Center, Manila, Philippines (REF: IRB-MMC #: 10-073; 31 August 2010).

Informed Consent Statement: Informed consent was obtained from all participants involved in the study.

Data Availability Statement: The data presented in this study are available on SRA (SAMN24775867-SAMN24775962, SAMN24115523-SAMN24115530). The bioinformatical commands to quantify the *TAF1* SVA (*CCCTCT*)$_n$ repeat length are described here: https://github.com/nanopol/xdp_sva/ (accessed on 10 December 2021).

Acknowledgments: The authors wish to thank the many patients and their families who volunteered and the efforts of the many clinical teams involved. The authors further acknowledge computational support from the OMICS computing cluster at the University of Luebeck.

Conflicts of Interest: C.K. serves as a medical advisor for genetic testing reports to CENTOGENE GmbH in the fields of movement disorders and dementia, excluding Parkinson's disease, and is a member of the Scientific Advisory Boards of Retromer Therapeutics and Klink. A.W. provides consultancy services around research projects for CENTOGENE GmbH. NB served as a consultant for BridgeBio, Biomarin and Centogene and is a member of the scientific advisory board of BridgeBio. He received speaker honoraria from Abbott and Zambon.

References

1. Lee, L.V.; Pascasio, F.M.; Fuentes, F.D.; Viterbo, G.H. Torsion dystonia in Panay, Philippines. *Adv. Neurol.* **1976**, *14*, 137–151.
2. Rosales, R.L. X-linked dystonia parkinsonism: Clinical phenotype, genetics and therapeutics. *J. Mov. Disord.* **2010**, *3*, 32–38. [CrossRef]
3. Lee, L.V.; Munoz, E.L.; Tan, K.T.; Reyes, M.T. Sex linked recessive dystonia parkinsonism of Panay, Philippines (XDP). *Mol. Pathol.* **2001**, *54*, 362–368. [PubMed]
4. Pauly, M.G.; Lopez, M.R.; Westenberger, A.; Saranza, G.; Bruggemann, N.; Weissbach, A.; Rosales, R.L.; Diesta, C.C.; Jamora, R.D.G.; Reyes, C.J.; et al. Expanding Data Collection for the MDSGene Database: X-linked Dystonia-Parkinsonism as Use Case Example. *Mov. Disord.* **2020**, *35*, 1933–1938. [CrossRef]
5. Bragg, D.C.; Mangkalaphiban, K.; Vaine, C.A.; Kulkarni, N.J.; Shin, D.; Yadav, R.; Dhakal, J.; Ton, M.L.; Cheng, A.; Russo, C.T.; et al. Disease onset in X-linked dystonia-parkinsonism correlates with expansion of a hexameric repeat within an SVA retrotransposon in TAF1. *Proc. Natl. Acad. Sci. USA* **2017**, *114*, E11020–E11028. [CrossRef] [PubMed]
6. Westenberger, A.; Reyes, C.J.; Saranza, G.; Dobricic, V.; Hanssen, H.; Domingo, A.; Laabs, B.H.; Schaake, S.; Pozojevic, J.; Rakovic, A.; et al. A hexanucleotide repeat modifies expressivity of X-linked dystonia parkinsonism. *Ann. Neurol.* **2019**, *85*, 812–822. [CrossRef] [PubMed]
7. Reyes, C.J.; Laabs, B.H.; Schaake, S.; Luth, T.; Ardicoglu, R.; Rakovic, A.; Grutz, K.; Alvarez-Fischer, D.; Jamora, R.D.; Rosales, R.L.; et al. Brain Regional Differences in Hexanucleotide Repeat Length in X-Linked Dystonia-Parkinsonism Using Nanopore Sequencing. *Neurol. Genet.* **2021**, *7*, e608. [CrossRef]
8. Domingo, A.; Westenberger, A.; Lee, L.V.; Braenne, I.; Liu, T.; Vater, I.; Rosales, R.; Jamora, R.D.; Pasco, P.M.; Cutiongco-Dela Paz, E.M.; et al. New insights into the genetics of X-linked dystonia-parkinsonism (XDP, DYT3). *Eur. J. Hum. Genet.* **2015**, *23*, 1334–1340. [CrossRef]

9. Makino, S.; Kaji, R.; Ando, S.; Tomizawa, M.; Yasuno, K.; Goto, S.; Matsumoto, S.; Tabuena, M.D.; Maranon, E.; Dantes, M.; et al. Reduced neuron-specific expression of the TAF1 gene is associated with X-linked dystonia-parkinsonism. *Am. J. Hum. Genet.* **2007**, *80*, 393–406. [CrossRef] [PubMed]
10. Rakovic, A.; Ziegler, J.; Martensson, C.U.; Prasuhn, J.; Shurkewitsch, K.; Konig, P.; Paulson, H.L.; Klein, C. PINK1-dependent mitophagy is driven by the UPS and can occur independently of LC3 conversion. *Cell Death Differ.* **2019**, *26*, 1428–1441. [CrossRef]
11. Aneichyk, T.; Hendriks, W.T.; Yadav, R.; Shin, D.; Gao, D.; Vaine, C.A.; Collins, R.L.; Domingo, A.; Currall, B.; Stortchevoi, A.; et al. Dissecting the Causal Mechanism of X-Linked Dystonia-Parkinsonism by Integrating Genome and Transcriptome Assembly. *Cell* **2018**, *172*, 897–909. [CrossRef] [PubMed]
12. Ewing, A.D.; Smits, N.; Sanchez-Luque, F.J.; Faivre, J.; Brennan, P.M.; Richardson, S.R.; Cheetham, S.W.; Faulkner, G.J. Nanopore Sequencing Enables Comprehensive Transposable Element Epigenomic Profiling. *Mol. Cell* **2020**, *80*, 915–928. [CrossRef] [PubMed]
13. Wang, H.; Xing, J.; Grover, D.; Hedges, D.J.; Han, K.; Walker, J.A.; Batzer, M.A. SVA elements: A hominid-specific retroposon family. *J. Mol. Biol.* **2005**, *354*, 994–1007. [CrossRef]
14. Watkins, W.S.; Feusier, J.E.; Thomas, J.; Goubert, C.; Mallick, S.; Jorde, L.B. The Simons Genome Diversity Project: A Global Analysis of Mobile Element Diversity. *Genome Biol. Evol.* **2020**, *12*, 779–794. [CrossRef] [PubMed]
15. Pfaff, A.L.; Bubb, V.J.; Quinn, J.P.; Koks, S. Reference SVA insertion polymorphisms are associated with Parkinson's Disease progression and differential gene expression. *npj Parkinsons Dis.* **2021**, *7*, 44. [CrossRef] [PubMed]
16. Montague, T.G.; Cruz, J.M.; Gagnon, J.A.; Church, G.M.; Valen, E. CHOPCHOP: A CRISPR/Cas9 and TALEN web tool for genome editing. *Nucleic Acids Res.* **2014**, *42*, W401–W407. [CrossRef]
17. Harris, R.S.; Cechova, M.; Makova, K.D. Noise-cancelling repeat finder: Uncovering tandem repeats in error-prone long-read sequencing data. *Bioinformatics* **2019**, *35*, 4809–4811. [CrossRef]
18. De Coster, W.; De Rijk, P.; De Roeck, A.; De Pooter, T.; D'Hert, S.; Strazisar, M.; Sleegers, K.; Van Broeckhoven, C. Structural variants identified by Oxford Nanopore PromethION sequencing of the human genome. *Genome Res.* **2019**, *29*, 1178–1187. [CrossRef]
19. Bragg, D.C.; Sharma, N.; Ozelius, L.J. X-Linked Dystonia-Parkinsonism: Recent advances. *Curr. Opin. Neurol.* **2019**, *32*, 604–609. [CrossRef]
20. Nethisinghe, S.; Kesavan, M.; Ging, H.; Labrum, R.; Polke, J.M.; Islam, S.; Garcia-Moreno, H.; Callaghan, M.F.; Cavalcanti, F.; Pook, M.A.; et al. Interruptions of the FXN GAA Repeat Tract Delay the Age at Onset of Friedreich's Ataxia in a Location Dependent Manner. *Int. J. Mol. Sci.* **2021**, *22*, 7507. [CrossRef]
21. Findlay Black, H.; Wright, G.E.B.; Collins, J.A.; Caron, N.; Kay, C.; Xia, Q.; Arning, L.; Bijlsma, E.K.; Squitieri, F.; Nguyen, H.P.; et al. Frequency of the loss of CAA interruption in the HTT CAG tract and implications for Huntington disease in the reduced penetrance range. *Genet. Med.* **2020**, *22*, 2108–2113. [CrossRef]
22. Giesselmann, P.; Brandl, B.; Raimondeau, E.; Bowen, R.; Rohrandt, C.; Tandon, R.; Kretzmer, H.; Assum, G.; Galonska, C.; Siebert, R.; et al. Analysis of short tandem repeat expansions and their methylation state with nanopore sequencing. *Nat. Biotechnol.* **2019**, *37*, 1478–1481. [CrossRef] [PubMed]
23. Gilpatrick, T.; Lee, I.; Graham, J.E.; Raimondeau, E.; Bowen, R.; Heron, A.; Downs, B.; Sukumar, S.; Sedlazeck, F.J.; Timp, W. Targeted nanopore sequencing with Cas9-guided adapter ligation. *Nat. Biotechnol.* **2020**, *38*, 433–438. [CrossRef]
24. Capponi, S.; Stöffler, N.; Penney, E.B.; Grütz, K.; Nizamuddin, S.; Vermunt, M.W.; Castelijns, B.; Fernandez-Cerado, C.; Legarda, G.P.; Velasco-Andrada, M.S.; et al. Dissection of TAF1 neuronal splicing and implications for neurodegeneration in X-linked dystonia-parkinsonism. *Brain Commun.* **2021**, *3*, fcab253. [CrossRef]
25. Valente, E.M.; Bhatia, K.P. Solving Mendelian Mysteries: The Non-coding Genome May Hold the Key. *Cell* **2018**, *172*, 889–891. [CrossRef]
26. Yates, P.A.; Burman, R.W.; Mummaneni, P.; Krussel, S.; Turker, M.S. Tandem B1 elements located in a mouse methylation center provide a target for de novo DNA methylation. *J. Biol. Chem.* **1999**, *274*, 36357–36361. [CrossRef] [PubMed]
27. Lee, I.; Razaghi, R.; Gilpatrick, T.; Molnar, M.; Gershman, A.; Sadowski, N.; Sedlazeck, F.J.; Hansen, K.D.; Simpson, J.T.; Timp, W. Simultaneous profiling of chromatin accessibility and methylation on human cell lines with nanopore sequencing. *Nat. Methods* **2020**, *17*, 1191–1199. [CrossRef]
28. Horvath, S.; Zhang, Y.; Langfelder, P.; Kahn, R.S.; Boks, M.P.; van Eijk, K.; van den Berg, L.H.; Ophoff, R.A. Aging effects on DNA methylation modules in human brain and blood tissue. *Genome Biol.* **2012**, *13*, R97. [CrossRef] [PubMed]
29. Smith, A.K.; Kilaru, V.; Klengel, T.; Mercer, K.B.; Bradley, B.; Conneely, K.N.; Ressler, K.J.; Binder, E.B. DNA extracted from saliva for methylation studies of psychiatric traits: Evidence tissue specificity and relatedness to brain. *Am. J. Med. Genet. Part B Neuropsychiatr. Genet.* **2015**, *168B*, 36–44. [CrossRef] [PubMed]
30. Braun, P.R.; Han, S.; Hing, B.; Nagahama, Y.; Gaul, L.N.; Heinzman, J.T.; Grossbach, A.J.; Close, L.; Dlouhy, B.J.; Howard, M.A., 3rd; et al. Genome-wide DNA methylation comparison between live human brain and peripheral tissues within individuals. *Transl. Psychiatry* **2019**, *9*, 47. [CrossRef]
31. Krause, C.; Schaake, S.; Grutz, K.; Sievert, H.; Reyes, C.J.; Konig, I.R.; Laabs, B.H.; Jamora, R.D.; Rosales, R.L.; Diesta, C.C.E.; et al. DNA Methylation as a Potential Molecular Mechanism in X-linked Dystonia-Parkinsonism. *Mov. Disord.* **2020**, *35*, 2220–2229. [CrossRef] [PubMed]

Article

Validity and Prognostic Value of a Polygenic Risk Score for Parkinson's Disease

Sebastian Koch [1], Björn-Hergen Laabs [2], Meike Kasten [3,4], Eva-Juliane Vollstedt [4], Jos Becktepe [5], Norbert Brüggemann [4,6], Andre Franke [7], Ulrike M. Krämer [6], Gregor Kuhlenbäumer [5], Wolfgang Lieb [8], Brit Mollenhauer [9,10], Miriam Neis [6,11], Claudia Trenkwalder [10,12], Eva Schäffer [5], Tatiana Usnich [4], Michael Wittig [7], Christine Klein [4], Inke R. König [2], Katja Lohmann [4], Michael Krawczak [1] and Amke Caliebe [1,*]

[1] Institute of Medical Informatics and Statistics, Kiel University, University Medical Center Schleswig-Holstein, Campus Kiel, 24105 Kiel, Germany; koch@medinfo.uni-kiel.de (S.K.); krawczak@medinfo.uni-kiel.de (M.K.)
[2] Institute of Medical Biometry and Statistics, University of Luebeck, University Medical Center Schleswig-Holstein, Campus Luebeck, 23562 Luebeck, Germany; b.laabs@uni-luebeck.de (B.-H.L.); inke.koenig@uni-luebeck.de (I.R.K.)
[3] Department of Psychiatry, University of Luebeck, 23538 Luebeck, Germany; meike.kasten@neuro.uni-luebeck.de
[4] Institute of Neurogenetics, University of Luebeck, University Medical Center Schleswig-Holstein, Campus Luebeck, 23538 Luebeck, Germany; jule.vollstedt@neuro.uni-luebeck.de (E.-J.V.); norbert.brueggemann@neuro.uni-luebeck.de (N.B.); tatiana.usnich@neuro.uni-luebeck.de (T.U.); christine.klein@neuro.uni-luebeck.de (C.K.); katja.lohmann@neuro.uni-luebeck.de (K.L.)
[5] Department of Neurology, Kiel University, 24105 Kiel, Germany; josstefen.becktepe@uksh.de (J.B.); g.kuhlenbaeumer@neurologie.uni-kiel.de (G.K.); eva.schaeffer@uksh.de (E.S.)
[6] Department of Neurology, University of Luebeck, 23562 Luebeck, Germany; ulrike.kraemer@neuro.uni-luebeck.de (U.M.K.); mi.neis@uni-luebeck.de (M.N.)
[7] Institute of Clinical Molecular Biology, Kiel University, 24105 Kiel, Germany; a.franke@mucosa.de (A.F.); m.wittig@mucosa.de (M.W.)
[8] Institute of Epidemiology and PopGen Biobank, Kiel University, University Medical Center Schleswig-Holstein, Campus Kiel, 24105 Kiel, Germany; wolfgang.lieb@epi.uni-kiel.de
[9] Department of Neurology, University Medical Center Goettingen, 37075 Goettingen, Germany; brit.mollenhauer@med.uni-goettingen.de
[10] Paracelsus-Elena-Klinik, 34128 Kassel, Germany; ctrenkwalder@gmx.de
[11] Department of Midwifery Science, University of Luebeck, 23562 Luebeck, Germany
[12] Department of Neurosurgery, University Medical Center Goettingen, 37075 Goettingen, Germany
* Correspondence: caliebe@medinfo.uni-kiel.de

Abstract: Idiopathic Parkinson's disease (PD) is a complex multifactorial disorder caused by the interplay of both genetic and non-genetic risk factors. Polygenic risk scores (PRSs) are one way to aggregate the effects of a large number of genetic variants upon the risk for a disease like PD in a single quantity. However, reassessment of the performance of a given PRS in independent data sets is a precondition for establishing the PRS as a valid tool to this end. We studied a previously proposed PRS for PD in a separate genetic data set, comprising 1914 PD cases and 4464 controls, and were able to replicate its ability to differentiate between cases and controls. We also assessed theoretically the prognostic value of the PD-PRS, i.e., its ability to predict the development of PD in later life for healthy individuals. As it turned out, the PD-PRS alone can be expected to perform poorly in this regard. Therefore, we conclude that the PD-PRS could serve as an important research tool, but that meaningful PRS-based prognosis of PD at an individual level is not feasible.

Keywords: Parkinson's disease; polygenic risk score; replication; validation; prognostic value; genetic risk

1. Introduction

Parkinson's disease (PD) is the second most common neurodegenerative disorder after Alzheimer's disease, with a particularly high prevalence seen in Europe and North America [1]. PD has a complex multifactorial etiology in which both environmental and genetic factors play a prominent role. The main risk factor for PD hitherto identified, however, is age, and both prevalence and incidence increase exponentially in later life.

While some 3–5% of PD cases are monogenic, recent genome-wide association studies (GWAS) revealed that idiopathic PD is highly polygenic [2–4]. Therefore, the development of polygenic risk scores (PRSs) as a means to summarize the effect of the genetic background upon an individual's disease risk in a single number appears meaningful for idiopathic PD. Several PRSs have been developed for PD affection status, age-at-onset and specific symptoms in studies of variable size and using different methodologies [2,5–10].

Although the construction of a PRS is rather straightforward using existing software, the validation of existing PRSs through an assessment of their performance in independent data sets has still been undertaken only rarely and, to our knowledge, not for PD. One aim of our study therefore was to investigate in more detail the discriminatory power of a PRS for PD previously published by Nalls et al. [2]. This PRS was developed based upon the largest meta-GWAS for the disease to date and comprises 1805 single nucleotide polymorphisms (SNPs). Our second aim was to assess the prognostic value of this PD-PRS. In fact, while PRSs usually differentiate well between cases and controls, their utility for disease prognostics has been a matter of intensive debate [11,12].

2. Materials and Methods

2.1. Samples

The samples analyzed in the present study originated from five German cohorts comprising a total of 1914 PD cases and 4464 controls after quality control (Table A1). The data sets were collated within the framework of DFG Research Unit 'ProtectMove' (FOR2488). The samples of two PD patient and control cohorts (Kiel PD, Luebeck PD) were recruited locally in Schleswig-Holstein, the northernmost federal state of Germany. EPIPARK is an additional prospective and longitudinal observational single-center study from Luebeck, focused upon the non-motor symptoms of PD patients [13]. DeNoPa is a prospective and longitudinal observational single-center study from Kassel in central Germany, aimed specifically at improving early diagnosis and prognosis of PD. Participants include early untreated PD patients and matched healthy controls [14]. The PopGen biobank [15,16] is a central research infrastructure, maintained by Kiel University, for the recruitment of case-control cohorts for defined diseases [15,16]. For the present study, PopGen contributed 661 PD patients and 3093 unaffected individuals from the broader Kiel area.

2.2. Genotyping, Genotype Imputation and Quality Control

Genomic DNA was extracted from peripheral blood leukocytes and genotyped using the Infinium Global Screening Array with Custom Content (GSA; Illumina Inc., San Diego, CA, USA) which targets 645,896 variants. Quality control was performed with PLINK 1.9, PLINK 2.0 and R package *plinkQC* [17–22].

At the SNP level, quality control was carried out with thresholds of 0.01 for the minor allele frequency (MAF), of 0.98 for the SNP call rate and of 10^{-50} for the software-issued p value of the Hardy–Weinberg equilibrium test. Some 431,738 variants passed quality control and were used for imputation with *SHAPEIT2* [23] and *IMPUTE2* [24], based upon the public part of the HRC reference panel (release 1.1, The European Genome-Phenome Archive, EGAS00001001710) [25]. Imputation yielded genotype data for a total of 39,106,911 variants and after the exclusion of variants with MAF < 0.01 or an info score < 0.7, some 7,804,284 variants remained for further analyses.

At the participant level, 6794 individuals were initially available from the five cohorts. Individuals with a call rate < 0.98 or with a heterozygosity value > 3 standard deviations different from the mean on the non-imputed data were removed. To exclude potential relatives and population outliers, linkage disequilibrium pruning was performed using a window size of 50 variants, shifted by five variants, and an r^2 threshold of 0.2, leaving 186,064 variants. Pairwise identity-by-descent (IBD) was then estimated and individuals were removed in a customized selection process (see Appendix A.1) until all pairwise IBD values were <0.1. For details on the identification of population outliers, see Appendix A.2 and Figure A1. In total, 416 individuals were removed leaving 6378 individuals (1914 cases, 4464 controls) for further analysis. Principal component analysis (PCA) plots of the samples from our study and from the 1000Genomes project can be found in Figure A2.

2.3. Analysis of Parkinson's Disease Polygenic Risk Score (PD-PRS)

We evaluated a PRS for PD published by Nalls et al. [2]. The list of the 1805 SNPs included in this PD-PRS, together with reference alleles and effect sizes, was kindly provided to us by the first author. Matching the SNPs to our imputed SNPs was done by reference to their chromosomal positions. Some 1743 of the PD-PRS SNPs were represented in our data set, and all of these SNPs were imputed (the 62 omitted SNPs are listed in Table A2).

The PD-PRS values were standardized by subtraction of the mean and division by the standard deviation of the PD-PRS among controls. This standardized version of the PRS will henceforth be used and also referred to as 'PD-PRS' as well. Density plots were created with base-R function *density*. Logistic regression analysis was performed treating the case-control status as outcome and the PD-PRS value as influence variable, adjusted for the first three PCs, sex and age-at-sampling. An additional logistic regression analysis, excluding age-at-sampling, was performed among cases from the lowest and highest age-at-onset quartiles, treating quartile affiliation as outcome. A two-sided significance level of 0.05 was adopted for the Wald test embedded into the logistic regression analysis.

Receiver operating characteristic (ROC) curves and corresponding areas under curve (AUCs) were calculated with R package *pROC* [26] and 95% confidence intervals for odds ratios were constructed with the *oddsratio.wald* function from the *epitools* package [27].

2.4. Identification of Most Relevant PD-PRS SNPs

We evaluated which SNPs of the PD-PRS were most relevant for distinguishing cases from controls by determining their influence upon the AUC. This was done in three steps.

1. The PD-PRS was repeatedly calculated, excluding one SNP each time, and determining the AUC of the PD-PRS without the SNP. These AUCs will be referred to as 'AUC-SNP' values.
2. SNPs were sequentially removed from the PD-PRS based upon the steepest decline of the AUC of the remaining SNPs, until the 95% confidence interval of the residual AUC included 0.5. This set of removed SNPs will be referred to as 'most relevant SNPs'.
3. The results from step 1 and step 2 were combined in a single plot, relating the AUC-SNP values of SNPs (y axis) to their AUC-SNP-based rank (x axis) and color-coding the set of most relevant SNPs from step 2 together with the set of 47 genome-wide significant SNPs identified by Nalls et al. [2] and included in our PD-PRS.

R package *biomaRt* and the hsapiens_gene_ensembl data set from Ensembl were used to identify genes that included at least one of the most relevant SNPs [28–30]. Coding and functional information on individual SNPs were obtained from dbSNP [31].

2.5. Prognostic Value of PD-PRS

The *coords* function from R package *pROC* [26] was used to derive appropriate PD-PRS thresholds from ROC curves, and to determine the corresponding values of sensitivity and specificity. Thresholds were calculated by maximizing a weighted Youden-Index:

$$\max(\text{costs} \cdot \text{sensitivity} + \text{specificity})$$

where 'costs' was defined as the relative severity of a false negative compared to a false positive result (i.e., classification or prediction as PD). Costs were varied from 1 to 5 in steps of 0.0001.

For fixed specificity and sensitivity, the positive and negative predictive values (ppv, npv) were computed with Bayes formula as

$$\text{ppv} = \frac{\text{sensitivity} \cdot \text{prevalence}}{\text{sensitivity} \cdot \text{prevalence} + (1 - \text{specificity}) \cdot (1 - \text{prevalence})}$$

$$\text{npv} = \frac{\text{specificity} \cdot (1 - \text{prevalence})}{\text{specificity} \cdot (1 - \text{prevalence}) + (1 - \text{sensitivity}) \cdot \text{prevalence}}$$

To evaluate the prognostic value of the PD-PRS, we had to include the residual lifetime incidence in the above formulae instead of the disease prevalence. To this end, we adopted the age-specific incidence and death rates $I_{[\text{interval}]}$ and $D_{[\text{interval}]}$ from the SIa strategy in [32]. The SIa strategy used only cases with at least two diagnoses of PD to avoid false positive diagnoses. $I_{[\text{interval}]}$ and $D_{[\text{interval}]}$ were given for 5-year age intervals, starting from [50–54] and ending with [95+]. Since the death rates were given as annual probabilities to die within a given interval, the probability to survive that interval can be approximated by $S_{[\text{interval}]} = (1 - D_{[\text{interval}]})^5$. For individuals from a given age interval [d,d+5], the residual lifetime incidence can then be computed as

$$I_{[d,\,95+]} = I_{[d,\,d+5]} + (I_{[d+6,\,d+11]} \cdot S_{[d,\,d+5]} \cdot (1 - I_{[d,\,d+5]})) + \ldots + (I_{[95+]} \cdot S_{[d,\,d+5]} \cdot \ldots \cdot S_{[90,\,94]} \cdot (1 - I_{[d,\,d+5]}) \cdot \ldots \cdot (1 - I_{[90,\,94]})).$$

The resulting residual lifetime incidence values are listed in Table A3.

3. Results

3.1. Validation of Published Parkinson's Disease Polygenic Risk Score (PD-PRS)

To independently validate the (standardized) PD-PRS proposed by Nalls et al. [2], we investigated the performance of this PRS in a separate data set comprising 1914 PD cases and 4464 controls (Table A1). The distribution of the PD-PRS clearly differed between the two groups (Figure 1A; Wald test $p < 10^{-5}$, Table 1). Nagelkerke's pseudo-R^2 from the logistic regression analysis equaled 0.35 when including PD-PRS, sex, age and the first three principal components (PCs), and 0.30 when the PD-PRS was not included (Table 1). The area under curve (AUC) for the receiver operating characteristic (ROC) curve (Figure 1B) was 0.65, which was comparable to the AUC obtained in the original study [2]. The disease odds ratios (ORs) for the 2nd to 10th deciles of the PRS distribution among controls ranged from 1.26 (2nd decile) to 6.10 (10th decile; 1st decile used as reference; Figure 2).

The PD-PRS was also able to distinguish well between cases from the 1st and 4th age-at-onset (AAO) quartile ($\leq$54 years vs. >70 years, Figure 3A, $p = 1.61 \times 10^{-5}$, Table 1). Nagelkerke's pseudo-R^2 from the logistic regression was 0.039 including PD-PRS, sex and the first three PCs, and 0.009 when the PD-PRS was not included. The AUC of the ROC equaled 0.59 (Figure 3B, Table 1) and was hence considerably smaller than the AUC obtained for distinguishing cases from controls.

3.2. Most Relevant SNPs in PD-PRS

We identified 422 SNPs as being the most relevant for distinguishing cases from controls, judged by their influence upon the AUC in a backward-selection process (see Methods). Of these SNPs, 287 are located within a gene. Table 2 lists the top 20 most relevant SNPs inside genes (for a complete list, see Table A4). Of all 1743 SNPs analyzed, some 47 had been genome-wide significant in the meta-GWAS by Nalls et al. [2]. Thirty-two of these (68%) were among the 422 most relevant SNPs identified here, and 25 of them (78%) were intra-genic. When all 1743 SNPs were ranked according to the AUC obtained when a given SNP was removed (Figure 4), the 422 most relevant SNPs occurred mostly on the left side of the graph meaning that the AUC is strongly reduced upon the removal of the SNP. The 32 most relevant and genome-wide significant SNPs, in particular, were found to cluster at the far left of the graph.

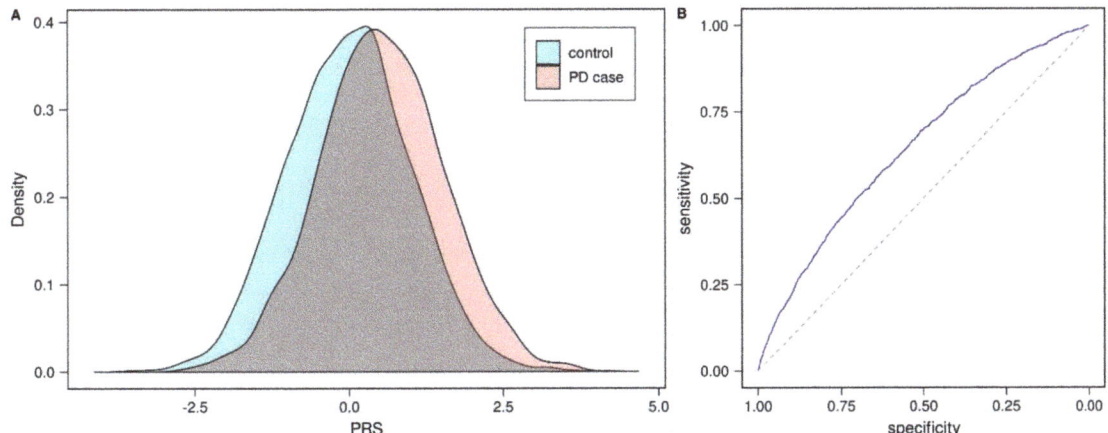

Figure 1. PD-PRS in PD cases and controls. (**A**) Density of PD-PRS in cases and controls. (**B**) ROC curve for PD-PRS as a predictor of case-control status. PRS: polygenic risk score, PD: Parkinson's disease, ROC: receiver operating characteristic.

Table 1. Comparative validation of PD-PRS.

Data Set	Samples (N)	SNPs (N)	AUC [95% CI]	Nagelkerke's Pseudo-R^2 [a]	p Value [b]	Nagelkerke's Pseudo-R^2 [c]
This study (case/control)	6378	1743	0.645 [0.630, 0.660]	0.348	$<10^{-5}$	0.298
Nalls training [d] (case/control)	11,243	1809	0.640 [0.630, 0.650]	n.a.	$<10^{-5}$	n.a.
Nalls validation [e] (case/control)	999	1805	0.692 [0.660, 0.725]	n.a.	$<10^{-5}$	n.a.
This study (AAO) [f]	836	1743	0.590 [0.551, 0.629]	0.039	1.6×10^{-5}	0.009

[a] From logistic regression analysis of PD case-control status (first line) and AAO 1st vs 4th quartile (fourth line), each time including PD-PRS, sex, age (only for the analysis of case-control status) and the first three PCs as independent variables. Nalls et al. [2] used a different approach to evaluate logistic regression models, hence a comparison of pseudo-R^2 is not meaningful. [b] p value for PD-PRS as an independent variable in the logistic regression analysis (Wald test). [c] Same logistic regression model as before, but without PD-PRS as an independent variable. [d] NeuroX-dbGaP data set (5851 cases, 5866 controls). [e] Harvard Biomarker Study (527 cases, 472 controls). [f] Samples belonging to the 1st and 4th AAO quartile among cases analyzed in this study. PD: Parkinson's disease, PRS: polygenic risk score, SNP: single nucleotide polymorphism, AUC: area under ROC curve, CI: confidence interval, AAO: age-at-onset, ROC: receiver operating characteristic, n.a.: not available.

3.3. Prognostic Value of PD-PRS

To investigate the prognostic value of the PD-PRS, an individual was defined as 'test-positive' if their PRS exceeded a given threshold of the PRS and 'test-negative' if not. Thus, sensitivity in this context means the probability that a person who develops PD in later life has a PRS above the threshold while specificity is the probability that a person who will not develop PD during their lifetime is test-negative. Since sensitivity is generally more important than specificity for screening tests, we considered different relative costs of false negative vs false positive test results when maximizing a weighted Youden index to determine the optimal PD-PRS threshold (Table 3). For costs of 1, i.e., when false positives and false negatives are deemed equally serious, the optimal PD-PRS threshold equaled 0.33, yielding a sensitivity of 0.58 and a specificity of 0.63. For costs of 5, the sensitivity equaled 1 and the specificity equaled 0.003 at an optimal PD-PRS threshold of −2.667 (Table 3, Figure 5A).

For fixed costs, the age-specific predictive values of the PD-PRS differed only little up to age interval [70–74], after which the positive predictive value (ppv) declined and the negative predictive value (npv) increased (Table 4, Figure 5B). Across all age groups and costs levels, the ppv was very low with a maximum of 0.027 up to 74 years at costs of 1. The minimum ppv was 0.005 for the highest age group (90+) at costs of 5. The npv varied between 0.988 (≤74 years, costs 1) and 1 (all age groups, costs 5).

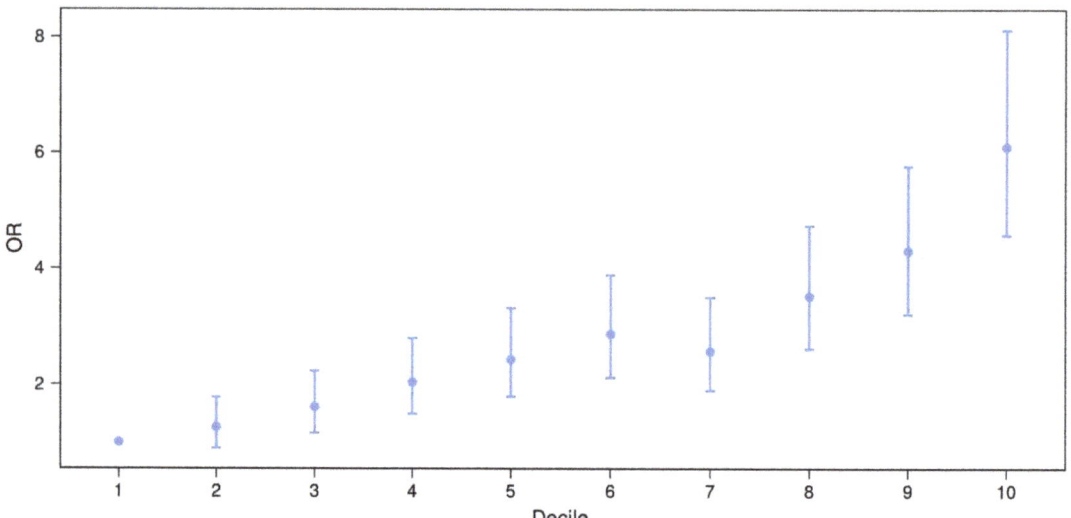

Figure 2. Disease OR for the 2nd to 10th deciles of the PD-PRS distribution among controls. (1st decile used as reference). Vertical bars demarcate 95% confidence intervals. OR: odds ratio, PD: Parkinson's disease, PRS: polygenic risk score.

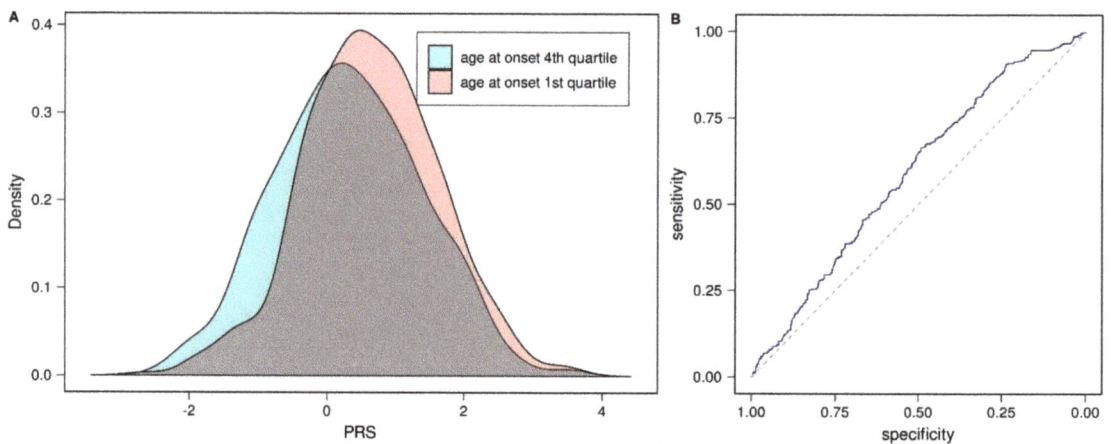

Figure 3. PD-PRS in early and late onset cases. (**A**) Density of PD-PRS in the 1st and 4th AAO quartile of cases. (**B**) ROC curve for PD-PRS as a predictor of 1st vs 4th AAO quartile. AAO: age-at-onset, PRS: polygenic risk score, PD: Parkinson's disease, ROC: receiver operating characteristic.

Table 2. Top 20 most relevant SNPs located within genes.

HGNC Symbol [1]	Chr	AUC	Start [2]	End [3]	SNP Position [4]	A1 [5]	A2 [6]	GS [7]	SNP Type
ENSG00000251095	4	0.643	90,472,507	90,647,654	90,626,111	G	A	yes	intron
SNCA	4	0.641	90,645,250	90,759,466	90,684,278	A	G	no	intron
HIP1R	12	0.640	123,319,000	123,347,507	123,326,598	G	T	yes	intron
TMEM175	4	0.639	926,175	952,444	951,947	T	C	yes	missense
SNCA	4	0.638	90,645,250	90,759,466	90,757,294	A	C	no	intron
ASH1L	1	0.637	155,305,059	155,532,598	155,437,711	G	A	no	intron
UBQLN4	1	0.634	156,005,092	156,023,585	156,007,988	G	A	no	intron
ENSG00000225342	12	0.633	40,579,811	40,617,605	40,614,434	C	T	yes	n.a.
LRRK2	12	0.633	40,590,546	40,763,087	40,614,434	C	T	yes	n.a.
STX1B	16	0.632	31,000,577	31,021,949	31,004,169	T	C	no	synonymous
INPP5F	10	0.631	121,485,609	121,588,652	121,536,327	G	A	yes	intron
CCSER1	4	0.631	91,048,686	92,523,064	91,164,040	C	T	no	intron
SLC2A13	12	0.630	40,148,823	40,499,891	40,388,109	C	T	no	intron
FBXL19	16	0.630	30,934,376	30,960,104	30,943,096	A	G	no	intron
ENSG00000251095	4	0.629	90,472,507	90,647,654	90,619,032	C	T	no	intron
CAB39L	13	0.629	49,882,786	50,018,262	49,927,732	T	C	yes	intron
STK39	2	0.628	168,810,530	169,104,651	168,979,290	C	T	no	intron
CCT3	1	0.628	156,278,759	156,337,664	156,300,731	T	C	no	intron
ENSG00000225342	12	0.627	40,579,811	40,617,605	40,614,656	A	G	no	n.a.
LRRK2	12	0.627	40,590,546	40,763,087	40,614,656	A	G	no	n.a.

[1] HGNC symbol or Ensemble gene ID if there is no HGNC symbol available. [2] Base pair position of start of gene. [3] Base pair position of end of gene. [4] Genomic position of SNP. [5] Major SNP allele. [6] Minor SNP allele. [7] Genome-wide significant (GS) in the meta-GWAS by Nalls et al. [2]. HGNC: HUGO Gene Nomenclature Committee, Chr: Chromosome, AUC: area under ROC curve, ROC: receiver operating characteristic, PRS: polygenic risk score, PD: Parkinson's disease, n.a.: not available.

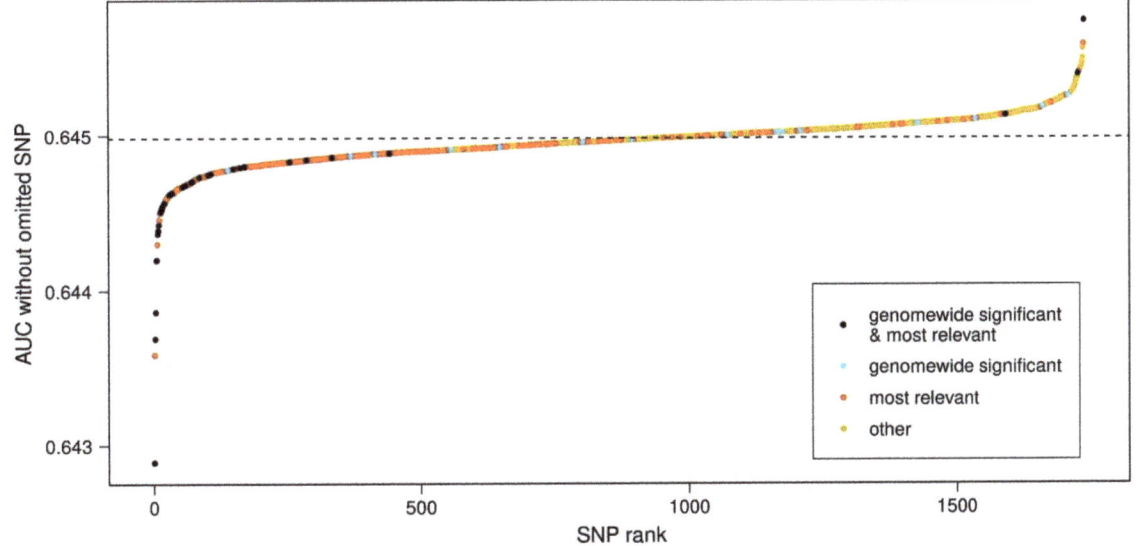

Figure 4. Influence of individual SNPs upon PD-PRS performance. For each of the 1743 PD-PRS SNPs, the AUC was calculated after removing the SNP from the PRS. SNPs were color-coded as either genome-wide significant in a meta-GWAS [2] (blue), as 'most relevant' in the present study (red), both of the former (black) or none of the former (yellow). SNP: single nucleotide polymorphism, PD: Parkinson's disease, PRS: polygenic risk score, AUC: area under ROC curve, ROC: receiver operating characteristic, GWAS: genome-wide association study.

Table 3. Prognostic value of PD-PRS.

	Costs				
	1	2	3	4	5
Sensitivity	0.581	0.921	0.981	0.999	1
[95% CI]	[0.479, 0.733]	[0.880, 0.981]	[0.973, 1]	[0.983, 1]	[0.996, 1]
Specificity	0.625	0.198	0.067	0.006	0.003
[95% CI]	[0.472, 0.725]	[0.075, 0.289]	[0.004, 0.096]	[0.002, 0.082]	[0.002, 0.034]
Threshold [1]	0.330	−0.868	−1.507	−2.533	−2.667

[1] Optimal threshold for PD-PRS as determined by maximizing a weighed Youden index. PD: Parkinson's disease, PRS: polygenic risk score, CI: confidence interval.

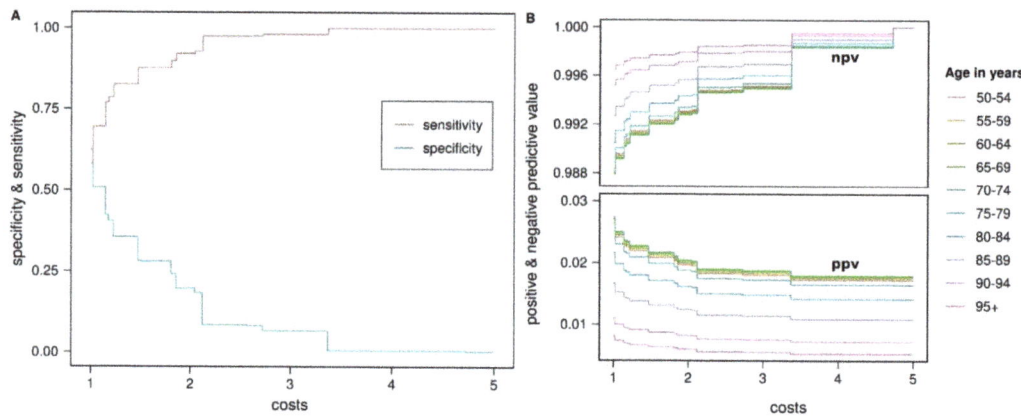

Figure 5. Prognostic value of PD-PRS. (**A**) Sensitivity and specificity of PD-PRS for the optimal threshold were determined by maximizing a weighted Youden index. The relative costs of false negative vs false positive results varied from 1 to 5. (**B**) ppv and npv were calculated from the costs-based sensitivity and specificity and the residual lifetime incidence (see Methods and Table A3) in 10 age groups. PRS: polygenic risk score, PD: Parkinson's disease, ppv: positive predictive value, npv: negative predictive value.

Table 4. Costs- and age-dependent PD-PRS predictive values.

		Costs									
		1		2		3		4		5	
		ppv	npv	ppv	npv	ppv	npv	ppv	npv	ppv	npv
Age group (Years)	50–54	0.026	0.988	0.020	0.993	0.018	0.995	0.017	0.998	0.017	1
	55–59	0.027	0.988	0.020	0.993	0.018	0.995	0.018	0.998	0.018	1
	60–64	0.027	0.988	0.020	0.993	0.019	0.995	0.018	0.998	0.018	1
	65–69	0.027	0.988	0.021	0.993	0.019	0.995	0.018	0.998	0.018	1
	70–74	0.027	0.988	0.020	0.993	0.019	0.995	0.018	0.998	0.018	1
	75–79	0.025	0.989	0.019	0.993	0.017	0.995	0.017	0.999	0.016	1
	80–84	0.022	0.990	0.016	0.994	0.015	0.996	0.014	0.999	0.014	1
	85–89	0.017	0.993	0.012	0.996	0.011	0.997	0.011	0.999	0.011	1
	90–94	0.011	0.995	0.008	0.997	0.008	0.998	0.007	0.999	0.007	1
	95+	0.008	0.996	0.006	0.998	0.005	0.999	0.005	1.000	0.005	1

PRS: polygenic risk score, PD: Parkinson's disease, ppv: positive predictive value, npv: negative predictive value.

4. Discussion

In the present study, we replicated the performance of the PD-PRS developed by Nalls et al. [2] in an independent data set. It turned out that the PD-PRS was clearly able to distinguish between cases and controls and that it was increased in cases of early age-at-onset. Individuals in the 10th PRS decile had an OR of around 6 of having PD as compared to individuals in the lowest decile. This is in line with the results by Nalls et al. [2] who reported ORs of 3.74 and 6.25 for the highest quartiles in their two data sets. The most relevant PRS SNPs identified in our study included many genome-wide significant SNPs from the Nalls et al. study [2], as was to be expected. In fact, of the 47 genome-wide significant SNPs, some 32 (68%) were found to be most relevant in the sense of our study. However, this is still only a small fraction (7.5%) of the total number of 422 most relevant SNPs, which highlights the polygenic background of PD with several low-effect variants and justifies the fact that not only genome-wide significant SNPs were originally included in the PRS.

In the recent past, the research community has become increasingly aware of the problem of non-replicability of research findings in independent data sets or with different methods [33]. This has been termed the "replication crisis" or "reproducibility crisis" [34,35]. Studies aiming at validating existing PRSs are still rare and, usually, new data set-specific PRSs are developed instead because this is easy with existing software. Nevertheless, PRS replication should be mandatory [36] and our replication of the results reported by Nalls et al. [2], in an independent data set, is reassuring. It supports the idea that this PD-PRS can be used to capture the contribution of the genetic background of an individual to their PD risk. The PD-PRS could hence be a valid instrument to adjust for the genetic background component in statistical models for PD. Moreover, it may also facilitate studies of the genetic overlap between different diseases or disease subtypes and of the interaction between genetic and environmental factors.

It has to be kept in mind, however, that PRSs only capture the effect of common genetic variants. Highly-penetrant rare or private variants as well as other types of variations such as copy number variants or indels are not represented [37]. Another drawback of PRSs is their dependency on the ancestry of populations [38]. The PD-PRS analyzed in the present study was both constructed and validated in populations of European ancestry, and transferability of the results to other ancestries cannot be taken for granted but has to be investigated in future studies. On a related note, it must be kept in mind that all PD-PRS SNPs considered in our study were imputed. This does not seem to have impaired our replication of the results of Nalls et al. [2], probably due to our stringent quality control. For populations, where a good imputation reference is lacking, consistent PRS performance may not be taken for granted.

Quality control in our study led to the exclusion of 62 of the original 1805 PD-PRS SNPs. The omitted SNPs showed on average a larger effect size in the original meta-analysis than the SNPs included in our PRS (Table A2). The former were excluded mostly (79%) because of very low MAF and the rest because the info score was below 0.70. Despite the higher effect sizes, it is therefore not clear if the additional usage of the 62 SNPs would enhance the performance of the PD-PRS because of low MAF and perhaps difficult imputation. The loss of variants from the score due to difficulties in imputation is a good argument for the adoption of the development of standardized PRSs based on reference variants which are available in common genotyping arrays. This would reduce the imputation problem.

Whereas PRSs deserve a role in etiological research and statistical modelling of diseases, their prognostic value is dubious [11,12,36]. PRSs are developed to differentiate between cases and controls. Although the level of differentiation achieved is reasonable at a group level, the obtained AUCs are usually insufficient for individual diagnostic or prognostic testing, where an AUC > 0.90 is required [11]. In this study, we evaluated the prognostic value of a specific PD-PRS and calculated its sensitivity and specificity as well as its predictive values for various assumptions about the relative importance of mis-prognoses. Our results were in accordance with the generally held view that a prognostic application of PRSs alone is not meaningful. The negative predictive values were high which means that people with a low PRS can be reasonably sure not to develop PD, at least not of the type considered in this study. However, the positive predictive values were only of the order of a few percent which means that the probability of a person with a high PRS developing the disease is quite low. Here, the comparison to a hypothetical test which gives everybody a negative test result is helpful: Assuming a lifetime incidence of 5% [39], the negative predictive value of this (nonsense) test would be 95%, i.e., quite similar to a test based solely on the PD-PRS.

There are three ways in which a prognostic test for PD, or any other disease, could potentially help to reduce incidence or severity: change of lifestyle factors, enhanced surveillance or preventive treatment. Of these, a change towards a healthier lifestyle is always meaningful, both from an individual and a population health perspective, and only a test with a positive predictive value much higher, for example, than that of the PD-PRS would mean an additional individual incentive for change. Moreover, with a low incidence and positive predictive value, frequent medical screening of individuals with a high PRS would mean spending valuable resources for individuals who have only a probability of a few percent to actually develop the disease in question. The same holds true for possible preventive treatment if such treatment were available in the first place. Apart from economic constraints, side-effects might result in a negative benefit-risk balance when the incidence of the disease in question is as low as for PD.

A limitation of our study has been that the predictive values were only calculated from theoretical models and were not based directly upon empirical observations. This is a general drawback when evaluating the prognostic value of PRSs because adequate long-term studies would be time-consuming, require large sample sizes and would hence be rather expensive. This notwithstanding, PRSs have to be externally validated and compared to other (clinical) risk models in a clinically meaningful prospective set-up [12,36] because this is a *conditio sine qua non* for the applicability in practice of any prognostic marker. Only a few studies have taken first steps in this direction [40–42], and most have found none or only little additional prognostic value of PRSs over and above clinical and demographic predictors. To our knowledge, no such study has been performed yet for PD, where the combination of a PRS with established prodromal markers [43] might be specifically worth investigating in future prospective studies.

5. Conclusions

The PD-PRS proposed by Nalls et al. [2] could be validated independently in German patients and controls, suggesting that the PRS may be a meaningful research tool to investigate and adjust for the polygenic component of PD. Individual risk prediction using the PD-PRS alone is, however, not meaningful.

Author Contributions: Conceptualization, A.C., C.K., I.R.K., S.K., M.K. (Michael Krawczak) and K.L.; methodology, A.C., I.R.K., S.K. and M.K. (Michael Krawczak); formal analysis, S.K.; investigation, A.C.; resources, J.B., N.B., A.F., G.K., U.M.K., B.-H.L., K.L., W.L., B.M., M.N., E.S., C.T., T.U. and M.W.; data curation, M.K. (Meike Kasten) and E.-J.V.; writing—original draft preparation, A.C. and S.K.; writing—review and editing, A.C., C.K., I.R.K., S.K., M.K. (Michael Krawczak) and K.L.; visualization, S.K.; supervision, A.C. and M.K. (Michael Krawczak); project administration, C.K.; funding acquisition, A.C. and C.K. All authors have read and agreed to the published version of the manuscript.

Funding: This research was funded by the German Research Foundation (FOR2488 to N.B., A.C., M.K. (Meike Kasten), M.K. (Michael Krawczak), C.K., I.R.K., K.L. and TR-CRC134 to U.M.K., M.K. (Meike Kasten), C.K.).

Institutional Review Board Statement: The study was conducted according to the guidelines of the Declaration of Helsinki, and approved by the Ethics Committees of the University of Lübeck, Germany (protocol code 16-039, date of approval 27 September 2019) and the P2N supervisory board, Kiel University, Germany (protocol code 2021-037, date of approval 16 September 2021).

Informed Consent Statement: Informed consent was obtained from all subjects involved in the study.

Data Availability Statement: The data that support the results of this study are available upon reasonable request from the corresponding author.

Acknowledgments: We thank Mike A. Nalls for providing us with the list of the 1805 SNPs included in their published PRS (together with reference alleles and effect sizes β).

Conflicts of Interest: C.K. serves as a medical advisor for genetic testing reports in the field of movement disorders and dementia, but excluding Parkinson's disease, to Centogene and as a member of the Scientific Advisory Board of Retromer Therapeutics. N.B. has previously served as a consultant for Centogene GmbH. The other authors declare no conflict of interest. The funders had no role in the design of the study; in the collection, analyses, or interpretation of data; in the writing of the manuscript, or in the decision to publish the results.

Appendix A

Appendix A.1. Removal of Related Individuals

Clusters of related individuals were generated such that each individual in a cluster had an IBD value ≥ 0.1 with at least one other individual in the cluster. Typical clusters were siblings or parent-child clusters but also larger clusters of extended families were found. A total of 238 disjunct clusters comprising 503 individuals were detected in our data set. For each cluster, the largest subset of unrelated individuals (all pairwise IBD values < 0.1) was next selected, and since cases were more valuable for our analysis than controls, the former were given double weight in the selection process. If two equally large subsets remained, the subset with the highest AAO for a case was selected because idiopathic PD typically has high AAO. If this was not possible, selection was in favor of the subset with the oldest control. Of the 503 individuals in clusters, 243 were kept for further analysis.

Appendix A.2. Removal of Population Outliers

Population outliers were removed in our study by two different approaches. In the first approach, our data set was merged with 2504 individuals from the 1000Genomes project (1000 Genomes Phase III, imputed). A PCA was then done with *PLINK 1.9* [21] at the default setting of 20 PCs. Next, a polygon was constructed around the European populations of the 1000Genomes data (CEU, FIN, GBR, IBS and TSI) to identify population outliers in our own data by considering PC1 and PC2. In more detail, the polygon was generated by first transforming the PC1:PC2-coordinates of the European individuals from 1000Genomes and of our samples into spatial data, using R package *sp* [44,45]. Ideally, a circle around each European 1000Genomes data point (sample) would represent the genetic neighborhood of the respective individual, and the union of these circles would be the region of probable European ancestry. However, that is technically difficult and therefore R package *rgeos* was used to calculate 20-polygonal approximations of circles with a width of 0.0005 around each data point [46] (Figure A1). The width of these circle-polygons was chosen such that the union of all circle-polygons was connected. The width roughly equaled 1/8 of the mean of the first PC and 1/4 of the mean of the second PC of the 1000Genomes European data. As a boundary of the union of the circle-polygons, a polygon was then computed with an additional distance of 0.0005 to the circle-polygons to smooth indentations. Finally, we gauged the samples from our data set against this boundary and every sample outside the boundary was removed.

As a second approach to remove population outliers, we applied the K nearest neighbor (KNN) method suggested in [47] using R packages *bigsnpr* and *bigparallelr* [48,49]. Utilizing a scree plot, three PCs were considered important and a threshold of 0.15 was used for the KNN statistics.

Table A1. Cohorts used in this study.

Cohort	N	N Cases	N Controls	N Female Cases	N Female Controls	Age-at-Sampling Cases [1]	Age-at-Sampling Controls [1]	Age-at-Onset Cases [1]
Kiel PD	184	184	0	59 (32%)	0	68 [61–76]	-	58 [48–68]
Luebeck PD	928	395	533	139 (35%)	323 (61%)	68 [57–75]	44 [35–48]	60 [51–68]
EPIPARK [13]	1271	525	746	205 (39%)	353 (47%)	69 [60–76]	67 [61–71]	60 [52–70]
DeNoPa [14]	241	149	92	52 (35%)	32 (35%)	67 [59–73]	67 [62–70]	67 [59–73]
Popgen [15,16]	3754	661	3093	262 (40%)	1527 (49%)	71 [66–77]	54 [41–65]	64 [56–71]

[1] Median and interquartile-range. PD: Parkinson's disease.

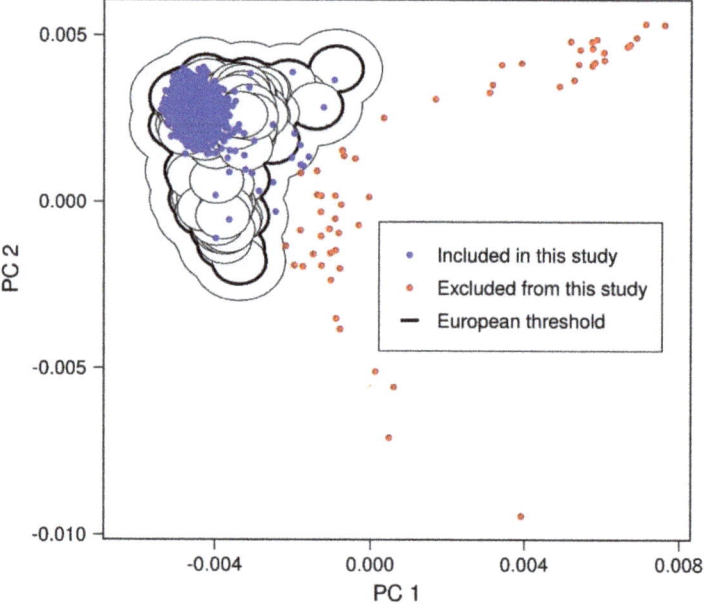

Figure A1. Identification of population outliers by PCA drawing upon 1000Genomes data. White circles represent polygonal circle approximations around European samples of the 1000Genomes project. The thick black line marks the union set, the thinner line marks the final boundary. Dots representing our samples are colored according to their inclusion in or exclusion from the study. Samples were excluded if they were outside the boundary. PC: principal component, PCA: principal component analysis.

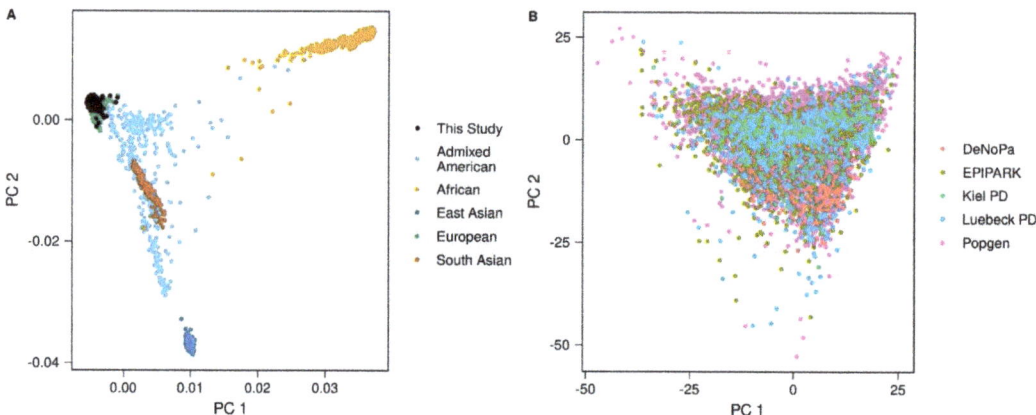

Figure A2. PCA plots after quality control. (**A**) Plot of the first two PCs from the 1000Genomes supra populations and the samples of this study. Our study samples were plotted on top, therefore obscuring part of the European samples from the 1000Genomes project. (**B**) Plot of the first two PCs from the cohorts included in our study (Table A1). PC: principal component, PCA: principal component analysis.

Table A2. SNPs omitted from PD-PRS.

SNP Location [1]	Beta [2]	GS [3]	MAF [4]
1:1,186,833	−0.4394	no	0.0178
1:145,716,763	0.0448	no	not imputed
1:154,837,939	0.2467	no	0.0052
1:155,205,634	0.7662	yes	0.0022
1:232,161,497	−0.2638	no	0.0087
1:62,675,673	0.317	no	0.0134
2:100,906,427	0.1534	no	0.0098
2:102,368,870	0.2332	no	0.0048
2:102,655,773	0.2056	no	0.0046
2:136,388,639	−0.0656	no	0.0513
2:191,364,828	0.2497	no	0.0079
2:63,783,507	0.173	no	0.0094
3:112,245,295	−0.1391	no	0.9907
3:48,406,286	0.0789	no	0.0398
3:96,921,359	0.1607	no	0.0069
3:97,799,541	0.1819	no	0.0062
4:133,792,853	0.1797	no	0.0057
4:77,645,873	−0.2104	no	0.0096
4:90,603,678	−0.203	no	0.0087
4:90,673,143	−0.3266	no	0.0032
4:90,810,340	0.3754	no	0.0062
4:90,955,553	0.2561	no	0.0052
4:90,967,340	0.2829	no	0.0081
4:91,033,047	0.3361	no	0.0078
4:91,278,545	0.3511	no	0.0022
5:112,288,617	0.2085	no	0.0076
5:141,311,896	0.1052	no	0.0434
5:177,972,560	0.1641	no	0.0080
5:60,150,889	0.1637	no	0.0069
6:109,972,453	0.1744	no	0.0071
6:27,483,385	0.1698	no	0.0072
6:32,036,055	−0.1716	no	0.0063
6:34,800,390	−0.2314	no	0.0029
6:48,781,938	0.2449	no	0.0087
7:6,070,199	0.1652	no	0.0096
9:116,138,770	0.2529	no	0.0042
9:139,566,889	−0.0812	no	0.1093
10:102,056,734	0.3817	no	0.0019
10:103,373,463	0.1323	no	0.0099
10:103,941,875	0.1667	no	0.0080
10:105,038,008	0.1579	no	0.0076
10:27,198,118	0.2103	no	0.0012
10:48,433,720	0.0481	no	0.1562
11:93,561,149	0.1769	no	0.0041
12:123,341,500	0.2448	no	0.0064
12:123,923,612	0.2771	no	0.0077
12:40,734,202	2.4354	yes	0.0001
12:72,179,446	0.2839	no	0.0156
14:103,351,731	0.1973	no	0.0046
16:429,926	0.2396	no	0.0077
16:71,451,526	0.2423	no	0.0065
17:43,516,175	−0.2917	no	0.0130
17:43,559,955	−0.2548	no	0.0098
17:43,857,449	−0.3906	no	0.0162
17:44,687,696	−0.5875	no	0.0172
17:44,914,558	−0.1824	no	0.0095
17:44,916,533	0.2253	no	0.0095
17:8,209,654	−0.1341	no	0.0131
19:11,084,467	0.2043	no	0.0083
19:38,222,914	0.1495	no	0.0085
19:39,756,425	−0.1751	no	0.0092
20:31,687,446	0.2054	no	0.0080
median [IQR] omitted 62 SNPs	0.207 [0.166, 0.262] [5]		0.0080 [0.0062, 0.0098]
median [IQR] 1743 SNPs used in this study	0.056 [0.042, 0.091] [5]		0.1916 [0.0102, 0.4407]

[1] Location of SNPs, given as chromosome:basepair position. [2] β from the meta-GWAS performed by Nalls et al. [2]. [3] Genome-wide significant (GS) in the meta-GWAS performed by Nalls et al. [2]. [4] MAF in our data set. [5] median and IQR of the absolute values of β. SNP: single nucleotide polymorphism, MAF: minor allele frequency, IQR: inter-quartile range, PRS: polygenic risk score, PD: Parkinson's disease.

Table A3. Incidence of PD in different age groups.

Age Interval in Years	Incidence [1]	Survival [2]	Residual Lifetime Incidence [3]
50–54	0.0002	0.994	0.017
55–59	0.0005	0.992	0.017
60–64	0.0009	0.987	0.018
65–69	0.0016	0.983	0.018
70–74	0.0034	0.974	0.018
75–79	0.0051	0.958	0.016
80–84	0.0067	0.929	0.014
85–89	0.0072	0.874	0.011
90–94	0.0056	0.782	0.007
95+	0.0052	0.654	0.005

[1] Probability to develop PD during age interval (from [32]). [2] Probability to survive a year from the respective age interval (from [32]). [3] Probability to develop PD in later life (see Methods section). PD: Parkinson's disease.

Table A4. Most relevant SNPs located within genes.

HGNC Symbol [1]	Chr	AUC	Start [2]	End [3]	SNP Position [4]	A1 [5]	A2 [6]	GS [7]
ENSG00000251095	4	0.643	90,472,507	90,647,654	90,626,111	G	A	yes
SNCA	4	0.641	9,0645,250	90,759,466	90,684,278	A	G	no
HIP1R	12	0.640	123,319,000	123,347,507	123,326,598	G	T	yes
TMEM175	4	0.639	926,175	952,444	951,947	T	C	yes
SNCA	4	0.638	90,645,250	90,759,466	90,757,294	A	C	no
ASH1L	1	0.637	155,305,059	155,532,598	155,437,711	G	A	no
UBQLN4	1	0.634	156,005,092	156,023,585	156,007,988	G	A	no
ENSG00000225342	12	0.633	40,579,811	40,617,605	40,614,434	C	T	yes
LRRK2	12	0.633	40,590,546	40,763,087	40,614,434	C	T	yes
STX1B	16	0.632	31,000,577	31,021,949	31,004,169	T	C	no
INPP5F	10	0.631	121,485,609	121,588,652	121,536,327	G	A	yes
CCSER1	4	0.631	91,048,686	92,523,064	91,164,040	C	T	no
SLC2A13	12	0.630	40,148,823	40,499,891	40,388,109	C	T	no
FBXL19	16	0.630	30,934,376	30,960,104	30,943,096	A	G	no
ENSG00000251095	4	0.629	90,472,507	90,647,654	90,619,032	C	T	no
CAB39L	13	0.629	49,882,786	50,018,262	49,927,732	T	C	yes
STK39	2	0.628	168,810,530	169,104,651	168,979,290	C	T	no
CCT3	1	0.628	156,278,759	156,337,664	156,300,731	T	C	no
ENSG00000225342	12	0.627	40,579,811	40,617,605	40,614,656	A	G	no
LRRK2	12	0.627	40,590,546	40,763,087	40,614,656	A	G	no
SH3GL2	9	0.627	17,579,080	17,797,127	17,726,888	C	T	no
LRRK2	12	0.626	40,590,546	40,763,087	40,713,899	T	C	no
ENSG00000251095	4	0.625	90,472,507	90,647,654	90,573,396	G	A	no
ASXL3	18	0.625	31,158,579	31,331,156	31,304,318	G	T	yes
SH3GL2	9	0.624	17,579,080	17,797,127	17,579,690	T	G	yes
ENSG00000259675	15	0.623	61,931,548	62,007,370	61,997,385	T	C	yes
RGS10	10	0.623	121,259,340	121,302,220	121,260,786	A	G	no
CASC16	16	0.622	52,586,002	52,686,017	52,636,242	C	A	yes
EPRS	1	0.621	220,141,943	220,220,000	220,163,026	C	A	no
BRIP1	17	0.621	59,758,627	59,940,882	59,918,091	A	G	no
PCGF3	4	0.620	699,537	764,428	758,444	C	T	no
ENSG00000249592	4	0.620	756,175	775,637	758,444	C	T	no
ENSG00000233799	4	0.620	758,275	758,862	758,444	C	T	no
NDUFAF2	5	0.620	60,240,956	60,448,853	60,297,500	A	G	no
DLG2	11	0.619	83,166,055	85,338,966	83,488,901	C	T	no
SEC16A	9	0.618	139,334,549	139,372,141	139,336,813	T	G	no
FCGR2A	1	0.617	161,475,220	161,493,803	161,478,859	T	C	no
SPTSSB	3	0.617	161,062,580	161,090,668	161,077,630	A	G	yes
DSCAM	21	0.616	41,382,926	42,219,065	41,452,034	C	T	no
GAK	4	0.616	843,064	926,161	893,712	C	T	no
CTSB	8	0.615	11,700,033	11,726,957	11,707,174	A	G	no

Table A4. Cont.

HGNC Symbol [1]	Chr	AUC	Start [2]	End [3]	SNP Position [4]	A1 [5]	A2 [6]	GS [7]
ASH1L	1	0.615	155,305,059	155,532,598	155,347,819	A	C	no
DCST1	1	0.614	155,006,300	155,023,406	155,014,968	T	G	no
LRSAM1	9	0.614	130,213,765	130,265,780	130,261,113	G	A	no
UBAP2	9	0.614	33,921,691	34,048,947	34,046,391	C	T	yes
GCH1	14	0.613	55,308,726	55,369,570	55,348,869	C	T	yes
PCGF2	17	0.613	36,890,150	36,906,070	36,896,751	G	A	no
SETD5	3	0.612	9,439,299	9,520,924	9,504,099	G	A	no
LRRK2	12	0.611	40,590,546	40,763,087	40,753,796	T	C	no
PRSS3	9	0.611	33,750,515	33,799,230	33,778,399	G	A	no
KANSL1	17	0.611	44,107,282	44,302,733	44,189,067	A	G	no
ENSG00000214871	7	0.610	23,210,760	23,234,503	23,232,659	T	C	no
NUPL2	7	0.610	23,221,446	23,240,630	23,232,659	T	C	no
SEC23IP	10	0.610	121,652,223	121,702,014	121,667,020	T	C	no
ENSG00000251095	4	0.610	90,472,507	90,647,654	90,538,467	A	G	no
SLC38A1	12	0.609	46,576,846	46,663,800	46,623,807	G	A	no
MED12L	3	0.609	150,803,484	151,154,860	151,112,968	C	A	no
NOD2	16	0.608	50,727,514	50,766,988	50,736,656	A	G	yes
UBTF	17	0.608	42,282,401	42,298,994	42,294,462	A	G	no
BTN2A2	6	0.608	26,383,324	26,395,102	26,389,926	C	T	no
PGS1	17	0.607	76,374,721	76,421,195	76,377,458	A	G	no
MRVI1	11	0.607	10,594,638	10,715,535	10,660,840	G	T	no
TMEM163	2	0.607	135,213,330	135,476,570	135,443,940	A	G	no
ENSG00000264031	17	0.606	27,887,565	28,034,108	27,897,585	T	C	no
TP53I13	17	0.606	27,893,070	27,900,175	27,897,585	T	C	no
ZNF165	6	0.606	28,048,753	28,057,341	28,054,198	A	G	no
PCGF3	4	0.606	699,537	764,428	733,630	G	A	no
PITPNM2	12	0.605	123,468,027	123,634,562	123,585,705	C	T	no
PCGF3	4	0.605	699,537	764,428	734,351	A	G	no
C10orf32-ASMT	10	0.605	104,614,029	104,661,656	104,635,103	G	A	no
AS3MT	10	0.605	104,629,273	104,661,656	104,635,103	G	A	no
ENSG00000232667	7	0.604	79,959,508	80,014,295	79,998,372	T	C	no
RNF141	11	0.604	10,533,225	10,562,777	10,558,777	A	G	yes
STK39	2	0.604	168,810,530	169,104,651	169,023,263	T	C	no
CCSER1	4	0.603	91,048,686	92,523,064	91,057,794	A	G	no
SEZ6L2	16	0.602	29,882,480	29,910,868	29,892,184	G	A	no
VSTM5	11	0.602	93,551,398	93,583,697	93,576,556	T	C	no
SPATA19	11	0.602	133,710,526	133,715,433	133,714,560	A	C	no
ENSG00000251095	4	0.601	90,472,507	90,647,654	90,606,518	T	G	no
H2AFX	11	0.600	118,964,564	118,966,177	118,965,479	G	A	no
MSTO1	1	0.599	155,579,979	155,718,153	155,698,425	C	T	no
MSTO2P	1	0.599	155,581,011	155,720,105	155,698,425	C	T	no
DAP3	1	0.599	155,657,751	155,708,801	155,698,425	C	T	no
GABRB1	4	0.599	46,995,740	47,428,461	47,372,139	A	C	no
TMEM163	2	0.599	135,213,330	135,476,570	135,464,616	A	G	yes
MFSD6	2	0.598	191,273,081	191,373,931	191,300,402	A	G	no
AMPD3	11	0.598	10,329,860	10,529,126	10,525,791	A	C	no
ADD1	4	0.598	2,845,584	2,931,803	2,901,349	A	G	no
NSF	17	0.597	44,668,035	44,834,830	44,808,902	G	A	no
HCAR1	12	0.597	123,104,824	123,215,390	123,124,138	T	C	no
NR1I3	1	0.597	161,199,456	161,208,092	161,205,966	G	T	no
GAK	4	0.596	843,064	926,161	903,249	G	A	no
EIF3K	19	0.595	39,109,735	39,127,595	39,116,961	A	G	no
BPTF	17	0.595	65,821,640	65,980,494	65,885,911	C	T	no
FBRSL1	12	0.595	133,066,137	133,161,774	133,081,895	C	T	no

Table A4. Cont.

HGNC Symbol [1]	Chr	AUC	Start [2]	End [3]	SNP Position [4]	A1 [5]	A2 [6]	GS [7]
ENSG00000260958	16	0.594	34,442,308	34,518,517	34,466,252	T	C	no
RIT2	18	0.594	40,323,192	40,695,657	40,673,380	A	G	yes
C10orf2	10	0.594	102,747,124	102,754,158	102,747,363	G	T	no
MYOC	1	0.593	171,604,557	171,621,823	171,612,267	G	A	no
XPO1	2	0.592	61,704,984	61,765,761	61,763,207	T	C	no
CRHR1	17	0.591	43,699,267	43,913,194	43,744,203	C	T	yes
ENSG00000263715	17	0.591	43,699,274	43,893,909	43,744,203	C	T	yes
PPP6R2	22	0.590	50,781,733	50,883,514	50,794,282	C	A	no
NRG1	8	0.590	31,496,902	32,622,548	31,942,557	G	A	no
NRG1-IT1	8	0.590	31,883,735	31,996,991	31,942,557	G	A	no
LTK	15	0.590	41,795,836	41,806,085	41,798,614	T	C	no
SAA1	11	0.589	18,287,721	18,291,524	18,290,067	G	T	no
KCNIP3	2	0.589	95,963,052	96,051,825	96,025,765	A	G	no
PCGF3	4	0.588	699,537	764,428	749,620	T	G	no
ART3	4	0.588	76,932,337	77,033,955	76,990,450	C	T	no
ARL15	5	0.588	53,179,775	53,606,412	53,537,742	G	A	no
ENSG00000272414	4	0.587	77,135,193	77,204,933	77,198,054	C	T	yes
FAM47E	4	0.587	77,172,874	77,232,282	77,198,054	C	T	yes
FAM47E-STBD1	4	0.587	77,172,886	77,232,752	77,198,054	C	T	yes
SCARB2	4	0.587	77,079,890	77,135,046	77,100,807	T	C	no
WNT3	17	0.587	44,839,872	44,910,520	44,868,187	G	A	no
DSCR9	21	0.586	38,580,804	38,594,037	38,593,620	G	T	no
MYLK3	16	0.586	46,740,891	46,824,319	46,778,070	G	A	no
ENSG00000251095	4	0.586	90,472,507	90,647,654	90,513,701	G	A	no
BST1	4	0.585	15,704,573	15,739,936	15,737,348	G	A	yes
C9orf129	9	0.585	96,080,481	96,108,696	96,087,807	C	T	no
MMRN1	4	0.584	90,800,683	90,875,780	90,804,532	C	T	no
MAPT-AS1	17	0.584	43,921,017	43,972,966	43,935,838	T	C	no
MCCC1	3	0.584	182,733,006	182,833,863	182,760,073	T	G	yes
MUC19	12	0.583	40,787,197	40,964,632	40,829,565	G	A	no
ENSG00000258167	12	0.583	40,789,655	40,837,649	40,829,565	G	A	no
CCNT2-AS1	2	0.583	135,493,034	135,676,280	135,500,179	G	A	no
XKR6	8	0.583	10,753,555	11,058,875	10,999,583	C	T	no
RCAN2	6	0.582	46,188,475	46,459,709	46,229,444	C	T	no
ITGA8	10	0.582	15,555,948	15,762,124	15,563,450	C	T	no
RANBP9	6	0.581	13,621,730	13,711,796	13,657,040	G	A	no
IGF2BP3	7	0.581	23,349,828	23,510,086	23,462,162	C	A	no
FAM47E	4	0.580	77,135,193	77,204,933	77,202,861	A	G	no
ENSG00000272414	4	0.580	77,172,874	77,232,282	77,202,861	A	G	no
FAM47E-STBD1	4	0.580	77,172,886	77,232,752	77,202,861	A	G	no
ENSG00000251095	4	0.579	90,472,507	90,647,654	90,594,987	G	A	no
SCARB2	4	0.578	77,079,890	77,135,046	77,111,032	C	T	no
ARHGAP27	17	0.578	43,471,275	43,511,787	43,472,507	A	G	no
ZYG11B	1	0.578	53,192,126	53,293,014	53,233,374	T	C	no
ENSG00000244128	3	0.577	164,924,748	165,373,211	165,020,212	A	G	no
PER1	17	0.577	8,043,790	8,059,824	8,051,639	A	G	no
KCNS3	2	0.577	18,059,114	18,542,882	18,132,092	C	T	no
HIBCH	2	0.576	191,054,461	191,208,919	191,071,057	G	A	no
RN7SL416P	7	0.576	100,127,987	100,128,282	100,128,114	G	A	no
YLPM1	14	0.575	75,230,069	75,322,244	75,234,329	G	A	no
FGFRL1	4	0.574	1,003,724	1,020,685	1,008,212	C	T	no
CRHR1	17	0.574	43,699,267	43,913,194	43,798,308	G	A	yes
ENSG00000263715	17	0.574	43,699,274	43,893,909	43,798,308	G	A	yes
HIP1R	12	0.574	123,319,000	123,347,507	123,334,442	C	T	no
MYO15B	17	0.573	73,584,139	73,622,929	73,587,257	A	G	no
PITPNM2	12	0.573	123,468,027	123,634,562	123,525,280	A	G	no
PREX2	8	0.573	68,864,353	69,149,265	69,029,244	C	A	no

Table A4. *Cont.*

HGNC Symbol [1]	Chr	AUC	Start [2]	End [3]	SNP Position [4]	A1 [5]	A2 [6]	GS [7]
ENSG00000255468	11	0.573	66,115,421	66,132,275	66,115,782	G	T	no
SIPA1L2	1	0.572	232,533,711	232,697,304	232,664,611	C	T	yes
AMPD3	11	0.571	10,329,860	10,529,126	10,475,856	G	A	no
PAM	5	0.571	102,089,685	102,366,809	102,363,402	C	T	no
IFT140	16	0.571	1,560,428	1,662,111	1,593,645	C	T	no
TMEM204	16	0.571	1,578,689	1,605,581	1,593,645	C	T	no
CLIP1	12	0.570	122,755,979	122,907,179	122,891,863	C	T	no
ABCB9	12	0.570	123,405,498	123,466,196	123,418,656	G	T	no
ZC3H7B	22	0.570	41,697,526	41,756,151	41,755,105	A	G	no
CRHR1	17	0.569	43,699,267	43,913,194	43,784,228	T	C	no
ENSG00000263715	17	0.569	43,699,274	43,893,909	43,784,228	T	C	no
LRRK2	12	0.569	40,590,546	40,763,087	40,730,463	C	T	no
ENSG00000235423	12	0.569	123,736,577	123,746,030	123,744,082	C	A	no
MSRA	8	0.568	9,911,778	10,286,401	10,280,818	A	C	no
LYVE1	11	0.568	10,578,513	10,633,236	10,628,883	G	A	no
MRVI1	11	0.568	10,594,638	10,715,535	10,628,883	G	A	no
FAM162A	3	0.568	122,103,023	122,131,181	122,109,601	T	C	no
MMRN1	4	0.567	90,800,683	90,875,780	90,868,355	T	C	no
ENSG00000236656	1	0.567	158,444,244	158,464,676	158,453,419	A	C	no
ENSG00000235495	2	0.567	67,792,736	67,911,209	67,806,472	A	G	no
DEFB119	20	0.566	29,964,967	29,978,406	29,971,435	G	A	no
NGEF	2	0.566	233,743,396	233,877,982	233,864,457	C	T	no
MGAT5	2	0.566	134,877,554	135,212,192	135,202,455	A	G	no
ASAH1	8	0.565	17,913,934	17,942,494	17,927,609	C	T	no
CPNE8	12	0.565	39,040,624	39,301,232	39,174,139	T	G	no
SEMA3G	3	0.565	52,467,069	52,479,101	52,468,940	T	C	no
PBRM1	3	0.564	52,579,368	52,719,933	52,649,748	A	G	no
HMBOX1	8	0.564	28,747,911	28922281	28,809,951	A	G	no
HMBOX1-IT1	8	0.564	28,807,193	28,813,472	28,809,951	A	G	no
SNCA	4	0.563	90,645,250	90,759,466	90,700,329	T	C	no
MAPT	17	0.563	43,971,748	44,105,700	44,071,851	G	A	no
ENSG00000258881	2	0.563	71,166,448	71,222,466	71,202,989	T	C	no
ENSG00000251095	4	0.562	90,472,507	90,647,654	90,627,967	G	A	no
CRHR1	17	0.562	43,699,267	43,913,194	43,901,665	T	C	no
ARHGEF7	13	0.562	111,766,906	111,958,084	111,863,720	C	T	no
GNPTAB	12	0.561	102,139,275	102,224,716	102,151,977	C	T	no
FAM220A	7	0.561	6,369,040	6,388,612	6,369,946	A	G	no
BRD2	6	0.561	32,936,437	32,949,282	32,941,506	C	T	no
ATG4D	19	0.561	10,654,571	10,664,094	10,663,997	C	T	no
KRI1	19	0.561	10,663,761	10,676,713	10,663,997	C	T	no
FBXO34	14	0.560	55,738,021	55,828,636	55,801,687	A	C	no
ENSG00000258455	14	0.560	55,792,552	55,806,219	55,801,687	A	C	no
CCDC101	16	0.560	28,565,236	28,603,111	28,566,158	G	T	no
C14orf159	14	0.560	91,526,677	91,691,976	91,682,844	T	C	no
KIF21A	12	0.560	39,687,030	39,837,192	39,738,666	G	A	no
PRRC2C	1	0.559	171,454,651	171,562,650	171,471,672	T	C	no
RNF141	11	0.559	10,533,225	10,562,777	10,560,447	A	C	no
SOX2-OT	3	0.559	180,707,558	181,554,668	180,797,921	T	G	no
SLC2A13	12	0.558	40,148,823	40,499,891	40,437,969	A	G	no
RPP14	3	0.558	58,291,974	58,310,422	58,292,485	G	A	no
DGKG	3	0.557	185,823,457	186,080,026	185,834,290	T	C	no
ENSG00000251364	11	0.557	7,448,497	7,533,746	7,532,175	T	G	no
OLFML1	11	0.557	7,506,619	7,532,608	7,532,175	T	G	no
ADAM15	1	0.557	155,023,042	155,035,252	155,033,317	T	C	no
TRHDE	12	0.556	72,481,046	73,059,422	72,714,601	G	T	no
GAK	4	0.556	843,064	926,161	852,939	G	A	no
CCDC134	22	0.555	42,196,683	42,222,303	42,216,326	A	G	no

Table A4. Cont.

HGNC Symbol [1]	Chr	AUC	Start [2]	End [3]	SNP Position [4]	A1 [5]	A2 [6]	GS [7]
LZTS2	10	0.555	10,275,6375	102,767,593	102,764,511	G	A	no
SLC44A2	19	0.555	10,713,133	10,755,235	10,730,352	G	A	no
FYN	6	0.554	111,981,535	112,194,655	112,164,313	G	A	no
RNF212	4	0.554	1,050,038	1,107,350	1,082,829	T	C	no
CCSER1	4	0.553	91,048,686	92,523,064	91,383,333	G	A	no
ZNF589	3	0.553	48,282,590	48,340,743	48,333,546	T	C	no
FGF14	13	0.553	102,372,134	103,054,124	102,996,713	A	G	no
FGF14-IT1	13	0.553	102,944,677	103,046,869	102,996,713	A	G	no
TFRC	3	0.552	195,754,054	195,809,060	195,775,449	C	T	no
MAEA	4	0.552	1,283,639	1,333,935	1,312,394	C	T	no
ANKRD11	16	0.551	89,334,038	89,556,969	89,369,869	A	G	no
ZZZ3	1	0.551	78,028,101	78,149,104	78,070,458	C	T	no
DNM3	1	0.551	171,810,621	172,387,606	171,845,192	G	T	no
LARP1B	4	0.550	128,982,423	129,144,086	129,107,049	T	C	no
STK39	2	0.550	168,810,530	169,104,651	169,071,190	G	T	no
NEXN	1	0.550	78,354,198	78,409,580	78,392,446	G	A	no
CD38	4	0.550	15,779,898	15,854,853	15,829,612	A	G	no
HAVCR1	5	0.549	156,456,424	156,486,130	156,479,424	A	C	no
SCAND3	6	0.549	28,539,407	28,583,989	28,547,283	T	C	no
APOM	6	0.548	31,620,193	31,625,987	31,622,606	C	A	no
TRIM37	17	0.548	57,059,999	57,184,282	57,111,269	A	C	no
OR9Q1	11	0.548	57,791,353	57,949,088	57,870,219	G	A	no
KIAA1841	2	0.547	61,293,006	61,391,960	61,347,469	C	T	no
TATDN2	3	0.547	10,289,707	10,322,902	10,300,941	A	G	no
ENSG00000272410	3	0.547	10,291,056	10,327,480	10,300,941	A	G	no
ZNF320	19	0.547	53,367,043	53,400,946	53,399,832	C	T	no
ENSG00000272657	21	0.546	35,445,892	35,732,332	35,677,897	G	A	no
ENSG00000214955	21	0.546	35,577,356	35,697,334	35,677,897	G	A	no
ITGAL	16	0.546	30,483,979	30,534,506	30,520,856	C	T	no
UNKL	16	0.546	1,413,206	1,464,752	1,436,510	G	A	no
FYN	6	0.545	111,981,535	112,194,655	112,122,373	C	T	no
SYBU	8	0.545	110,586,207	110,704,020	110,644,774	T	C	no
AGMO	7	0.545	15,239,943	15,601,640	15,262,499	G	T	no
MED12L	3	0.544	150,803,484	151,154,860	151,133,211	G	A	no
SYNDIG1	20	0.544	24,449,835	24,647,252	24,645,939	G	A	no
MYO7A	11	0.544	76,839,310	76,926,284	76,920,983	A	G	no
CAPRIN2	12	0.543	30,862,486	30,907,885	30,895,251	T	C	no
BRSK2	11	0.543	1,411,129	1,483,919	1,478,565	T	C	no
ARID2	12	0.542	46,123,448	46,301,823	46,134,812	T	C	no
RALYL	8	0.542	85,095,022	85,834,079	85,772,129	A	G	no
HCAR1	12	0.542	123,104,824	123,215,390	123,189,794	T	C	no
ENSG00000256249	12	0.542	123,171,672	123,200,526	123,189,794	T	C	no
SPPL2B	19	0.541	2,328,614	2,355,099	2,341,047	C	T	yes
RNF165	18	0.541	43,906,772	44,043,103	44,040,660	T	C	no
HSF5	17	0.541	56,497,528	56,565,745	56,507,063	C	T	no
ENO3	17	0.540	4,851,387	4,860,426	4,858,206	A	G	no
WBP1L	10	0.539	104,503,727	104,576,021	104,562,212	C	T	no
ERC2	3	0.538	55,542,336	56,502,391	56,014,781	A	G	no
MYO1H	12	0.538	109,785,708	109,893,328	109,846,466	G	T	no
MAEA	4	0.538	1,283,639	1,333,935	1,311,933	G	T	no
ENSG00000244036	7	0.538	129,593,074	129,666,391	129,663,496	C	T	no
ZC3HC1	7	0.538	129,658,126	129,691,291	129,663,496	C	T	no
CSMD1	8	0.537	2,792,875	4,852,494	3,078,351	A	G	no
ENSG00000259848	2	0.537	95,533,231	95,613,086	95,555,581	T	C	no
POU2F3	11	0.536	120,107,349	120,190,653	120,178,753	T	G	no
HLA-DOA	6	0.536	32,971,955	32,977,389	32,973,303	T	C	no
TMPO	12	0.536	98,909,290	98,944,157	98,939,838	C	A	no
MTF2	1	0.536	93,544,792	93,604,638	93,570,368	G	A	no
SLC16A10	6	0.535	111,408,781	111,552,397	111,489,059	G	T	no

Table A4. *Cont.*

HGNC Symbol [1]	Chr	AUC	Start [2]	End [3]	SNP Position [4]	A1 [5]	A2 [6]	GS [7]
ENSG00000250003	5	0.535	38,025,799	38,184,034	38,046,354	G	A	no
ENSG00000225981	7	0.534	1,499,573	1,503,644	1,502,497	C	T	no
LRRK2	12	0.534	4,059,0546	40,763,087	40,707,861	C	T	no
TRAPPC13	5	0.533	64,920,543	64,962,060	64,952,500	C	T	no
METTL13	1	0.533	171,750,788	171,783,163	171,772,453	T	G	no
ENSG00000259675	15	0.533	61,931,548	62,007,370	62,005,917	C	A	no
AIRE	21	0.532	45,705,721	45,718,531	45,708,277	C	T	no
ENSG00000272305	3	0.532	53,003,135	53,133,469	53,087,621	A	G	no
C6orf10	6	0.531	32,256,303	32,339,684	32,303,848	G	A	no
HLA-DQA2	6	0.530	32,709,119	32,714,992	32,712,666	C	T	no
XPO1	2	0.530	61,704,984	61,765,761	61,763,170	C	T	no
HLA-DQB1	6	0.529	32,627,244	32,636,160	32,634,646	T	C	no
LRRK2	12	0.529	40,579,811	40,617,605	40,607,566	G	A	no
ENSG00000225342	12	0.529	40,590,546	40,763,087	40,607,566	G	A	no
C1orf167	1	0.529	11,821,844	11,849,642	11,827,776	A	G	no
ENSG00000249988	4	0.528	14,166,079	14,244,437	14,167,196	A	G	no
LAMA2	6	0.528	129,204,342	129,837,714	129,537,858	G	A	no
SOX6	11	0.528	15,987,995	16,761,138	16,158,420	G	A	no
CCDC69	5	0.527	150,560,613	150,603,706	150,566,196	C	T	no
ENSG00000223343	3	0.527	49,022,482	49,027,421	49,025,101	A	C	no
MAP4K4	2	0.527	102,313,312	102,511,149	102,468,624	A	G	no
KLHL7	7	0.526	23,145,353	23,217,533	23,208,043	G	A	no
ENSG00000253194	6	0.526	119,255,950	119,352,706	119,322,992	C	T	no
FAM184A	6	0.526	119,280,928	119,470,552	119,322,992	C	T	no
QRICH1	3	0.525	49,067,140	49,131,796	49,083,566	G	A	no
SYT17	16	0.525	19,179,293	19,279,652	19,279,380	T	C	no
CCDC62	12	0.524	123,258,874	123,312,075	123,296,204	G	A	no
SHC4	15	0.524	49,115,932	49,255,641	49,174,661	C	T	no
PNKD	2	0.523	219,135,115	219,211,516	219,142,491	C	T	no
TMBIM1	2	0.523	219,138,915	219,157,309	219,142,491	C	T	no
DIP2C	10	0.523	320,130	735,683	570,172	T	C	no
SCCPDH	1	0.523	246,887,349	246,931,439	246,893,948	C	T	no
IP6K1	3	0.522	49,761,727	49,823,975	49,808,007	A	G	no
FAM167A	8	0.522	11,278,972	11,332,224	11,309,780	G	A	no
ADCY5	3	0.521	123,001,143	123,168,605	123,143,272	G	A	no
PCGF3	4	0.521	699,537	764,428	701,896	A	G	no
RPRD2	1	0.520	150,335,567	150,449,042	150,438,362	A	C	no
CARM1	19	0.520	10,982,189	11,033,453	11,025,817	G	A	no
ENSG00000251246	1	0.519	155,036,224	155,059,283	155,055,863	G	A	no
EFNA3	1	0.519	155,036,224	155,060,014	155,055,863	G	A	no
MMS22L	6	0.519	97,590,037	97,731,093	97,662,784	G	A	no
C12orf40	12	0.519	40,019,969	40,302,102	40,042,940	C	T	no
C3orf84	3	0.518	49,215,065	49,229,291	49,220,504	A	C	no
MMRN1	4	0.518	90,800,683	90,875,780	90,859,279	G	A	no
RILPL2	12	0.517	123,899,936	123,921,264	123,912,213	T	C	no
CHAT	10	0.517	50,817,141	50,901,925	50,821,191	G	T	no
TMEM161B	5	0.517	87,485,450	87,565,293	87,513,775	C	T	no
BIN3	8	0.517	22,477,931	22,526,661	22,525,980	T	C	yes
TRPM4	19	0.516	49,660,998	49,715,093	49,695,007	A	G	no
USP8	15	0.516	50,716,577	50,793,280	50,741,068	A	C	no
BCAR3	1	0.516	94,027,347	94,312,706	94,038,847	G	A	no
TNXB	6	0.516	32,008,931	32,083,111	32,062,687	G	A	no

[1] HGNC symbol or Ensemble gene ID if there is no HGNC symbol available. [2] Base pair position of start of gene. [3] Base pair position of end of gene. [4] Genomic position of SNP. [5] Major SNP allele. [6] Minor SNP allele. [7] Genome-wide significant in the meta-GWAS by Nalls et al. [2]. HGNC: HUGO Gene Nomenclature Committee, Chr: Chromosome, AUC: area under ROC curve, ROC: receiver operating characteristic, PRS: polygenic risk score, PD: Parkinson's disease, n.a.: not available.

References

1. Kalia, L.V.; Lang, A.E. Parkinson's disease. *Lancet* **2015**, *386*, 896–912. [CrossRef]
2. Nalls, M.A.; Blauwendraat, C.; Vallerga, C.L.; Heilbron, K.; Bandres-Ciga, S.; Chang, D.; Tan, M.; Kia, D.A.; Noyce, A.J.; Xue, A.; et al. Identification of novel risk loci, causal insights, and heritable risk for Parkinson's disease: A meta-analysis of genome-wide association studies. *Lancet Neurol.* **2019**, *18*, 1091–1102. [CrossRef]
3. Chang, D.; Nalls, M.A.; Hallgrimsdottir, I.B.; Hunkapiller, J.; van der Brug, M.; Cai, F.; International Parkinson's Disease Genomics Consortium; 23andMe Research Team; Kerchner, G.A.; Ayalon, G.; et al. A meta-analysis of genome-wide association studies identifies 17 new Parkinson's disease risk loci. *Nat. Genet.* **2017**, *49*, 1511–1516. [CrossRef] [PubMed]
4. Bloem, B.R.; Okun, M.S.; Klein, C. Parkinson's disease. *Lancet* **2021**, *397*, 2284–2303. [CrossRef]
5. Nalls, M.A.; Pankratz, N.; Lill, C.M.; Do, C.B.; Hernandez, D.G.; Saad, M.; DeStefano, A.L.; Kara, E.; Bras, J.; Sharma, M.; et al. Large-scale meta-analysis of genome-wide association data identifies six new risk loci for Parkinson's disease. *Nat. Genet.* **2014**, *46*, 989–993. [CrossRef] [PubMed]
6. Ibanez, L.; Dube, U.; Saef, B.; Budde, J.; Black, K.; Medvedeva, A.; Del-Aguila, J.L.; Davis, A.A.; Perlmutter, J.S.; Harari, O.; et al. Parkinson disease polygenic risk score is associated with Parkinson disease status and age at onset but not with α-synuclein cerebrospinal fluid levels. *BMC Neurol.* **2017**, *17*, 198. [CrossRef]
7. Li, W.W.; Fan, D.Y.; Shen, Y.Y.; Zhou, F.Y.; Chen, Y.; Wang, Y.R.; Yang, H.; Mei, J.; Li, L.; Xu, Z.Q.; et al. Association of the polygenic risk score with the incidence risk of Parkinson's disease and cerebrospinal fluid α-synuclein in a Chinese cohort. *Neurotox. Res.* **2019**, *36*, 515–522. [CrossRef]
8. Escott-Price, V.; Sims, R.; Bannister, C.; Harold, D.; Vronskaya, M.; Majounie, E.; Badarinarayan, N.; Morgan, K.; Passmore, P.; Holmes, C.; et al. Common polygenic variation enhances risk prediction for Alzheimer's disease. *Brain* **2015**, *138*, 3673–3684. [CrossRef]
9. Jacobs, B.M.; Belete, D.; Bestwick, J.; Blauwendraat, C.; Bandres-Ciga, S.; Heilbron, K.; Dobson, R.; Nalls, M.A.; Singleton, A.; Hardy, J.; et al. Parkinson's disease determinants, prediction and gene-environment interactions in the UK Biobank. *J. Neurol. Neurosurg. Psychiatry* **2020**, *91*, 1046–1054. [CrossRef] [PubMed]
10. Paul, K.C.; Schulz, J.; Bronstein, J.M.; Lill, C.M.; Ritz, B.R. Association of polygenic risk score with cognitive decline and motor progression in Parkinson disease. *JAMA Neurol.* **2018**, *75*, 360–366. [CrossRef]
11. Wald, N.J.; Old, R. The illusion of polygenic disease risk prediction. *Genet. Med.* **2019**. [CrossRef] [PubMed]
12. Caliebe, A.; Heinzel, S.; Schmidtke, J.; Krawczak, M. Genorakel polygene Risikoscores: Möglichkeiten und Grenzen. *Dtsch. Arztebl. Int.* **2021**, *118*, A410.
13. Kasten, M.; Hagenah, J.; Graf, J.; Lorwin, A.; Vollstedt, E.J.; Peters, E.; Katalinic, A.; Raspe, H.; Klein, C. Cohort Profile: A population-based cohort to study non-motor symptoms in parkinsonism (EPIPARK). *Int. J. Epidemiol.* **2013**, *42*, 128–128k. [CrossRef] [PubMed]
14. Mollenhauer, B.; Trautmann, E.; Sixel-Doring, F.; Wicke, T.; Ebentheuer, J.; Schaumburg, M.; Lang, E.; Focke, N.K.; Kumar, K.R.; Lohmann, K.; et al. Nonmotor and diagnostic findings in subjects with de novo Parkinson disease of the DeNoPa cohort. *Neurology* **2013**, *81*, 1226–1234. [CrossRef]
15. Lieb, W.; Jacobs, G.; Wolf, A.; Richter, G.; Gaede, K.I.; Schwarz, J.; Arnold, N.; Bohm, R.; Buyx, A.; Cascorbi, I.; et al. Linking pre-existing biorepositories for medical research: The PopGen 2.0 Network. *J. Community Genet.* **2019**, *10*, 523–530. [CrossRef]
16. Krawczak, M.; Nikolaus, S.; von Eberstein, H.; Croucher, P.J.; El Mokhtari, N.E.; Schreiber, S. PopGen: Population-based recruitment of patients and controls for the analysis of complex genotype-phenotype relationships. *Community Genet.* **2006**, *9*, 55–61. [CrossRef]
17. Meyer, H. plinkQC: Genotype Quality Control with 'PLINK'. R Package Version 0.3.4. 2021. Available online: https://cran.r-project.org/web/packages/plinkQC/index.html (accessed on 15 October 2021).
18. Chang, C.C.; Chow, C.C.; Tellier, L.C.; Vattikuti, S.; Purcell, S.M.; Lee, J.J. Second-generation PLINK: Rising to the challenge of larger and richer datasets. *Gigascience* **2015**, *4*, 7. [CrossRef]
19. Wigginton, J.E.; Cutler, D.J.; Abecasis, G.R. A note on exact tests of Hardy-Weinberg equilibrium. *Am. J. Hum. Genet.* **2005**, *76*, 887–893. [CrossRef]
20. Purcell, S.; Neale, B.; Todd-Brown, K.; Thomas, L.; Ferreira, M.A.; Bender, D.; Maller, J.; Sklar, P.; de Bakker, P.I.; Daly, M.J.; et al. PLINK: A tool set for whole-genome association and population-based linkage analyses. *Am. J. Hum. Genet.* **2007**, *81*, 559–575. [CrossRef]
21. Purcell, S.; Chang, C. PLINK 1.9. Available online: https://www.cog-genomics.org/plink (accessed on 22 November 2021).
22. Purcell, S.; Chang, C. PLINK 2.0. Available online: https://www.cog-genomics.org/plink/2.0 (accessed on 22 November 2021).
23. O'Connell, J.; Gurdasani, D.; Delaneau, O.; Pirastu, N.; Ulivi, S.; Cocca, M.; Traglia, M.; Huang, J.; Huffman, J.E.; Rudan, I.; et al. A general approach for haplotype phasing across the full spectrum of relatedness. *PLoS Genet.* **2014**, *10*, e1004234. [CrossRef] [PubMed]
24. Howie, B.N.; Donnelly, P.; Marchini, J. A flexible and accurate genotype imputation method for the next generation of genome-wide association studies. *PLoS Genet.* **2009**, *5*, e1000529. [CrossRef]
25. McCarthy, S.; Das, S.; Kretzschmar, W.; Delaneau, O.; Wood, A.R.; Teumer, A.; Kang, H.M.; Fuchsberger, C.; Danecek, P.; Sharp, K.; et al. A reference panel of 64,976 haplotypes for genotype imputation. *Nat. Genet.* **2016**, *48*, 1279–1283. [CrossRef] [PubMed]

26. Robin, X.; Turck, N.; Hainard, A.; Tiberti, N.; Lisacek, F.; Sanchez, J.C.; Muller, M. pROC: An open-source package for R and S+ to analyze and compare ROC curves. *BMC Bioinform.* **2011**, *12*, 77. [CrossRef]
27. Aragon, T. Epitools: Epidemiology Tools. R Package Version 0.5-10.1. 2012. Available online: https://cran.r-project.org/web/packages/epitools/index.html (accessed on 22 November 2021).
28. Durinck, S.; Moreau, Y.; Kasprzyk, A.; Davis, S.; De Moor, B.; Brazma, A.; Huber, W. BioMart and Bioconductor: A powerful link between biological databases and microarray data analysis. *Bioinformatics* **2005**, *21*, 3439–3440. [CrossRef] [PubMed]
29. Durinck, S.; Spellman, P.T.; Birney, E.; Huber, W. Mapping identifiers for the integration of genomic datasets with the R/Bioconductor package biomaRt. *Nat. Protoc.* **2009**, *4*, 1184–1191. [CrossRef]
30. Howe, K.L.; Achuthan, P.; Allen, J.; Allen, J.; Alvarez-Jarreta, J.; Amode, M.R.; Armean, I.M.; Azov, A.G.; Bennett, R.; Bhai, J.; et al. Ensembl 2021. *Nucleic Acids Res.* **2021**, *49*, D884–D891. [CrossRef]
31. Sherry, S.T.; Ward, M.H.; Kholodov, M.; Baker, J.; Phan, L.; Smigielski, E.M.; Sirotkin, K. dbSNP: The NCBI database of genetic variation. *Nucleic Acids Res.* **2001**, *29*, 308–311. [CrossRef] [PubMed]
32. Nerius, M.; Fink, A.; Doblhammer, G. Parkinson's disease in Germany: Prevalence and incidence based on health claims data. *Acta Neurol. Scand.* **2017**, *136*, 386–392. [CrossRef]
33. Hoffmann, S.; Schonbrodt, F.; Elsas, R.; Wilson, R.; Strasser, U.; Boulesteix, A.L. The multiplicity of analysis strategies jeopardizes replicability: Lessons learned across disciplines. *R. Soc. Open Sci.* **2021**, *8*, 201925. [CrossRef]
34. Baker, M. 1500 scientists lift the lid on reproducibility. *Nature* **2016**, *533*, 452–454. [CrossRef]
35. Loken, E.; Gelman, A. Measurement error and the replication crisis. *Science* **2017**, *355*, 584–585. [CrossRef] [PubMed]
36. Janssens, A. Validity of polygenic risk scores: Are we measuring what we think we are? *Hum. Mol. Genet.* **2019**, *28*, R143–R150. [CrossRef]
37. Fullerton, J.M.; Nurnberger, J.I. Polygenic risk scores in psychiatry: Will they be useful for clinicians? *F1000Research* **2019**, *8*. [CrossRef] [PubMed]
38. Martin, A.R.; Kanai, M.; Kamatani, Y.; Okada, Y.; Neale, B.M.; Daly, M.J. Clinical use of current polygenic risk scores may exacerbate health disparities. *Nat. Genet.* **2019**, *51*, 584–591. [CrossRef]
39. Altenbuchinger, M.; Weihs, A.; Quackenbush, J.; Grabe, H.J.; Zacharias, H.U. Gaussian and Mixed Graphical Models as (multi-)omics data analysis tools. *Biochim. Biophys. Acta Gene Regul. Mech.* **2020**, *1863*, 194418. [CrossRef]
40. Elliott, J.; Bodinier, B.; Bond, T.A.; Chadeau-Hyam, M.; Evangelou, E.; Moons, K.G.M.; Dehghan, A.; Muller, D.C.; Elliott, P.; Tzoulaki, I. Predictive accuracy of a polygenic risk score-enhanced prediction model vs a clinical risk score for coronary artery disease. *JAMA* **2020**, *323*, 636–645. [CrossRef]
41. Landi, I.; Kaji, D.A.; Cotter, L.; Van Vleck, T.; Belbin, G.; Preuss, M.; Loos, R.J.F.; Kenny, E.; Glicksberg, B.S.; Beckmann, N.D.; et al. Prognostic value of polygenic risk scores for adults with psychosis. *Nat. Med.* **2021**, *27*, 1576–1581. [CrossRef]
42. Yanes, T.; Young, M.A.; Meiser, B.; James, P.A. Clinical applications of polygenic breast cancer risk: A critical review and perspectives of an emerging field. *Breast Cancer Res.* **2020**, *22*, 21. [CrossRef] [PubMed]
43. Heinzel, S.; Berg, D.; Gasser, T.; Chen, H.; Yao, C.; Postuma, R.B.; Disease, M.D.S.T.F.o.t.D.o.P.s. Update of the MDS research criteria for prodromal Parkinson's disease. *Mov. Disord.* **2019**, *34*, 1464–1470. [CrossRef]
44. Pebesma, E.; Bivand, R. Classes and Methods for Spatial Data in R. *R. News* **2005**, *5*, 9–13.
45. Bivand, R.; Pebesma, E.; Gómez Rubio, V. *Applied Spatial Data Analysis With R*; Springer: New York, NY, USA, 2013.
46. Bivand, R.; Rundel, C. Rgeos: Interface to Geometry Engine-Open Source (GEOS). R Package Version 0.5-8. 2021. Available online: https://cran.r-project.org/web/packages/rgeos/index.html (accessed on 22 November 2021).
47. Prive, F.; Luu, K.; Blum, M.G.B.; McGrath, J.J.; Vilhjalmsson, B.J. Efficient toolkit implementing best practices for principal component analysis of population genetic data. *Bioinformatics* **2020**, *36*, 4449–4457. [CrossRef] [PubMed]
48. Prive, F.; Aschard, H.; Ziyatdinov, A.; Blum, M.G.B. Efficient analysis of large-scale genome-wide data with two R packages: Bigstatsr and bigsnpr. *Bioinformatics* **2018**, *34*, 2781–2787. [CrossRef] [PubMed]
49. Privé, F. Bigparallelr: Easy Parallel Tools. R Package Version 0.3.1. 2021. Available online: https://rdrr.io/cran/bigparallelr/man/bigparallelr-package.html (accessed on 22 November 2021).

2. mtDNA deletions, and
3. an overall reduction of mtDNA copy numbers [38].

Both, point mutations and mtDNA deletions are subject to clonal expansion. As mitochondria replicate independently from the cell cycle and distribute randomly to the daughter cells after mitosis, the degree of heteroplasmy can widely differ within a given tissue [39]. If a certain heteroplasmy threshold is exceeded, mitochondrial homeostasis can be impaired, subsequently leading to impairments similar to those seen in primary mitochondrial disease and ultimately to cell death [40] (see Figure 2). There is strong experimental evidence that genetic variations in mtDNA increase with age, which also translates to our pathophysiological understanding of the development of neurodegenerative diseases [41]. It has also been shown that the mtDNA mutation rate accelerates with higher age which is especially relevant to postmitotic neurons [42].

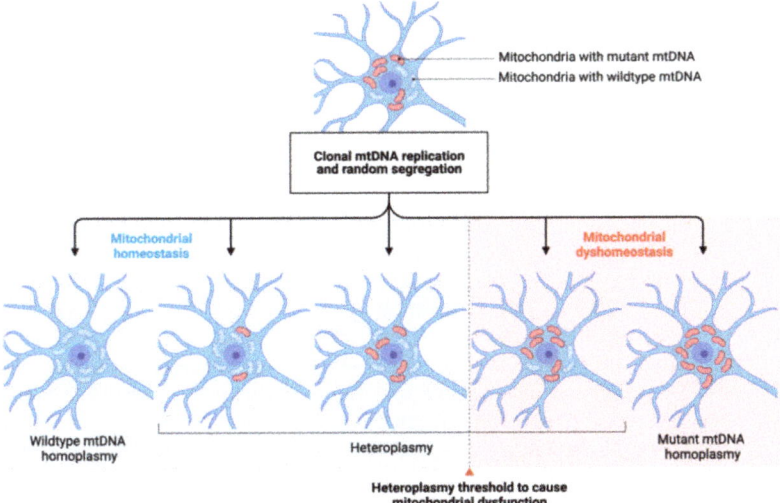

Figure 2. The concept of mtDNA heteroplasmy and its contribution to mitochondrial dysfunction in PD. The arbitrary occurrence of mtDNA mutations (depicted by red-colored mitochondria) and their respective clonal expansion leads to different degrees of heteroplasmy in a given neuronal cell. Cell-specific heteroplasmy thresholds (dashed line) to cause mitochondrial dysfunction are widely unknown. By, e.g., shifting the degree of heteroplasmy towards a higher ratio of wildtype/mutated mtDNA, mitochondrial homeostasis might be restored. mtDNA: mitochondrial DNA.

1.5.2. Inherited and Somatic mtDNA Point Mutations and Their Role in the Pathophysiology of Parkinson's Disease

Inherited mtDNA point mutations are of negligible relevance for the vast majority of PD cases. However, Shoffner et al. (1993) described a point mutation (m.1555A>C., *MT-RNR1*) within the 12S-rRNA gene in a pedigree of maternally transmitted hearing loss and levodopa-responsive parkinsonism [43]. In another pedigree, the heteroplasmic mtDNA point mutation m.1095A>C in the *MT-RNR1* gene has been identified as a potential cause of PD [44]. The latter is especially intriguing as it impairs the complex I function of the ETC, a pathophysiological hallmark often observed in PD [44]. These findings were supported by the later identification of additional mtDNA missense mutations present in nearly all mtDNA-encoded subunits of complex I [45]. Collectively, inherited mtDNA variants, referred to as mtDNA haplogroups, are associated with a lower or higher risk of developing PD [46–48]. Many of these reports need additional experimental validation. In summary, there is no direct evidence to suggest that inherited mtDNA point mutations are a primary cause of PD [49]. One study examined the combined mutational burden of somatic mtDNA point mutations in all genes encoding complex I subunits in postmortem

PD brain tissue [37]. The authors concluded that there was no significant difference in the overall number of mtDNA point mutations in PD patients and controls.

In contrast, this work revealed relatively low levels of somatic mtDNA point mutations within the *MT-ND5* gene exclusively observed in idiopathic PD patients. However, the heteroplasmy thresholds were generally less than 1%, where no functional consequences would be expected [37]. The conflicting experimental data so far also points toward potential challenges for developing mitochondria-targeted gene therapies: Somatic mutations (deletions and/or point mutations) occur randomly by means of heteroplasmy and localization within the mitochondrial genome [32]. As the pathophysiological role of inherited mtDNA mutations is still under debate, the arbitrary occurrence of somatic mtDNA mutations complicates the design of mitochondrial genome editing techniques for this purpose. In addition, it is unclear how to define functionally relevant cumulative thresholds based on the simultaneous presence of different kinds and respective frequencies of somatic mtDNA mutations.

1.5.3. The Role of Mitochondrial DNA Deletions and Copy Number Variations in PD

The most frequent deletion in human mtDNA encompasses ca. 5 kbp. This deletion includes most of the complexes of the ETC, leading to an overall bioenergetic deficit [43]. While it is not entirely understood how mtDNA deletions occur, several hypotheses were suggested: In most cases, mtDNA deletions occur randomly and appear to undergo clonal expansion. Another theory suggests that critical pathways for mtDNA replication and quality control are impaired in neurodegenerative disorders [40]. The maintenance of mtDNA requires a variety of nDNA-encoded gene products. The proteins involved in mtDNA replication have been termed replisome [50]. The mtDNA replisome consists of the mtDNA polymerase γ (a complex of *POLG* and *POLG2* gene products), the mitochondrial transcription factor (*TFAM*), the DNA helicase twinkle (*TWNK*), and the mitochondrial single-stranded binding protein (*mtSSB*) [50]. Remarkably, variants in *POLG*, *TWNK*, and *TFAM* are not only known as a monogenic cause of primary mitochondrial disorders (occasionally presenting with parkinsonism) but can also increase the risk for PD [51]. Based on the known function of the mitochondrial replisome, mutations in these three genes can result in mtDNA deletions and decreased mtDNA copy numbers [49]. All three genes show high expression levels in neuroanatomical key structures involved in PD disease development such as the substantia nigra (SN) [51]. Postmortem studies also revealed lower levels of mtDNA transcription factor *TFAM* in the SN of PD patients [52]. In this study, *TFAM* and *TFB2M* levels correlated with decreased expression levels of complex I. Noteworthy, decreased mtDNA copy numbers showed a cell-specific distribution in PD [53]. In contrast to dopaminergic neurons of the SN, cholinergic neurons isolated from PD brains were associated with a higher mtDNA copy number [54]. It is also worth mentioning that many of the known monogenic PD genes (e.g., *PRKN* or *LRRK2*) have been linked to altered mtDNA maintenance [55–57]. However, future studies are needed to fully understand the interconnectedness of mtDNA maintenance and their impact on monogenic and idiopathic PD.

Even though mtDNA alterations have been observed in physiological aging, the increased amount of mtDNA rearrangements and deletions in PD patients suggest a certain disease specificity [58]. Accordingly, SN-related mtDNA deletions and copy number variations are more common in PD than in patients with other neurodegenerative diseases (e.g., Alzheimer's disease, AD) [41]. PD patients are thus more likely to accumulate mtDNA mutations, particularly in dopaminergic neurons. Therefore, regulation of mtDNA deletions and copy number variations seems to be a potential mechanism to protect SN neurons from cell death or apoptosis.

Additional experimental evidence originates from animal models. A conditional *TFAM* knock-out mouse (MitoPark mouse) is characterized by respiratory chain deficiencies and low neuronal cell counts including progressive loss of dopaminergic neurons in the SN [59–61]. In another mouse model, mutant *TWNK* has been expressed in CNS neurons, leading to an increase of age-related mtDNA deletions and dopaminergic neurodegeneration [62]. These mice suffer from levodopa-responsive motor impairment and show phenotypic features of premature aging. This data stresses that the integrity of the nuclear and the mitochondrial genome is critical for the survival of dopaminergic neurons.

The deepened understanding of mtDNA defects in PD may offer the opportunity for targeted therapies: mtDNA deletions in individual SN neurons can activate compensatory mechanisms mainly by triggering mitochondrial biogenesis [63,64]. These mechanisms increase the number of mtDNA copies, the formation of cristae networks, and dopamine synthesis. The compensatory response could be impaired or dysregulated by nDNA variants in the genes mentioned above and may impact PD onset and progression [35]. By employing compensatory mechanisms, individual neurons can overcome the harmful effects of mtDNA mutations below a certain threshold [64]. The increase of mtDNA copy numbers with a corresponding rise in wild-type mtDNA might therefore prevent respiratory chain defects in people with a high mtDNA deletion burden. Therefore, the enhancement of mitochondrial biogenesis could be specifically targeted by gene therapy to combat the unspecific accumulation of mtDNA mutations in PD patients [65].

2. Main Body
2.1. Defining Neuroanatomical Treatment Targets

The treatment of mitochondrial dysfunction in PD comes with unique challenges and opportunities. To date, it is unclear which brain regions are especially vulnerable to mitochondrial dysfunction. Most post-mortem studies focus on the SN as a target region of PD pathophysiology [66]. The neurodegeneration of the dopaminergic neurons in the SN has been postulated as a histopathological hallmark of PD [67]. Besides, the co-occurrence of impaired dopamine metabolism, the emergence of oxidative stress, and mitochondrial dysfunction have been proposed as a vicious cycle, self-amplifying the molecular roots of neurodegenerative processes in this disorder [68,69]. However, mitochondrial dysfunction is not restricted to dopaminergic neurons but also affects other neuronal and non-neuronal cell populations. This idea is additionally supported by our current understanding of PD as a network disease, affecting widespread areas of the human brain [70]. Whether spatially non-specific drug delivery to the CNS or targeted intraparenchymal drug delivery in predefined neuroanatomical regions (e.g., the basal ganglia) will be the most promising approach in the future, is still under debate [15]. However, the concept of spatial drug delivery does not only concern distinct neuroanatomically defined regions [71]. There is a close metabolic interconnectedness between glial cells and neurons, and mitochondrial dysfunction most likely extends to several CNS cellular subpopulations [72]. The spatial complexity is not the only challenge concerning drug delivery. Different treatment strategies are discussed in the following. Based on the complex intracellular compartmentalization of mammalian cells, it is vital to consider whether gene therapeutic approaches target the nucleus or the mitochondrial matrix. These different levels of spatial complexity substantially aggravate anyhow preexisting challenges for CNS-based drug delivery (e.g., overcoming the blood-brain barrier (BBB)) [73].

In general, two common approaches for drug delivery are available: direct and indirect CNS delivery [74]. Direct CNS drug delivery describes the administration route via intraparenchymal application (e.g., by stereotactically placed catheters, similar to the procedure for electrode implantation for deep brain stimulation) [74]. This term also extends to intrathecal, intracerebroventricular, and subpial administration. In contrast, indirect CNS drug delivery describes the administration via an intravenous infusion [75]. An additional supportive method is the transient opening of the BBB by focused ultrasound which has, however, not yet been clinically evaluated [76]. The concept of cellular tropism (by means of tissue-/cell type-specificity) can be achieved by employing different Adeno-associated virus (AAV) serotypes and respective transgene designs (e.g., by using cell-specific promoters) [77]. If the development of targeted liposomes can achieve identical (pre-)clinical efficacy to AAV-based methods, will be the subject of future studies [78].

AAVs are great candidates for gene therapy because of their low risk of insertional mutation and long-term persistence within cells [79]. Currently, AAV-based approaches are the central concept for gene therapy in preclinical and human use. AAVs are non-pathogenic in humans but may induce immunological host responses, potentially hindering their long-term use in a given patient. AAV-based vectors can be fine-tuned by specific capsid and promotor designs. Research has shown that distinct virus strains show a relatively specific tissue tropism [79].

The payload of gene therapies can be specifically designed to meet molecular needs: This can be achieved by the overexpression of genes by gene replacement (e.g., for loss-of-function mutations), the silencing of genes by small hairpin RNA (shRNA), or small interfering RNA (siRNA) (e.g., for gain-of-function mutations), site-directed genome editing (in general suitable for a variety of mutation types), and the modulation of gene expression by microRNAs (miRNAs) or modified genome editing technologies [17,80]. However, most of these approaches have not made their way into clinical use. In addition, for safe and effective gene therapy, specific gene regulatory elements can be chosen to achieve cell-specificity (e.g., by cell-specific promoter sequences) [79]. When choosing a target or disorder to pursue gene therapy, it is crucial to consider the possibility of successfully delivering an AAV vector into the CNS.

2.2. Treatment Strategies

The development of gene therapies for neurological disorders is a highly dynamic field of research, and the last years have shown impressive advances. However, gene therapeutic approaches targeting mitochondrial dysfunction in neurodegenerative diseases are currently sparse, even in pre-clinical phases. This may be due to the significant challenges mitochondrial gene therapy encounters in vivo, many caused by the complex mitochondrial biology [81]. In general, the combination of different gene therapies would likely be the most efficient treatment strategy, mainly due to the interwovenness of the nuclear and mitochondrial genome, unique characteristics of mtDNA, such as high mtDNA copy numbers, heteroplasmy, and the mtDNA-specific genetic code. These limitations must be addressed before mitochondrial gene therapy can be used effectively in the context of PD. In this review, we will describe specific challenges for mitochondrial gene therapy and will focus on four potential therapeutic strategies:

1. gene replacement/correction of monogenic PD genes,
2. gene replacement of nuclear-encoded mitochondrial genes,
3. allotopic expression of mtDNA-encoded genes, and
4. mtDNA genome editing.

2.2.1. Gene Therapies of Monogenic Parkinson's Disease Genes to Treat Mitochondrial Dysfunction

The discovery of monogenic PD genes has led to in-depth insights into relevant disease mechanisms, broadening our molecular understanding of idiopathic PD [82]. Previously, mitochondrial dysfunction has already been implicated in idiopathic PD cases based on environmental studies [83]. The discovery of the *PRKN* and *PINK1* genes has grounded the concept of mitochondrial dysfunction on a genetic basis [4].

Mutations in both genes are inherited in an autosomal recessive fashion. Truncated or missense variants of the *PRKN* or *PINK1* gene have been shown to result in a loss-of-function or complete inactivation of their respective gene products. Later studies have demonstrated that *PRKN* and *PINK1* work together in a shared pathway and are mainly responsible for mitochondrial quality control by removing dysfunctional mitochondria (a process called mitophagy) [84]. Dysfunctional Parkin or PINK1 leads to impaired clearance of damaged mitochondria [85,86]. Intracellularly, damaged mitochondria can present with any aspects of mitochondrial dysfunction, including OXPHOS deficiency and impaired mtDNA maintenance [43]. Another monogenic PD gene that has been directly linked to mitochondrial dysfunction is *DJ-1* [87,88].

Even though the precise function of DJ-1 remains unclear, it is thought of as an oxidative stress sensor and works synergistically together with *PRKN* and *PINK1*. *PRKN*, *PINK1*, and *DJ-1* are the most prominent examples of monogenic PD genes directly leading to mitochondrial dysfunction [89,90]. However, most of the other identified monogenic PD genes (e.g., *SNCA* or *LRRK2*) have also been experimentally associated with mitochondrial dysfunction [91–94]. Whether mitochondrial dysfunction is a cause or consequence of neurodegenerative processes in other monogenic PD forms will need additional experimental validation.

Nonetheless, overexpression or silencing (depending on the relevant mutation type) of other monogenic PD genes can present viable treatment targets for improving mitochondrial dysfunction in monogenic PD [4]. Many of these treatment strategies may also extend to idiopathic PD. We kindly refer the reader to the review by Bloem et al. [1].

Previous studies have shown that *PRKN* overexpression can protect against cellular insults directed against mitochondria [95–98]. For example, the overexpression of wild-type *PRKN* in transgenic mice models reduced 1-Methyl-4-phenyl-1,2,3,6-tetrahydropyridine (MPTP)-induced (a known inhibitor of complex I) mitochondrial damage and prevented dopaminergic neurodegeneration [99]. The intranigral AAV-based delivery of wild-type *PRKN* prevented motor impairments and dopaminergic cell loss in a chronic MPTP minipump mouse model [100]. In Drosophila flies, the knockout of the *PINK1* homolog can lead to male sterility and progressive muscle wasting [101]. Here, defects in mitochondrial structure and increased sensitivity to oxidative stress can be observed. The Drosophila *PINK1*-KO phenotype can be rescued by overexpression of human *PINK1* [95]. The overexpression of *PRKN* has been shown to rescue mutated *PINK1* phenotypes, most likely by Miro-mediated phosphorylation and subsequent proteasomal degradation of dysfunctional mitochondria [102,103]. Furthermore, overexpression of *PINK1* has also been shown to rescue the α-synuclein-induced phenotype in Drosophila [104,105]. Additional evidence can be derived from siRNA experiments. Here, *PINK1*-silencing caused neuronal toxicity, which has been aggravated by MPTP administration in mice [106]. The wild-type but not the mutated form of *PINK1* protected neurons against MPTP-mediated cell death. The AAV-mediated expression of *PRKN* or *DJ-1* can protect mitochondria of dopaminergic neurons, even when *PINK1* is absent [106]. An overview of monogenic PD genes and their respective link to mitochondrial dysfunction is highlighted in Table 1.

Table 1. Established causative genes of monogenic PD and their respective association with mitochondrial dysfunction.

Gene Name	Mode of Inheritance	Parkinson's Disease Phenotype	Mitochondrial Involvement in Disease Pathophysiology–Key Mechanisms	References
ATP13A2 (PARK9)	AR	Atypical PD, Kufor-Rakeb syndrome	Impaired mitochondrial clearance, mitochondrial dysfunction due to zinc dyshomeostasis	Ramirez et al., 2006 Grunewald et al., 2012 Park et al., 2014
DJ-1 (PARK7)	AR	Early-onset PD	Reduced anti-oxidative stress mechanisms	Bonifati et al., 2003 Takahashi-Niki et al., 2004
FBXO7 (PARK15)	AR	Atypical PD	Aggravated protein aggregation in mitochondria, impaired mitophagy	Shojaee et al., 2008 Zhou et al., 2018

Table 1. Cont.

Gene Name	Mode of Inheritance	Parkinson's Disease Phenotype	Mitochondrial Involvement in Disease Pathophysiology–Key Mechanisms	References
GBA	AD	resembling IPD with more rapid cognitive and motor progression, dementia with Lewy bodies	Impaired mitophagy	Sidransky et al., 2009 Barkhuizen et al., 2016 Zhao et al., 2016 Gegg et al., 2016 Moren et al., 2019
LRRK2 (PARK8)	AD	resembling IPD	Disturbance in mitochondrial ATP and ROS production, impaired mitochondrial dynamics and mitophagy, mitochondrial DNA damage	Zimprich et al., 2004 Mancini et al., 2020
PINK1 (PARK6)	AR	Early-onset PD	Defective mitochondrial quality control	Valente et al., 2004 Ge et al., 2020
PLA2G6 (PARK14)	AR	Atypical PD, NBIA type 2B, Infantile neuroaxonal dystrophy 1	Maintenance of mitochondrial function, impaired mitophagy	Paisan-Ruiz et al., 2009 Chiu et al., 2017 Chiu et al., 2019
PRKN (PARK2)	AR	Early-onset PD	Defective mitochondrial quality control	Kitada et al., 1998 Ge et al., 2020
SNCA (PARK1)	AD	May be atypical (higher frequency of cognitive/ psychiatric symptoms)	Mitochondrial toxicity, fragmented mitochondria	Polymeropoulos et al., 1997 Singleton et al., 2003 Chartier-Harlin et al., 2004
VPS35 (PARK17)	AD	resembling IPD	Regulation of mitochondrial dynamics and homeostasis	Vilarino-Guell et al., 2011 Zimprich et al., 2011 Cutillo et al., 2020

AD: autosomal dominant. AR: autosomal recessive. IPD: idiopathic Parkinson's disease. NBIA: neurodegeneration with brain iron accumulation. PD: Parkinson's disease. The table has been adapted and modified from Prasuhn et al. [8].

2.2.2. Gene Repair and Enhancement of Nuclear-Encoded Mitochondrial Genes

AAV-mediated gene therapies have been tested in different models of primary mitochondrial diseases [107]. Insights derived from primary mitochondrial disorders elucidated potential gene therapy targets to treat mitochondrial dysfunction in PD. These strategies extend to the repair or enhancement of non-classical monogenic PD genes. In this context, we have already discussed "mtDNA maintenance disorders", which can clinically present with parkinsonian features, and stressed the role of increased somatic mtDNA mutations in the onset and progression of PD [10]. Gene therapy-based replacement (if there is a disease-causing mutation in patients present) or enhancements (by overexpression of genes even in the absence of a disease-causing mutation therein) of the mtDNA replisome may be helpful to improve mtDNA maintenance. Based on previous studies, the most promising genes for this approach are POLG, POLG2, TWNK, and TFAM [51]. Experimental evidence can be derived from the transfection of TFAM. Here, PD-derived nigral cybrid cell lines (cell lines that incorporate the nuclear genome from one cell with the mitochondrial genome from another cell) can restore mitochondrial bioenergetics by overexpression of TFAM [108]. However, it is unlikely that all PD patients show a marked increase in mtDNA damage at a given time. If nDNA mutations in the given genes are present in a patient, different gene therapies treatment strategies (e.g., gene editing by CRISPR/Cas9, Clustered Regularly Interspaced Short Palindromic Repeats/CRISPR-associated protein 9) could also be experimentally evaluated [80].

Many potential target genes can be identified from primary mitochondrial disorders caused by nDNA-encoded genes [107]. The large number of nDNA defects leading to primary mitochondrial diseases helps to understand the many-faceted nature of mitochondrial dysfunction [5]. Even though about 300 nDNA genes have been suggested to be associated with primary mitochondrial disorders, their gene products only account for a distinct subset of the overall mitochondrial proteome [107]. However, it is necessary to prioritize treatment targets, and the nDNA mutations causing primary mitochondrial diseases represent a reasonable starting point for treating other disorders. In addition, functional data are available for many causative nDNA genes and have been linked to distinct partial aspects of mitochondrial dysfunction. For example, *DNM1L*, *GDAP1*, *MFF*, *MFN2*, *MSTO1*, *OPA1*, *STAT2*, *TRAK1*, and *YME1L1* cause primary mitochondrial diseases mainly by impacting mitochondrial dynamics [107]. Disruption of mitochondrial dynamics has already been proposed in the pathophysiology of PD [109]; therefore, it is reasonable to assume that influencing the aforementioned genes may help treat this aspect by altering their respective gene expression. Prioritizing treatment targets helps substantially to streamline the drug development pipeline. In combination with high-throughput methods, potentially positive treatment effects can be validated in a reasonable time frame [34]. Figure 3 provides an overview of mtDNA- and nDNA-encoded genes causative for primary mitochondrial diseases as potential therapeutic targets in PD.

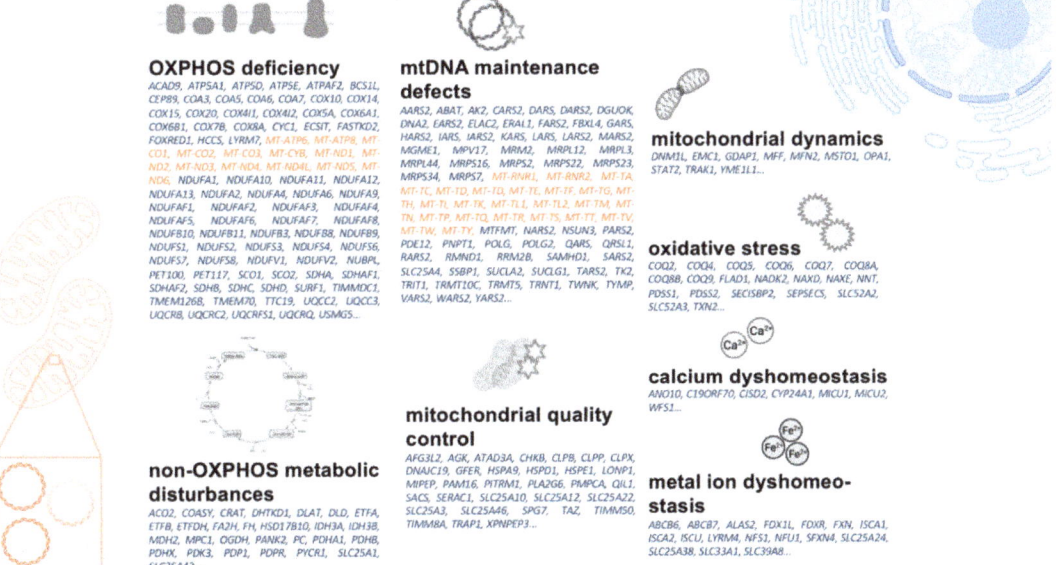

Figure 3. Overview on mtDNA- (orange) and nDNA- (blue) encoded genes causative for primary mitochondrial disorders. The listed genes are ordered based on pathophysiological concepts their respective disorders share with aspects of mitochondrial dysfunction in PD. mtDNA: mitochondrial DNA. nDNA: nuclear DNA. OXPHOS: oxidative phosphorylation. PD: Parkinson's disease.

2.2.3. Allotopic Expression of mtDNA-Encoded Mitochondrial Genes

Mutations within the mtDNA appear to accumulate randomly in PD patients over time. The lack of mtDNA quality control and repair systems leads to the assumption that "allotopic expression" of mtDNA can be an approach in treating mitochondrial dysfunction [110]. The term "allotopic" means that mtDNA-encoded genes are either transiently expressed in the nucleus or permanently inserted in the non-coding regions of the nuclear genome [111]. In general, the allotopic expression strategy was developed to treat primary mitochondrial diseases caused by mtDNA mutations. The design of allotopically expressed mtDNA genes must adhere to different design standards: a mitochondrial targeting sequence is necessary so that the encoded protein is directed to the mitochondria. Additionally, differences in the codons used by the nuclear and mitochondrial genomes and differing codon preferences between the nuclear-cytosolic and mitochondrial translation systems must be considered [112]. mtDNA-encoded genes include 13 ETC complex subunits (for complex I, III, IV, and V), 22 mitochondria-specific tRNAs, and two mitochondrial rRNAs [24]. Based on the random emergence of mtDNA variants in PD, all mtDNA-encoded genes could potentially represent valuable treatment targets. However, it is likely that OXPHOS deficiencies, as the hallmark and common end route of mitochondrial dysfunction, should be prioritized for drug development [113].

The concept of allotopic expression gene therapies is currently being tested in humans. A phase I/II clinical trial aims to treat Leber's hereditary optical neuropathy (LHON) by intravitreal injection and allotopic overexpression of the mtDNA-encoded NADH:ubiquinone oxidoreductase (complex I) (NCT04912843). The NADH:ubiquinone oxidoreductase can also be a viable treatment target in PD, where complex I deficiency has been repeatedly reported [44]. Whether a combined allotopic expression of mtDNA-encoded ETC complexes can provide additional therapeutic benefits needs additional experimental validation. Low vector capacities may additionally hinder the combined allotopic expression of mtDNA genes [79]. There are still a few challenges to accelerate the application of allotopic gene expression. These include necessary improvements in nuclear gene expression, mitochondrial import of cytosolic proteins, posttranslational protein modifications, and functional integration in mitochondrial protein complexes [114–116].

2.2.4. Mitochondrial DNA Genome Editing and Heteroplasmy Shifting

Most pathogenic mtDNA mutations require a critical threshold to cause harm to cells. This aspect has been employed as a potential treatment paradigm named heteroplasmy shifting [117,118]. The underlying idea is to decrease the cumulative amount of mutated mtDNA below a disease-causing heteroplasmy threshold. To achieve this goal, various genome editing methods have been modified to alter mtDNA sequences in a targeted and predictable manner [119]. These methods included mitochondria-targeted restriction endonucleases, zinc finger endonucleases (ZFNs), transcription activator-like effectors nucleases (TALENs), and the CRISPR/Cas9 methodology [118]. Many of these methods have already been employed in the preclinical treatment of primary mitochondrial diseases. For example, the restriction endonuclease SmaI decreased the mutation load in cybrid cell lines with the m.8399T>G mutation causing neuropathy, ataxia, and retinitis pigmentosa (NARP syndrome) [120]. Subsequent functional analyses revealed an increase in ATP levels following genome editing in these cell lines. These findings have also been confirmed in heteroplasmic mouse models of primary mitochondrial diseases following AAV transfection of restriction endonucleases [121].

However, suitable restriction sites are limited in mtDNA, so more flexible approaches have been designed. These limitations can potentially be overcome by introducing programmable nucleases like ZFNs [122] or TALENs [123,124]. These methods, widely known from nuclear genome editing, have been specifically modified to be employed in mtDNA genome editing (mtZFNs and mitoTALENs). Even though these methods have been successfully applied in the preclinical evaluation for primary mitochondrial disorders, the clinical use can be limited by inducing rapid mtDNA depletion in humans [125]. This can mainly be caused by the lack of suitable mtDNA repair mechanisms following the double-strand breaks introduced by these two methods [27]. Interestingly, the rise of CRISPR/Cas9 technology for nuclear genome editing faces significant challenges in mitochondrial genome editing: the import of sgRNA (single guide RNA, the relevant functional component of CRISPR/Cas9 for site-directed specificity) is generally limited by poor RNA import capabilities of mitochondria [119]. Even though mitochondrial genome editing and subsequent heteroplasmy shifting will likely be a viable approach for treating distinct primary mitochondrial disorders (caused by single-site mtDNA point mutations), the applicability in PD is unlikely. Because of the random appearance of multiple mtDNA point mutations, a specific site targeted design will be nearly impossible. However, extensive research is needed to evaluate whether single mtDNA point mutations in PD occur with a higher frequency to define potential treatment targets. Our knowledge in this area is still limited at the moment.

2.3. Special Considerations for Mitochondrial Gene Therapy

Allotopic expression of mtDNA faces significant challenges: each allotopic expressed mtDNA gene needs to be imported to the mitochondria via a complex mitochondria import machinery consisting of various proteins [126]. This is achieved most easily by attaching a mitochondrial targeting sequence to the allotopic expressed mtDNA gene [127]. It effectively leverages the mitochondrial import machinery to bring the respective protein to the correct mitochondrial compartment. Unfortunately, the import of allotopic expressed mtDNA-encoded proteins may not be as simple of a solution as it initially appears. Recent research has shown that the overproduction of nDNA-encoded and allotopically expressed mtDNA proteins can itself cause mitochondrial dysfunction [128]. Producing defective or misfolded mitochondrial protein precursors from the nuclear genome can cause a toxic build-up in the cytosol. This unphysiological high expression level of mitochondrial proteins can be named "mitochondrial protein import stress". The build-up accumulation of misfolded protein in the mitochondria can cause severe disruption of OXPHOS, proteotoxic stress, and mtDNA depletion [129]. Mitochondrial protein import stress can provide a tremendous challenge for gene therapies targeted against mitochondrial dysfunction [130]. For example, the subunits of ETC complexes to ensure efficient OXPHOS is highly regulated and kept in a balanced equilibrium state. By tipping over this fine-tuned balance, disassembled ETC complexes can lead to impaired OXPHOS and bioenergetic depletion [131]. The subsequent increase of ROS by impaired OXPHOS can further damage mitochondrial and overall cellular structures paving the way into a vicious cycle. Based on the additional presence of heteroplasmy, this situation can become highly complex and unpredictable concerning the design of mitochondrial gene therapies.

In summary, we have discussed several challenges for the experimental and clinical evaluation of mitochondrial gene therapies. It is necessary to identify PD patients with clear-cut mitochondrial dysfunction. There are currently no established methods for patient stratification. Most likely, not all patients with PD primarily suffer from disease-relevant mitochondrial dysfunction at any given time. Identifying a window of opportunity for treatment will be one of the significant challenges for successful clinical trial designs. Extensive longitudinal data is needed, in particular in the prodromal stage of PD. However, reliable biomarkers to achieve this goal have not yet been conclusively established [132]. A promising approach can derive from enhanced insights into the individual disease genetics (e.g., by presymptomatic genetic testing or polygenetic risk scoring) [133]. However, blood- or neuroimaging-based assessments of mitochondrial dysfunction can substantially enhance our current understanding of mitochondrial dysfunction's temporal and spatial dynamics in vivo [8]. Current knowledge gaps of essential aspects of mitochondrial biology need to be closed for the rational design of gene therapies. This aspect extends to unclear elements of the mitochondrial import machinery, our incomplete understanding of mitochondrial protein import stress, and unknown tissue-specific mtDNA heteroplasmy thresholds. Treatments targeting the mitochondrial genome require specific genome editing techniques. Drug delivery to the mitochondrial matrix can be substantially hindered by the subcellular compartmentalization and respective physical barriers to be overcome (BBB, cell membranes, OMM, and IMM).

3. Conclusions

Mitochondria-targeted gene therapies may offer potential possibilities in the treatment of PD. The recent progress in gene therapy-based treatment strategies for primary mitochondrial disorders is relevant to understanding the potential use in PD and other neurodegenerative diseases [134]. However, particular challenges need to be overcome and additional research is required to broaden our understanding of mitochondrial biology in PD. Delivery methods must consider the specific properties of the specific mitochondrial proteins, including their location within the mitochondrial organelle, and how they will be targeted there without overwhelming the mitochondrial import machinery. General advancements of gene therapy (e.g., genome editing technologies) will benefit the development of innovative treatment strategies. Prioritization of drug targets and sophisticated design strategies are needed to ensure the subsequent success of gene therapy in clinical trials. These challenges are not insurmountable, but remarkable knowledge gaps need to be closed before PD patients may benefit from such potentially disease-modifying treatments.

Author Contributions: J.P. performed the literature review and drafted the first version of the manuscript. N.B. reviewed and complemented the manuscript. All authors have read and agreed to the published version of the manuscript.

Funding: J.P. received funding from the Parkinson's Foundation, the Deutsche Parkinsongesellschaft, and the Deutsche Forschungsgemeinschaft via the Clinician Scientist School Lübeck (DFG-GEPRIS 413535489). N.B. received funding from the Deutsche Forschungs-Gemeinschaft (BR4328.2-1 [FOR2488], GRK1957).

Institutional Review Board Statement: Not applicable.

Informed Consent Statement: Not applicable.

Data Availability Statement: Not applicable.

Acknowledgments: We acknowledge financial support by Land Schleswig-Holstein within the funding program Open Access Publkationsfonds.

Conflicts of Interest: The authors declare no conflict of interest.

References

1. Bloem, B.R.; Okun, M.S.; Klein, C. Parkinson's disease. *Lancet* **2021**, *397*, 2284–2303. [CrossRef]
2. Vijiaratnam, N.; Simuni, T.; Bandmann, O.; Morris, H.R.; Foltynie, T. Progress towards therapies for disease modification in Parkinson's disease. *Lancet Neurol.* **2021**, *20*, 559–572. [CrossRef]
3. Vazquez-Velez, G.E.; Zoghbi, H.Y. Parkinson's Disease Genetics and Pathophysiology. *Annu. Rev. Neurosci.* **2021**, *44*, 87–108. [CrossRef] [PubMed]
4. Borsche, M.; Pereira, S.L.; Klein, C.; Grunewald, A. Mitochondria and Parkinson's Disease: Clinical, Molecular, and Translational Aspects. *J. Parkinsons Dis.* **2021**, *11*, 45–60. [CrossRef] [PubMed]
5. Trinh, D.; Israwi, A.R.; Arathoon, L.R.; Gleave, J.A.; Nash, J.E. The multi-faceted role of mitochondria in the pathology of Parkinson's disease. *J. Neurochem.* **2021**, *156*, 715–752. [CrossRef]
6. Goncalves, F.B.; Morais, V.A. PINK1: A Bridge between Mitochondria and Parkinson's Disease. *Life* **2021**, *11*, 371. [CrossRef]
7. Camilleri, A.; Vassallo, N. The centrality of mitochondria in the pathogenesis and treatment of Parkinson's disease. *CNS Neurosci. Ther.* **2014**, *20*, 591–602. [CrossRef]
8. Prasuhn, J.; Davis, R.L.; Kumar, K.R. Targeting Mitochondrial Impairment in Parkinson's Disease: Challenges and Opportunities. *Front. Cell Dev. Biol.* **2020**, *8*, 615461. [CrossRef]
9. Zanin, M.; Santos, B.F.R.; Antony, P.M.A.; Berenguer-Escuder, C.; Larsen, S.B.; Hanss, Z.; Barbuti, P.A.; Baumuratov, A.S.; Grossmann, D.; Capelle, C.M.; et al. Mitochondria interaction networks show altered topological patterns in Parkinson's disease. *NPJ Syst. Biol. Appl.* **2020**, *6*, 38. [CrossRef]
10. Illes, A.; Balicza, P.; Gal, A.; Pentelenyi, K.; Csaban, D.; Gezsi, A.; Molnar, V.; Molnar, M.J. Hereditary Parkinson's disease as a new clinical manifestation of the damaged POLG gene. *Orv. Hetil.* **2020**, *161*, 821–828. [CrossRef]
11. Coxhead, J.; Kurzawa-Akanbi, M.; Hussain, R.; Pyle, A.; Chinnery, P.; Hudson, G. Somatic mtDNA variation is an important component of Parkinson's disease. *Neurobiol. Aging* **2016**, *38*, 217.e1–217.e6. [CrossRef] [PubMed]
12. Finsterer, J. Parkinson's syndrome and Parkinson's disease in mitochondrial disorders. *Mov. Disord.* **2011**, *26*, 784–791. [CrossRef] [PubMed]
13. Al Shahrani, M.; Heales, S.; Hargreaves, I.; Orford, M. Oxidative Stress: Mechanistic Insights into Inherited Mitochondrial Disorders and Parkinson's Disease. *J. Clin. Med.* **2017**, *6*, 100. [CrossRef]
14. Mohammad, R. Key considerations in formulation development for gene therapy. *Drug Discov. Today* **2021**, in press. [CrossRef]
15. Merola, A.; Van Laar, A.; Lonser, R.; Bankiewicz, K. Gene therapy for Parkinson's disease: Contemporary practice and emerging concepts. *Expert Rev. Neurother.* **2020**, *20*, 577–590. [CrossRef] [PubMed]

16. Hwu, P.W.; Kiening, K.; Anselm, I.; Compton, D.R.; Nakajima, T.; Opladen, T.; Pearl, P.L.; Roubertie, A.; Roujeau, T.; Muramatsu, S.I. Gene therapy in the putamen for curing AADC deficiency and Parkinson's disease. *EMBO Mol. Med.* **2021**, *13*, e14712. [CrossRef]
17. Behl, T.; Kaur, I.; Kumar, A.; Mehta, V.; Zengin, G.; Arora, S. Gene Therapy in the Management of Parkinson's Disease: Potential of GDNF as a Promising Therapeutic Strategy. *Curr. Gene Ther.* **2020**, *20*, 207–222. [CrossRef]
18. Niethammer, M.; Tang, C.C.; LeWitt, P.A.; Rezai, A.R.; Leehey, M.A.; Ojemann, S.G.; Flaherty, A.W.; Eskandar, E.N.; Kostyk, S.K.; Sarkar, A.; et al. Long-term follow-up of a randomized AAV2-GAD gene therapy trial for Parkinson's disease. *JCI Insight* **2017**, *2*, e90133. [CrossRef]
19. Hitti, F.L.; Yang, A.I.; Gonzalez-Alegre, P.; Baltuch, G.H. Human gene therapy approaches for the treatment of Parkinson's disease: An overview of current and completed clinical trials. *Parkinsonism Relat. Disord.* **2019**, *66*, 16–24. [CrossRef]
20. Murley, A.; Nunnari, J. The Emerging Network of Mitochondria-Organelle Contacts. *Mol. Cell* **2016**, *61*, 648–653. [CrossRef]
21. Cocco, T.; Pacelli, C.; Sgobbo, P.; Villani, G. Control of OXPHOS efficiency by complex I in brain mitochondria. *Neurobiol. Aging* **2009**, *30*, 622–629. [CrossRef]
22. Wang, Z.; Guo, W.; Kuang, X.; Hou, S.; Liu, H. Nanopreparations for mitochondria targeting drug delivery system: Current strategies and future prospective. *Asian J. Pharm. Sci.* **2017**, *12*, 498–508. [CrossRef] [PubMed]
23. Perier, C.; Vila, M. Mitochondrial biology and Parkinson's disease. *Cold Spring Harb. Perspect. Med.* **2012**, *2*, a009332. [CrossRef]
24. Rath, S.; Sharma, R.; Gupta, R.; Ast, T.; Chan, C.; Durham, T.J.; Goodman, R.P.; Grabarek, Z.; Haas, M.E.; Hung, W.H.W.; et al. MitoCarta3.0: An updated mitochondrial proteome now with sub-organelle localization and pathway annotations. *Nucleic Acids Res.* **2021**, *49*, D1541–D1547. [CrossRef] [PubMed]
25. Heuer, B. Mitochondrial DNA: Unraveling the "other" genome. *J. Am. Assoc. Nurse Pract.* **2021**, *33*, 673–675. [CrossRef]
26. Ammal Kaidery, N.; Thomas, B. Current perspective of mitochondrial biology in Parkinson's disease. *Neurochem. Int.* **2018**, *117*, 91–113. [CrossRef]
27. Fontana, G.A.; Gahlon, H.L. Mechanisms of replication and repair in mitochondrial DNA deletion formation. *Nucleic Acids Res.* **2020**, *48*, 11244–11258. [CrossRef]
28. D'Erchia, A.M.; Atlante, A.; Gadaleta, G.; Pavesi, G.; Chiara, M.; De Virgilio, C.; Manzari, C.; Mastropasqua, F.; Prazzoli, G.M.; Picardi, E.; et al. Tissue-specific mtDNA abundance from exome data and its correlation with mitochondrial transcription, mass and respiratory activity. *Mitochondrion* **2015**, *20*, 13–21. [CrossRef]
29. Jenuth, J.P.; Peterson, A.C.; Shoubridge, E.A. Tissue-specific selection for different mtDNA genotypes in heteroplasmic mice. *Nat. Genet.* **1997**, *16*, 93–95. [CrossRef]
30. Phillips, A.F.; Millet, A.R.; Tigano, M.; Dubois, S.M.; Crimmins, H.; Babin, L.; Charpentier, M.; Piganeau, M.; Brunet, E.; Sfeir, A. Single-Molecule Analysis of mtDNA Replication Uncovers the Basis of the Common Deletion. *Mol. Cell* **2017**, *65*, 527–538.e6. [CrossRef]
31. Royrvik, E.C.; Johnston, I.G. MtDNA sequence features associated with 'selfish genomes' predict tissue-specific segregation and reversion. *Nucleic Acids Res.* **2020**, *48*, 8290–8301. [CrossRef] [PubMed]
32. Kirches, E. Do mtDNA Mutations Participate in the Pathogenesis of Sporadic Parkinson's Disease? *Curr. Genom.* **2009**, *10*, 585–593. [CrossRef] [PubMed]
33. Parakatselaki, M.E.; Ladoukakis, E.D. mtDNA Heteroplasmy: Origin, Detection, Significance, and Evolutionary Consequences. *Life* **2021**, *11*, 633. [CrossRef] [PubMed]
34. Nadanaciva, S.; Murray, J.; Wilson, C.; Gebhard, D.F.; Will, Y. High-throughput assays for assessing mitochondrial dysfunction caused by compounds that impair mtDNA-encoded protein levels in eukaryotic cells. *Curr. Protoc. Toxicol.* **2011**, *3*, 11. [CrossRef] [PubMed]
35. Nandakumar, P.; Tian, C.; O'Connell, J.; 23AndMe Research Team; Hinds, D.; Paterson, A.D.; Sondheimer, N. Nuclear genome-wide associations with mitochondrial heteroplasmy. *Sci. Adv.* **2021**, *7*, eabe7520. [CrossRef]
36. Chinnery, P.F.; Taylor, D.J.; Brown, D.T.; Manners, D.; Styles, P.; Lodi, R. Very low levels of the mtDNA A3243G mutation associated with mitochondrial dysfunction In Vivo. *Ann. Neurol.* **2000**, *47*, 381–384. [CrossRef]
37. Arthur, C.R.; Morton, S.L.; Dunham, L.D.; Keeney, P.M.; Bennett, J.P., Jr. Parkinson's disease brain mitochondria have impaired respirasome assembly, age-related increases in distribution of oxidative damage to mtDNA and no differences in heteroplasmic mtDNA mutation abundance. *Mol. Neurodegener.* **2009**, *4*, 37. [CrossRef]
38. Buneeva, O.; Fedchenko, V.; Kopylov, A.; Medvedev, A. Mitochondrial Dysfunction in Parkinson's Disease: Focus on Mitochondrial DNA. *Biomedicines* **2020**, *8*, 591. [CrossRef]
39. Pereira, C.V.; Gitschlag, B.L.; Patel, M.R. Cellular mechanisms of mtDNA heteroplasmy dynamics. *Crit. Rev. Biochem. Mol. Biol.* **2021**, *56*, 510–525. [CrossRef]
40. Ramon, J.; Vila-Julia, F.; Molina-Granada, D.; Molina-Berenguer, M.; Melia, M.J.; Garcia-Arumi, E.; Torres-Torronteras, J.; Camara, Y.; Marti, R. Therapy Prospects for Mitochondrial DNA Maintenance Disorders. *Int. J. Mol. Sci.* **2021**, *22*, 6447. [CrossRef]
41. Kumar, R.; Harila, S.; Parambi, D.G.T.; Kanthlal, S.K.; Rahman, M.A.; Alexiou, A.; Batiha, G.E.; Mathew, B. The Role of Mitochondrial Genes in Neurodegenerative Disorders. *Curr. Neuropharmacol.* **2021**, in press. [CrossRef] [PubMed]
42. Antonyova, V.; Kejik, Z.; Brogyanyi, T.; Kaplanek, R.; Pajkova, M.; Talianova, V.; Hromadka, R.; Masarik, M.; Sykora, D.; Miksatkova, L.; et al. Role of mtDNA disturbances in the pathogenesis of Alzheimer's and Parkinson's disease. *DNA Repair* **2020**, *91–92*, 102871. [CrossRef] [PubMed]

43. Martin-Jimenez, R.; Lurette, O.; Hebert-Chatelain, E. Damage in Mitochondrial DNA Associated with Parkinson's Disease. *DNA Cell Biol.* **2020**, *39*, 1421–1430. [CrossRef]
44. Flones, I.H.; Fernandez-Vizarra, E.; Lykouri, M.; Brakedal, B.; Skeie, G.O.; Miletic, H.; Lilleng, P.K.; Alves, G.; Tysnes, O.B.; Haugarvoll, K.; et al. Neuronal complex I deficiency occurs throughout the Parkinson's disease brain, but is not associated with neurodegeneration or mitochondrial DNA damage. *Acta Neuropathol.* **2018**, *135*, 409–425. [CrossRef]
45. Smigrodzki, R.; Parks, J.; Parker, W.D. High frequency of mitochondrial complex I mutations in Parkinson's disease and aging. *Neurobiol. Aging* **2004**, *25*, 1273–1281. [CrossRef]
46. Saha, T.; Roy, S.; Chakraborty, R.; Biswas, A.; Das, S.K.; Ray, K.; Ray, J.; Sengupta, M. Mitochondrial DNA Haplogroups and Three Independent Polymorphisms have no Association with the Risk of Parkinson's Disease in East Indian Population. *Neurol. India* **2021**, *69*, 461–465. [CrossRef] [PubMed]
47. Wu, H.M.; Li, T.; Wang, Z.F.; Huang, S.S.; Shao, Z.Q.; Wang, K.; Zhong, H.Q.; Chen, S.F.; Zhang, X.; Zhu, J.H. Mitochondrial DNA variants modulate genetic susceptibility to Parkinson's disease in Han Chinese. *Neurobiol. Dis.* **2018**, *114*, 17–23. [CrossRef]
48. Gaweda-Walerych, K.; Maruszak, A.; Safranow, K.; Bialecka, M.; Klodowska-Duda, G.; Czyzewski, K.; Slawek, J.; Rudzinska, M.; Styczynska, M.; Opala, G.; et al. Mitochondrial DNA haplogroups and subhaplogroups are associated with Parkinson's disease risk in a Polish PD cohort. *J. Neural. Transm.* **2008**, *115*, 1521–1526. [CrossRef]
49. Muller-Nedebock, A.C.; Brennan, R.R.; Venter, M.; Pienaar, I.S.; van der Westhuizen, F.H.; Elson, J.L.; Ross, O.A.; Bardien, S. The unresolved role of mitochondrial DNA in Parkinson's disease: An overview of published studies, their limitations, and future prospects. *Neurochem. Int.* **2019**, *129*, 104495. [CrossRef]
50. Oliveira, M.T.; Pontes, C.B.; Ciesielski, G.L. Roles of the mitochondrial replisome in mitochondrial DNA deletion formation. *Genet Mol. Biol.* **2020**, *43*, e20190069. [CrossRef]
51. Muller-Nedebock, A.C.; van der Westhuizen, F.H.; Koks, S.; Bardien, S. Nuclear Genes Associated with Mitochondrial DNA Processes as Contributors to Parkinson's Disease Risk. *Mov. Disord.* **2021**, *36*, 815–831. [CrossRef] [PubMed]
52. Yakubovskaya, E.; Chen, Z.; Carrodeguas, J.A.; Kisker, C.; Bogenhagen, D.F. Functional human mitochondrial DNA polymerase gamma forms a heterotrimer. *J. Biol. Chem.* **2006**, *281*, 374–382. [CrossRef] [PubMed]
53. Lin, M.T.; Cantuti-Castelvetri, I.; Zheng, K.; Jackson, K.E.; Tan, Y.B.; Arzberger, T.; Lees, A.J.; Betensky, R.A.; Beal, M.F.; Simon, D.K. Somatic mitochondrial DNA mutations in early Parkinson and incidental Lewy body disease. *Ann. Neurol.* **2012**, *71*, 850–854. [CrossRef] [PubMed]
54. Bury, A.G.; Pyle, A.; Elson, J.L.; Greaves, L.; Morris, C.M.; Hudson, G.; Pienaar, I.S. Mitochondrial DNA changes in pedunculopontine cholinergic neurons in Parkinson disease. *Ann. Neurol.* **2017**, *82*, 1016–1021. [CrossRef]
55. Pickrell, A.M.; Huang, C.H.; Kennedy, S.R.; Ordureau, A.; Sideris, D.P.; Hoekstra, J.G.; Harper, J.W.; Youle, R.J. Endogenous Parkin Preserves Dopaminergic Substantia Nigral Neurons following Mitochondrial DNA Mutagenic Stress. *Neuron* **2015**, *87*, 371–381. [CrossRef]
56. Podlesniy, P.; Puigros, M.; Serra, N.; Fernandez-Santiago, R.; Ezquerra, M.; Tolosa, E.; Trullas, R. Accumulation of mitochondrial 7S DNA in idiopathic and LRRK2 associated Parkinson's disease. *EBioMedicine* **2019**, *48*, 554–567. [CrossRef]
57. Gonzalez-Hunt, C.P.; Thacker, E.A.; Toste, C.M.; Boularand, S.; Deprets, S.; Dubois, L.; Sanders, L.H. Mitochondrial DNA damage as a potential biomarker of LRRK2 kinase activity in LRRK2 Parkinson's disease. *Sci. Rep.* **2020**, *10*, 17293. [CrossRef]
58. Lujan, S.A.; Longley, M.J.; Humble, M.H.; Lavender, C.A.; Burkholder, A.; Blakely, E.L.; Alston, C.L.; Gorman, G.S.; Turnbull, D.M.; McFarland, R.; et al. Ultrasensitive deletion detection links mitochondrial DNA replication, disease, and aging. *Genome Biol.* **2020**, *21*, 248. [CrossRef]
59. Langley, M.R.; Ghaisas, S.; Palanisamy, B.N.; Ay, M.; Jin, H.; Anantharam, V.; Kanthasamy, A.; Kanthasamy, A.G. Characterization of nonmotor behavioral impairments and their neurochemical mechanisms in the MitoPark mouse model of progressive neurodegeneration in Parkinson's disease. *Exp. Neurol.* **2021**, *341*, 113716. [CrossRef]
60. Beckstead, M.J.; Howell, R.D. Progressive parkinsonism due to mitochondrial impairment: Lessons from the MitoPark mouse model. *Exp. Neurol.* **2021**, *341*, 113707. [CrossRef]
61. Grauer, S.M.; Hodgson, R.; Hyde, L.A. MitoPark mice, an animal model of Parkinson's disease, show enhanced prepulse inhibition of acoustic startle and no loss of gating in response to the adenosine A(2A) antagonist SCH 412348. *Psychopharmacology* **2014**, *231*, 1325–1337. [CrossRef] [PubMed]
62. Song, L.; Shan, Y.; Lloyd, K.C.; Cortopassi, G.A. Mutant Twinkle increases dopaminergic neurodegeneration, mtDNA deletions and modulates Parkin expression. *Hum. Mol. Genet* **2012**, *21*, 5147–5158. [CrossRef] [PubMed]
63. Dolle, C.; Flones, I.; Nido, G.S.; Miletic, H.; Osuagwu, N.; Kristoffersen, S.; Lilleng, P.K.; Larsen, J.P.; Tysnes, O.B.; Haugarvoll, K.; et al. Defective mitochondrial DNA homeostasis in the substantia nigra in Parkinson disease. *Nat. Commun.* **2016**, *7*, 13548. [CrossRef]
64. Perier, C.; Bender, A.; Garcia-Arumi, E.; Melia, M.J.; Bove, J.; Laub, C.; Klopstock, T.; Elstner, M.; Mounsey, R.B.; Teismann, P.; et al. Accumulation of mitochondrial DNA deletions within dopaminergic neurons triggers neuroprotective mechanisms. *Brain* **2013**, *136*, 2369–2378. [CrossRef] [PubMed]
65. Chen, Y.; Jiang, Y.; Yang, Y.; Huang, X.; Sun, C. SIRT1 Protects Dopaminergic Neurons in Parkinson's Disease Models via PGC-1alpha-Mediated Mitochondrial Biogenesis. *Neurotox Res.* **2021**, *39*, 1393–1404. [CrossRef]
66. Reeve, A.K.; Grady, J.P.; Cosgrave, E.M.; Bennison, E.; Chen, C.; Hepplewhite, P.D.; Morris, C.M. Mitochondrial dysfunction within the synapses of substantia nigra neurons in Parkinson's disease. *NPJ Parkinsons Dis.* **2018**, *4*, 9. [CrossRef]

67. Vaillancourt, D.E.; Mitchell, T. Parkinson's disease progression in the substantia nigra: Location, location, location. *Brain* **2020**, *143*, 2628–2630. [CrossRef]
68. Margabandhu, G.; Vanisree, A.J. Dopamine, a key factor of mitochondrial damage and neuronal toxicity on rotenone exposure and also parkinsonic motor dysfunction-Impact of asiaticoside with a probable vesicular involvement. *J. Chem. Neuroanat.* **2020**, *106*, 101788. [CrossRef]
69. Jana, S.; Sinha, M.; Chanda, D.; Roy, T.; Banerjee, K.; Munshi, S.; Patro, B.S.; Chakrabarti, S. Mitochondrial dysfunction mediated by quinone oxidation products of dopamine: Implications in dopamine cytotoxicity and pathogenesis of Parkinson's disease. *Biochem. Biophys. Acta* **2011**, *1812*, 663–673. [CrossRef]
70. Bockova, M.; Vytvarova, E.; Lamos, M.; Klimes, P.; Jurak, P.; Halamek, J.; Goldemundova, S.; Balaz, M.; Rektor, I. Cortical network organization reflects clinical response to subthalamic nucleus deep brain stimulation in Parkinson's disease. *Hum. Brain Mapp.* **2021**, in press. [CrossRef]
71. Pinto, M.; Pickrell, A.M.; Moraes, C.T. Regional susceptibilities to mitochondrial dysfunctions in the CNS. *Biol. Chem.* **2012**, *393*, 275–281. [CrossRef] [PubMed]
72. Miyazaki, I.; Asanuma, M. Neuron-Astrocyte Interactions in Parkinson's Disease. *Cells* **2020**, *9*, 2623. [CrossRef] [PubMed]
73. Andrews, M.R. Gene therapy in the CNS-one size does not fit all. *Gene Ther.* **2021**, *28*, 393–395. [CrossRef] [PubMed]
74. Broadstock, M.; Yanez-Munoz, R.J. Challenges for gene therapy of CNS disorders and implications for Parkinson's disease therapies. *Hum. Gene Ther.* **2012**, *23*, 340–343. [CrossRef] [PubMed]
75. Gombash, S.E.; Foust, K.D. Systemic Gene Therapy for Targeting the CNS. *Methods Mol. Biol.* **2016**, *1382*, 231–237. [CrossRef]
76. Meng, Y.; Hynynen, K.; Lipsman, N. Applications of focused ultrasound in the brain: From thermoablation to drug delivery. *Nat. Rev. Neurol.* **2021**, *17*, 7–22. [CrossRef]
77. Bjorklund, T.; Davidsson, M. Next-Generation Gene Therapy for Parkinson's Disease Using Engineered Viral Vectors. *J. Parkinsons Dis.* **2021**, *11*, S209–S217. [CrossRef]
78. Jayant, R.D.; Sosa, D.; Kaushik, A.; Atluri, V.; Vashist, A.; Tomitaka, A.; Nair, M. Current status of non-viral gene therapy for CNS disorders. *Expert Opin. Drug Deliv.* **2016**, *13*, 1433–1445. [CrossRef]
79. Hocquemiller, M.; Giersch, L.; Audrain, M.; Parker, S.; Cartier, N. Adeno-Associated Virus-Based Gene Therapy for CNS Diseases. *Hum. Gene Ther.* **2016**, *27*, 478–496. [CrossRef]
80. Safari, F.; Hatam, G.; Behbahani, A.B.; Rezaei, V.; Barekati-Mowahed, M.; Petramfar, P.; Khademi, F. CRISPR System: A High-throughput Toolbox for Research and Treatment of Parkinson's Disease. *Cell Mol. Neurobiol.* **2020**, *40*, 477–493. [CrossRef]
81. Slone, J.; Huang, T. The special considerations of gene therapy for mitochondrial diseases. *NPJ Genom. Med.* **2020**, *5*, 7. [CrossRef] [PubMed]
82. Day, J.O.; Mullin, S. The Genetics of Parkinson's Disease and Implications for Clinical Practice. *Genes* **2021**, *12*, 1006. [CrossRef]
83. Greenamyre, J.T.; Betarbet, R.; Sherer, T.B. The rotenone model of Parkinson's disease: Genes, environment and mitochondria. *Parkinsonism Relat. Disord.* **2003**, *9* (Suppl. S2), S59–S64. [CrossRef]
84. Nicoletti, V.; Palermo, G.; Del Prete, E.; Mancuso, M.; Ceravolo, R. Understanding the Multiple Role of Mitochondria in Parkinson's Disease and Related Disorders: Lesson from Genetics and Protein-Interaction Network. *Front. Cell Dev. Biol.* **2021**, *9*, 636506. [CrossRef] [PubMed]
85. Lizama, B.N.; Chu, C.T. Neuronal autophagy and mitophagy in Parkinson's disease. *Mol. Aspects Med.* **2021**, *12*, 100972. [CrossRef] [PubMed]
86. Wang, X.L.; Feng, S.T.; Wang, Y.T.; Yuan, Y.H.; Li, Z.P.; Chen, N.H.; Wang, Z.Z.; Zhang, Y. Mitophagy, a Form of Selective Autophagy, Plays an Essential Role in Mitochondrial Dynamics of Parkinson's Disease. *Cell Mol. Neurobiol.* **2021**, in press. [CrossRef]
87. Buneeva, O.A.; Medvedev, A.E. DJ-1 Protein and Its Role in the Development of Parkinson's Disease: Studies on Experimental Models. *Biochemistry* **2021**, *86*, 627–640. [CrossRef]
88. Repici, M.; Giorgini, F. DJ-1 in Parkinson's Disease: Clinical Insights and Therapeutic Perspectives. *J. Clin. Med.* **2019**, *8*, 1377. [CrossRef]
89. van der Vlag, M.; Havekes, R.; Heckman, P.R.A. The contribution of Parkin, PINK1 and DJ-1 genes to selective neuronal degeneration in Parkinson's disease. *Eur. J. Neurosci.* **2020**, *52*, 3256–3268. [CrossRef]
90. Creed, R.B.; Menalled, L.; Casey, B.; Dave, K.D.; Janssens, H.B.; Veinbergs, I.; van der Hart, M.; Rassoulpour, A.; Goldberg, M.S. Basal and Evoked Neurotransmitter Levels in Parkin, DJ-1, PINK1 and LRRK2 Knockout Rat Striatum. *Neuroscience* **2019**, *409*, 169–179. [CrossRef]
91. Li, W.; Fu, Y.; Halliday, G.M.; Sue, C.M. PARK Genes Link Mitochondrial Dysfunction and Alpha-Synuclein Pathology in Sporadic Parkinson's Disease. *Front. Cell Dev. Biol.* **2021**, *9*, 612476. [CrossRef] [PubMed]
92. Risiglione, P.; Zinghirino, F.; Di Rosa, M.C.; Magri, A.; Messina, A. Alpha-Synuclein and Mitochondrial Dysfunction in Parkinson's Disease: The Emerging Role of VDAC. *Biomolecules* **2021**, *11*, 718. [CrossRef] [PubMed]
93. Mortiboys, H.; Furmston, R.; Bronstad, G.; Aasly, J.; Elliott, C.; Bandmann, O. UDCA exerts beneficial effect on mitochondrial dysfunction in LRRK2(G2019S) carriers and in vivo. *Neurology* **2015**, *85*, 846–852. [CrossRef] [PubMed]
94. Yang, S.; Xia, C.; Li, S.; Du, L.; Zhang, L.; Hu, Y. Mitochondrial dysfunction driven by the LRRK2-mediated pathway is associated with loss of Purkinje cells and motor coordination deficits in diabetic rat model. *Cell Death Dis.* **2014**, *5*, e1217. [CrossRef] [PubMed]

95. Si, H.; Ma, P.; Liang, Q.; Yin, Y.; Wang, P.; Zhang, Q.; Wang, S.; Deng, H. Overexpression of pink1 or parkin in indirect flight muscles promotes mitochondrial proteostasis and extends lifespan in Drosophila melanogaster. *PLoS ONE* **2019**, *14*, e0225214. [CrossRef]
96. Bonilla-Porras, A.R.; Arevalo-Arbelaez, A.; Alzate-Restrepo, J.F.; Velez-Pardo, C.; Jimenez-Del-Rio, M. PARKIN overexpression in human mesenchymal stromal cells from Wharton's jelly suppresses 6-hydroxydopamine-induced apoptosis: Potential therapeutic strategy in Parkinson's disease. *Cytotherapy* **2018**, *20*, 45–61. [CrossRef]
97. Rana, A.; Rera, M.; Walker, D.W. Parkin overexpression during aging reduces proteotoxicity, alters mitochondrial dynamics, and extends lifespan. *Proc. Natl. Acad. Sci. USA* **2013**, *110*, 8638–8643. [CrossRef]
98. Liu, B.; Traini, R.; Killinger, B.; Schneider, B.; Moszczynska, A. Overexpression of parkin in the rat nigrostriatal dopamine system protects against methamphetamine neurotoxicity. *Exp. Neurol.* **2013**, *247*, 359–372. [CrossRef]
99. Aguiar, A.S., Jr.; Tristao, F.S.; Amar, M.; Chevarin, C.; Lanfumey, L.; Mongeau, R.; Corti, O.; Prediger, R.D.; Raisman-Vozari, R. Parkin-knockout mice did not display increased vulnerability to intranasal administration of 1-methyl-4-phenyl-1,2,3,6-tetrahydropyridine (MPTP). *Neurotox Res.* **2013**, *24*, 280–287. [CrossRef]
100. Yasuda, T.; Hayakawa, H.; Nihira, T.; Ren, Y.R.; Nakata, Y.; Nagai, M.; Hattori, N.; Miyake, K.; Takada, M.; Shimada, T.; et al. Parkin-mediated protection of dopaminergic neurons in a chronic MPTP-minipump mouse model of Parkinson disease. *J. Neuropathol. Exp. Neurol.* **2011**, *70*, 686–697. [CrossRef]
101. Julienne, H.; Buhl, E.; Leslie, D.S.; Hodge, J.J.L. Drosophila PINK1 and parkin loss-of-function mutants display a range of non-motor Parkinson's disease phenotypes. *Neurobiol. Dis.* **2017**, *104*, 15–23. [CrossRef] [PubMed]
102. Kane, L.A.; Youle, R.J. PINK1 and Parkin flag Miro to direct mitochondrial traffic. *Cell* **2011**, *147*, 721–723. [CrossRef]
103. Wang, X.; Winter, D.; Ashrafi, G.; Schlehe, J.; Wong, Y.L.; Selkoe, D.; Rice, S.; Steen, J.; LaVoie, M.J.; Schwarz, T.L. PINK1 and Parkin target Miro for phosphorylation and degradation to arrest mitochondrial motility. *Cell* **2011**, *147*, 893–906. [CrossRef] [PubMed]
104. Todd, A.M.; Staveley, B.E. Expression of Pink1 with alpha-synuclein in the dopaminergic neurons of Drosophila leads to increases in both lifespan and healthspan. *Genet. Mol. Res.* **2012**, *11*, 1497–1502. [CrossRef]
105. Todd, A.M.; Staveley, B.E. Pink1 suppresses alpha-synuclein-induced phenotypes in a Drosophila model of Parkinson's disease. *Genome* **2008**, *51*, 1040–1046. [CrossRef]
106. Haque, M.E.; Mount, M.P.; Safarpour, F.; Abdel-Messih, E.; Callaghan, S.; Mazerolle, C.; Kitada, T.; Slack, R.S.; Wallace, V.; Shen, J.; et al. Inactivation of Pink1 gene in vivo sensitizes dopamine-producing neurons to 1-methyl-4-phenyl-1,2,3,6-tetrahydropyridine (MPTP) and can be rescued by autosomal recessive Parkinson disease genes, Parkin or DJ-1. *J. Biol. Chem.* **2012**, *287*, 23162–23170. [CrossRef]
107. Wallace, D.C. Mitochondrial genetic medicine. *Nat. Genet.* **2018**, *50*, 1642–1649. [CrossRef]
108. Keeney, P.M.; Quigley, C.K.; Dunham, L.D.; Papageorge, C.M.; Iyer, S.; Thomas, R.R.; Schwarz, K.M.; Trimmer, P.A.; Khan, S.M.; Portell, F.R.; et al. Mitochondrial gene therapy augments mitochondrial physiology in a Parkinson's disease cell model. *Hum. Gene Ther.* **2009**, *20*, 897–907. [CrossRef]
109. Santos, D.; Cardoso, S.M. Mitochondrial dynamics and neuronal fate in Parkinson's disease. *Mitochondrion* **2012**, *12*, 428–437. [CrossRef] [PubMed]
110. Zullo, S.J. Gene therapy of mitochondrial DNA mutations: A brief, biased history of allotopic expression in mammalian cells. *Semin. Neurol.* **2001**, *21*, 327–335. [CrossRef]
111. Liufu, T.; Wang, Z. Treatment for mitochondrial diseases. *Rev. Neurosci.* **2021**, *32*, 35–47. [CrossRef] [PubMed]
112. Lewis, C.J.; Dixit, B.; Batiuk, E.; Hall, C.J.; O'Connor, M.S.; Boominathan, A. Codon optimization is an essential parameter for the efficient allotopic expression of mtDNA genes. *Redox Biol.* **2020**, *30*, 101429. [CrossRef] [PubMed]
113. Chen, C.; McDonald, D.; Blain, A.; Sachdeva, A.; Bone, L.; Smith, A.L.M.; Warren, C.; Pickett, S.J.; Hudson, G.; Filby, A.; et al. Imaging mass cytometry reveals generalised deficiency in OXPHOS complexes in Parkinson's disease. *NPJ Parkinsons Dis.* **2021**, *7*, 39. [CrossRef] [PubMed]
114. Oca-Cossio, J.; Kenyon, L.; Hao, H.; Moraes, C.T. Limitations of allotopic expression of mitochondrial genes in mammalian cells. *Genetics* **2003**, *165*, 707–720. [CrossRef] [PubMed]
115. Figueroa-Martinez, F.; Vazquez-Acevedo, M.; Cortes-Hernandez, P.; Garcia-Trejo, J.J.; Davidson, E.; King, M.P.; Gonzalez-Halphen, D. What limits the allotopic expression of nucleus-encoded mitochondrial genes? The case of the chimeric Cox3 and Atp6 genes. *Mitochondrion* **2011**, *11*, 147–154. [CrossRef]
116. Mukhopadhyay, A.; Zullo, S.J.; Weiner, H. Factors that might affect the allotopic replacement of a damaged mitochondrial DNA-encoded protein. *Rejuvenation Res.* **2006**, *9*, 182–190. [CrossRef]
117. Naeem, M.M.; Sondheimer, N. Heteroplasmy Shifting as Therapy for Mitochondrial Disorders. *Adv. Exp. Med. Biol.* **2019**, *1158*, 257–267. [CrossRef]
118. Jackson, C.B.; Turnbull, D.M.; Minczuk, M.; Gammage, P.A. Therapeutic Manipulation of mtDNA Heteroplasmy: A Shifting Perspective. *Trends Mol. Med.* **2020**, *26*, 698–709. [CrossRef]
119. Gammage, P.A.; Moraes, C.T.; Minczuk, M. Mitochondrial Genome Engineering: The Revolution May Not Be CRISPR-Ized. *Trends Genet.* **2018**, *34*, 101–110. [CrossRef]

120. Tanaka, M.; Borgeld, H.J.; Zhang, J.; Muramatsu, S.; Gong, J.S.; Yoneda, M.; Maruyama, W.; Naoi, M.; Ibi, T.; Sahashi, K.; et al. Gene therapy for mitochondrial disease by delivering restriction endonuclease SmaI into mitochondria. *J. Biomed. Sci.* **2002**, *9*, 534–541. [CrossRef]
121. Bacman, S.R.; Williams, S.L.; Duan, D.; Moraes, C.T. Manipulation of mtDNA heteroplasmy in all striated muscles of newborn mice by AAV9-mediated delivery of a mitochondria-targeted restriction endonuclease. *Gene Ther.* **2012**, *19*, 1101–1106. [CrossRef] [PubMed]
122. Minczuk, M.; Papworth, M.A.; Miller, J.C.; Murphy, M.P.; Klug, A. Development of a single-chain, quasi-dimeric zinc-finger nuclease for the selective degradation of mutated human mitochondrial DNA. *Nucleic Acids Res.* **2008**, *36*, 3926–3938. [CrossRef] [PubMed]
123. Bacman, S.R.; Kauppila, J.H.K.; Pereira, C.V.; Nissanka, N.; Miranda, M.; Pinto, M.; Williams, S.L.; Larsson, N.G.; Stewart, J.B.; Moraes, C.T. MitoTALEN reduces mutant mtDNA load and restores tRNA(Ala) levels in a mouse model of heteroplasmic mtDNA mutation. *Nat. Med.* **2018**, *24*, 1696–1700. [CrossRef] [PubMed]
124. Hashimoto, M.; Bacman, S.R.; Peralta, S.; Falk, M.J.; Chomyn, A.; Chan, D.C.; Williams, S.L.; Moraes, C.T. MitoTALEN: A General Approach to Reduce Mutant mtDNA Loads and Restore Oxidative Phosphorylation Function in Mitochondrial Diseases. *Mol. Ther.* **2015**, *23*, 1592–1599. [CrossRef] [PubMed]
125. Falkenberg, M.; Hirano, M. Editing the Mitochondrial Genome. *N. Engl. J. Med.* **2020**, *383*, 1489–1491. [CrossRef]
126. Wiedemann, N.; Pfanner, N. Mitochondrial Machineries for Protein Import and Assembly. *Annu. Rev. Biochem.* **2017**, *86*, 685–714. [CrossRef] [PubMed]
127. Omura, T. Mitochondria-targeting sequence, a multi-role sorting sequence recognized at all steps of protein import into mitochondria. *J. Biochem.* **1998**, *123*, 1010–1016. [CrossRef] [PubMed]
128. Priesnitz, C.; Becker, T. Pathways to balance mitochondrial translation and protein import. *Genes Dev.* **2018**, *32*, 1285–1296. [CrossRef] [PubMed]
129. Ruan, L.; Zhou, C.; Jin, E.; Kucharavy, A.; Zhang, Y.; Wen, Z.; Florens, L.; Li, R. Cytosolic proteostasis through importing of misfolded proteins into mitochondria. *Nature* **2017**, *543*, 443–446. [CrossRef]
130. Mohanraj, K.; Nowicka, U.; Chacinska, A. Mitochondrial control of cellular protein homeostasis. *Biochem. J.* **2020**, *477*, 3033–3054. [CrossRef]
131. Perales-Clemente, E.; Fernandez-Silva, P.; Acin-Perez, R.; Perez-Martos, A.; Enriquez, J.A. Allotopic expression of mitochondrial-encoded genes in mammals: Achieved goal, undemonstrated mechanism or impossible task? *Nucleic Acids Res.* **2011**, *39*, 225–234. [CrossRef]
132. Postuma, R.B.; Berg, D.; Stern, M.; Poewe, W.; Olanow, C.W.; Oertel, W.; Obeso, J.; Marek, K.; Litvan, I.; Lang, A.E.; et al. MDS clinical diagnostic criteria for Parkinson's disease. *Mov. Disord.* **2015**, *30*, 1591–1601. [CrossRef] [PubMed]
133. Prasuhn, J.; Kasten, M.; Vos, M.; Konig, I.R.; Schmid, S.M.; Wilms, B.; Klein, C.; Bruggemann, N. The Use of Vitamin K2 in Patients with Parkinson's Disease and Mitochondrial Dysfunction (PD-K2): A Theranostic Pilot Study in a Placebo-Controlled Parallel Group Design. *Front. Neurol.* **2020**, *11*, 592104. [CrossRef] [PubMed]
134. Coutinho, E.; Batista, C.; Sousa, F.; Queiroz, J.; Costa, D. Mitochondrial Gene Therapy: Advances in Mitochondrial Gene Cloning, Plasmid Production, and Nanosystems Targeted to Mitochondria. *Mol. Pharm.* **2017**, *14*, 626–638. [CrossRef] [PubMed]

Article

Microarray Genotyping Identifies New Loci Associated with Dementia in Parkinson's Disease

Sungyang Jo [1], Kye Won Park [2], Yun Su Hwang [1], Seung Hyun Lee [1], Ho-Sung Ryu [3] and Sun Ju Chung [1,*]

1. Department of Neurology, Asan Medical Center, University of Ulsan College of Medicine, Seoul 05505, Korea; sungyangjo@gmail.com (S.J.); ghkddbstn1@naver.com (Y.S.H.); doors327@naver.com (S.H.L.)
2. Department of Neurology, Uijeongbu Eulji Medical Center, Eulji University School of Medicine, Uijeongbu-si 11759, Gyeonggi-do, Korea; karabach88@gmail.com
3. Department of Neurology, Kyungpook National University Hospital, Daegu 41944, Korea; ryuhosung138@gmail.com
* Correspondence: sjchung@amc.seoul.kr; Tel.: +82-2-3010-3988

Abstract: Dementia is one of the most disabling nonmotor symptoms of Parkinson's disease (PD). However, the risk factors contributing to its development remain unclear. To investigate genetic variants associated with dementia in PD, we performed microarray genotyping based on a customized platform utilizing variants identified in previous genetic studies. Microarray genotyping was performed in 313 PD patients with dementia, 321 PD patients without dementia, and 635 healthy controls. The primary analysis was performed using a multiple logistic regression model adjusted for age and sex. *SNCA* single nucleotide polymorphism (SNP) rs11931074 was determined to be most significantly associated with PD (odds ratio = 0.66, 95% confidence interval = 0.56–0.78, $p = 7.75 \times 10^{-7}$). In the analysis performed for patients with PD only, *MUL1* SNP rs3738128 (odds ratio = 2.52, 95% confidence interval = 1.68–3.79, $p = 8.75 \times 10^{-6}$) was found to be most significantly associated with dementia in PD. SNPs in *ZHX2* and *ERP29* were also associated with dementia in PD. This microarray genomic study identified new loci of *MUL1* associated with dementia in PD, suggesting an essential role of mitochondrial dysfunction in the development of dementia in patients with PD.

Keywords: genome-wide association study; Parkinson's disease; dementia; cognition

1. Introduction

Parkinson's disease (PD) is the second most prevalent neurodegenerative disease globally, affecting more than six million people worldwide [1]. The diagnosis of PD is based on specific motor symptoms, including bradykinesia, rigidity, tremor, or gait disturbance [2]. However, patients with PD suffer from various nonmotor symptoms, such as fatigue, pain, sleep disturbance, dementia, depression, anxiety, and autonomic dysfunction [3]. Dementia is one of the disabling nonmotor symptoms that substantially impairs the quality of life of patients with PD, increasing caregiver burden and economic costs [4]. The prevalence of dementia is high, with up to 75% of patients with PD developing dementia within 10 years from diagnosis [5,6]. However, the determining factors involved in the development of dementia in patients with PD are still unclear.

Genome-wide association studies (GWAS) have widened our understanding of the genetics of PD and have identified more than 90 genetic loci that are associated with the development of PD [7–13]. However, a majority of the previously conducted GWAS have focused on the susceptibility of PD, and GWAS specifically investigating motor or non-motor presentations—including dementia—of PD have been limited. In a recent GWAS, we reported that *RYR2* and other genetic loci are associated with cognitive impairment in PD; however, the assessment of cognitive function was based only on Mini-Mental Status Examination (MMSE) and the Montreal Cognitive Assessment (MoCA) scores [14]. Many recent GWAS have reported genomic variants, including *GBA*, *APOE*, *SNCA*, and *CNTN1*

that are associated with dementia with Lewy bodies (DLB) [15–18]. Although Parkinson's disease dementia (PDD) and DLB share clinical, neurochemical, and morphological features, no consensus has been established yet with respect to the consideration of the two extremes on the one continuous spectrum of Lewy body disease [19]. Interestingly, in a large multinational cohort of patients with PD, PDD, and DLB, parkinsonism and dementia showed two distinct association profiles with the 3' or 5' regions of the *SNCA* gene, suggesting that PD, PDD, and DLB have distinct genetic etiologies. Therefore, further studies undertaking genome-wide investigations are necessary to identify distinct genetic variants associated with the development of dementia in patients with PD, independent of DLB.

In this study, we employed a novel customized microarray platform to comprehensively investigate the genetic variants associated with dementia in patients with PD.

2. Materials and Methods

2.1. Study Population

We prospectively enrolled patients with PDD, patients with PD without dementia (PD-ND), and healthy controls at Asan Medical Center, Seoul, Korea. All participants were ethnic Koreans. The diagnosis of PD was based on the UK Brain Bank criteria [2] and the diagnosis of PDD was based on the criteria proposed by the Movement Disorder Society Task Force [20]. Healthy controls were recruited from among the spouses of the patients, and the inclusion criterion included the absence of neurological diseases including PD or dementia. Blood samples were collected from all participants for genetic tests, and patient information including that related to age, sex, and educational qualification (number of years of education) was collected at the time of sampling. Mini-Mental Status Examination (MMSE) was performed for the screening of cognitive function. For patients with PD, age at disease onset, age at diagnosis of dementia if applicable, age at latest follow-up, and the latest MMSE scores were obtained.

2.2. Development of Microarray Genotyping Platform

We designed a microarray genotyping platform that contained genetic variants with biological plausibility for PDD, suggested by our previous GWAS or other previous genetic studies. The platform included: (1) Genetic variants that showed a high level of association (p-value < 10^{-4}) with PD in our previous GWAS performed using ethnicity-specific Korean Chip (K-CHIP). K-CHIP was designed by the Center for Genome Science, Korea National Institute of Health (4845–301, 3000–3031) (www.cdc.go.kr) [14,21]. K-CHIP consists of an imputation GWAS grid (505,000 Asian-based grid with minor allele frequency (MAF) > 5% in Asians); exome content (84,000 Korean-based grid with MAF > 5% in Koreans, 149,000 coding single-nucleotide polymorphisms, and insertions and deletions determined based on data derived from 2000 whole-exome sequences and 400 whole-genome sequences with MAF > 0.1%); new exome/loss of function contents (44,000 variants); expression quantitative trait loci (17,000 variants); genes associated with absorption, distribution, metabolism, and excretion; and other miscellaneous variants. (2) Genetic variants that showed significant association with PD in previous GWAS [7–13]. (3) Genetic mutations that were reported to be a cause of monogenic familial PD with Mendelian inheritance (https://www.omim.org/). (4) Genetic variants that showed significant association with DLB in previous GWAS [15,16,22]. (5) Genetic variants that showed significant association with Alzheimer's disease in previous GWAS [23–26]. (6) Genetic variants associated with neuroinflammation in previous GWAS [11,27,28].

Annotation of the variants was performed using the nspEff tool to confirm the distribution of the gene effect [29]. From a total of 219,065 variants, we excluded 109,804 "novel—not recommended and neutral" markers for the score data, because the performance or efficacy of genotyping might be low (Table S1). The final selection was performed by excluding duplicate markers, markers not included in the 1000 genome project phase 3 data, markers with a minor allele frequency of zero in East Asian GWAS data, and proxy single nucleotide polymorphisms (SNPs) (tagging $r^2 > 0.8$) (Table S2). The final candidate markers consisted of 74,224 markers (Table S3).

2.3. Sample Quality Control

Samples with a low call rate and high heterozygosity were excluded. Samples that deviated from the whole sample were excluded from the analysis by an assessment performed using multidimensional scaling. We also excluded excessive singleton, samples with gender discrepancies, and cryptic first-degree relatives using the PLINK program (version 1.90, NIH–NIDDK Laboratory of Biological Modeling, Bethesda, MD, USA).

2.4. SNP Quality Control

We performed an SNPolisher analysis to exclude low-quality SNPs. SNPs with call rates over 95% in both cases and controls were included. SNPs with p-value $> 10^{-4}$ in a Hardy–Weinberg equilibrium test were excluded. We excluded SNPs with minor allele frequency < 1% in both cases and controls. We performed cluster quality control for every SNP with $p < 0.001$ using linkage disequilibrium within 150 kilobases through visual inspection.

2.5. Statistical Analysis

We compared the demographics and clinical characteristics of patients with PDD, those with PD-ND, and healthy controls using Kruskal-Wallis tests for continuous variables, which did not meet the assumption of the homogeneity of variance, as well as with chi-squared test for categorical variables. Post hoc analysis was performed using Dunnett's post hoc tests and Bonferroni correction.

The association between the genetic variants and PD or PDD was analyzed using a multiple logistic regression model after adjusting for age, sex, and education years. For each genetic variant, we calculated the odds ratios (OR), 95% confidence interval (CI), and two-tailed p-value. Bonferroni correction was applied to adjust for multiple comparisons. Manhattan plots and quantile–quantile plots (Q-Q plots) were constructed for p-values of all genotyped variants that passed quality control.

Statistical analysis was performed using R (version 3.1.2, Free Software Foundation, Inc., Boston, MA, USA), the PLINK program (version 1.90, NIH–NIDDK Laboratory of Biological Modeling, Bethesda, MD, USA), Haploview (version 4.2, Daly Lab at the Broad Institute, Cambridge, MA, USA), and LocusZoom (version 1.4, University of Michigan, Department of Biostatistics, Center for Statistical Genetics, Ann Arbor, MI, USA).

3. Results

3.1. Clinical Characteristics

We enrolled 318 patients with PDD, 326 patients with PD-ND, and 648 healthy controls. After quality control assessment, 5 patients with PDD, 5 patients with PD-ND, and 13 healthy controls were excluded. The final study population included 313 patients with PDD, 321 patients with PD-ND, and 635 healthy controls. The ages noted at the latest follow-up for patients with PD or those noted at study enrollment for healthy controls were significantly different among the three groups (median 76.0 vs. 75.0 vs. 68.0, $p < 0.001$) (Table 1). In the post hoc analysis, ages noted at the latest follow-up were significantly lower among healthy controls than those among patients with PDD or PD-ND (all $p < 0.001$). The ages at disease onset and the disease durations were not significantly different between patients with PDD and PD-ND. The median disease duration was 12.0 years for both PDD and PD-ND groups. The percentage of females was significantly higher in the PDD group compared to that in the healthy controls (57.2% vs. 44.9%, $p = 0.0007$ in post hoc analysis). The number of education years (total years of academic education) was significantly lower in the PDD group than that in the PD-ND or healthy control group (both $p < 0.001$ in post hoc analysis).

Table 1. Baseline clinical characteristics of the study subjects.

Characteristics	PD Dementia (N = 313)	PD without Dementia (N = 321)	Controls (N = 635)	p-Value
Age at onset, years	64.0 (57.0–68.0)	63.0 (57.0–68.0)	-	0.449
Age at latest follow-up, years	76.0 (72.0–81.0)	75.0 (72.0–80.0)	68.0 (64.0–72.0) [a,b]	<0.001
Disease duration, years	12.0 (9.0–17.0)	12.0 (9.0–16.0)	-	0.896
Female, N (%)	179 (57.2%)	166 (51.7%)	285 (44.9%) [a]	0.001
Education, years	6.0 (2.0–12.0)	12.0 (6.0–16.0) [c]	12.0 (9.0–16.0) [a]	<0.001
Latest MMSE	17.0 (13.0–20.0)	27.0 (26.0–29.0) [c]	28.0 (26.0–29.0) [a]	<0.001
Age at dementia, years	73.0 (69.0–78.0)	-	-	

PD, Parkinson's disease; MMSE, Mini-Mental Status Examination. [a] Significant difference compared with PD dementia using Dunn's post hoc test. [b] Significant difference compared with PD without dementia using Dunn's post hoc test. [c] Significant difference compared with healthy controls using Dunn's post hoc test.

3.2. Genetic Association with Susceptibility to PD

The 41,534 genetic variants that passed quality control were genotyped and analyzed. Multiple logistic regression with additive coding schemes was performed to compare genetic variants between patients with PD (both, patients with PDD and those with PD-ND) and healthy controls after adjusting for age and sex. Q-Q plots were generated for the diagnosis of patients with PD in comparison with healthy controls (Figure S1). The Manhattan plot is depicted in Figure 1. Among the top 10 genetic variants associated with PD, five SNPs were observed in the loci of *SNCA* (rs11931074, rs12642514, rs75876872, rs80184884, and rs75231811) (Table 2), and two *SNCA* SNPs (rs11931074 and rs12642514) showed statistical significance after Bonferroni correction (Figure 1). Among the *SNCA* SNPs, SNP rs11931074 was most significantly associated with PD (OR = 0.66, 95% CI = 0.56–0.78, $p = 7.75 \times 10^{-7}$). *SPHK1* SNP rs2247856 (OR = 0.65, 95% CI = 0.53–0.80, $p = 4.35 \times 10^{-5}$) and *FYN* SNP rs7772036 (OR = 0.72, 95% CI = 0.61–0.85, $p = 9.74 \times 10^{-5}$) were also associated with PD.

3.3. Genetic Association with Dementia in PD

We compared genetic variants between PDD and PD-ND using multiple logistic regression with additive coding schemes after adjusting for age, sex, and education years. Q–Q plots were generated for the diagnosis of PDD compared with PD-ND (Figure S2). The respective Manhattan plot is depicted in Figure 2. Among the top 10 SNPs associated with PDD, two SNPs were observed in the loci of *MUL1* (rs3738128 and rs12566937) (Table 3). *MUL1* SNP rs3738128 (OR = 2.52, 95% CI = 1.68–3.79, $p = 8.75 \times 10^{6}$) was most significantly associated with dementia in PD. In linkage analysis, *MUL1* SNP rs12566937 showed moderate linkage disequilibrium with *MUL1* SNP rs3738128, which was associated with the lowest p-value (Figure 3). SNPs in *ZHX2* (OR = 0.56 95% CI = 0.43–0.74, $p = 3.65 \times 10^{-5}$) and *ERP29* (OR = 3.05, 95% CI = 1.77–5.27, $p = 6.41 \times 10^{-5}$) were also associated with dementia in PD. However, following Bonferroni correction, none of the SNPs showed statistical significance.

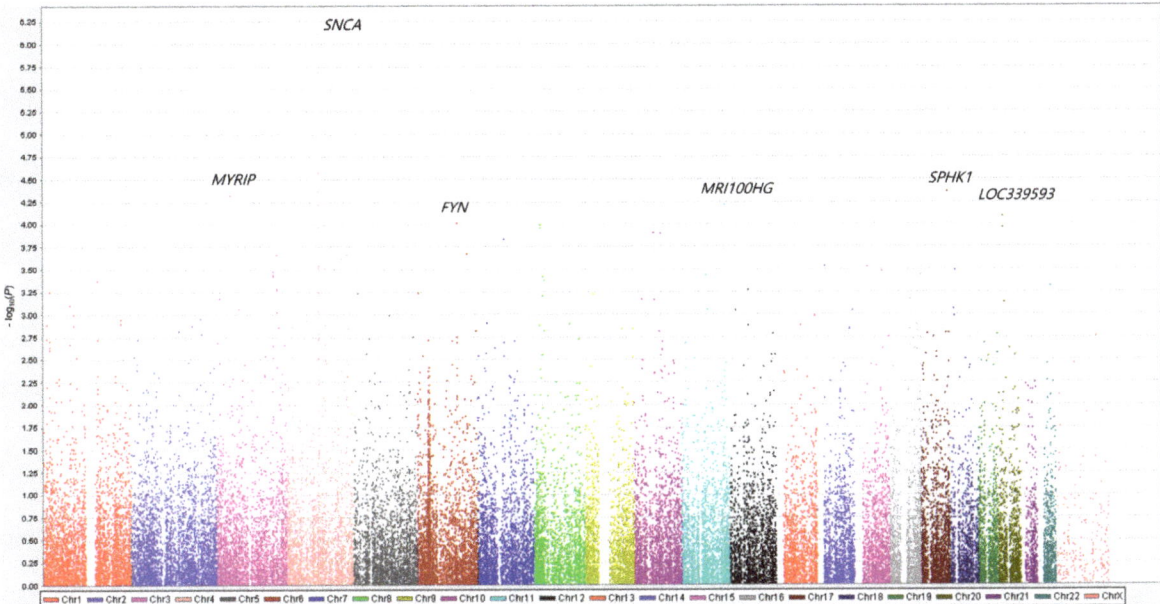

Figure 1. Manhattan plots for Parkinson's disease (PD). The genes nearest to the top 10 significant variants are labeled. The *x*-axis represents the base pair position of the variants from chromosome 1 to chromosome 22. The *SNCA* loci showed a statistically significant association with PD after Bonferroni correction. *SNCA* SNP rs11931074 was most significantly associated with PD (OR = 0.66, 95% CI = 0.56–0.78, $p = 7.75 \times 10^{-7}$). The *SPHK1* and *FYN* loci were also associated with PD.

Table 2. Top 10 genetic variants associated with Parkinson's disease in the order of statistical significance.

Gene	SNP	Chr	Position	Region Relative to the Gene	Allele (Minor/Major)	Minor Allele Frequency (Case/Control)	OR (95% CI)	*p*-Value
SNCA, GPRIN3	rs11931074	4	89718364	intron, downstream, upstream	G/C	0.37/0.46	0.66 (0.56, 0.78)	7.75×10^{-7}
SNCA, GPRIN3	rs12642514	4	89708246	intron, downstream, upstream	A/C	0.36/0.46	0.66 (0.58, 0.79)	2.08×10^{-6}
SNCA	rs356191	4	89766969	Intron	A/G	0.06/0.10	0.52 (0.38, 0.70)	2.64×10^{-5}
SNCA, GPRIN3	rs80184884	4	89705068	intron, downstream, upstream	G/A	0.06/0.10	0.52 (0.38 0.71)	4.24×10^{-5}
SPHK1	rs2247856	17	76385474	missense, UTR-5, exon	A/G	0.16/0.22	0.65 (0.53, 0.80)	4.35×10^{-5}
MYRIP	rs6599077	3	40055127	Intron	A/G	0.43/0.35	1.42(1.20, 1.68)	4.81×10^{-5}
MRI100HG	rs577924	11	122264399	Intron	C/T	0.43/0.35	1.41 (1.19, 1.67)	6.05×10^{-5}
SNCA, GPRIN3	rs75876872	4	89705795	intron, downstream, upstream	G/A	0.05/0.08	0.49 (0.35, 0.69)	6.07×10^{-5}
LOC339593	rs1473702	20	11253884	intron, downstream	C/T	0.51/0.44	1.38 (1.18, 1.62)	8.05×10^{-5}
FYN	rs7772036	6	111739596	Intron	G/A	0.32/0.39	0.72 (0.61, 0.85)	9.74×10^{-5}

Chr, chromosome; OR, odds ratio; CI, confidence interval.

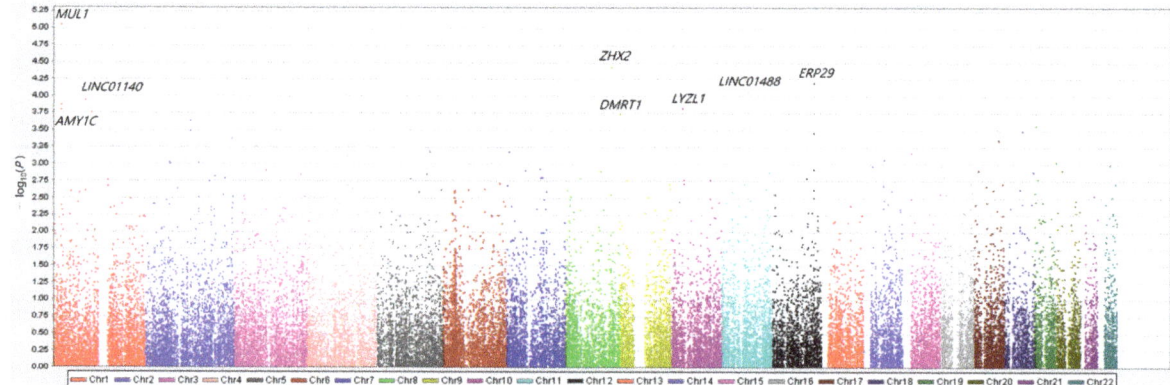

Figure 2. Manhattan plots for dementia in Parkinson's disease (PD). The genes nearest to the top 10 significant variants are labeled. The *x*-axis represents the base pair position of the variants from chromosome 1 to chromosome 22. The *MUL1* loci was most significantly associated with dementia in PD. *MUL1* SNP rs3738128 (OR = 2.52, 95% CI = 1.68–3.79, $p = 8.75 \times 10^{-6}$) was most significantly associated with dementia in PD. The *ZHX2* and *ERP29* loci were also associated with dementia in PD.

Table 3. Top 10 genetic variants associated with dementia in Parkinson's disease in the order of statistical significance.

Gene	SNP	Chr	Position	Region Relative to the Gene	Allele (Minor/Major)	Minor Allele Frequency (Case/Control)	OR (95% CI)	*p*-Value
MUL1	rs3738128	1	20499992	UTR-3	G/C	0.07/0.11	2.52 (1.68, 3.79)	8.75×10^{-6}
ZHX2	rs11779459	8	122968311	Intron	T/C	0.34/0.29	0.56 (0.43, 0.74)	3.65×10^{-5}
ERP29, NAA25	rs4767293	12	112025492	downstream	A/G	0.04/0.06	3.05 (1.77, 5.27)	6.41×10^{-5}
LINC01488	rs7395791	11	69448148	upstream, downstream	A/G	0.56/0.50	0.61 (0.47, 0.78)	8.44×10^{-5}
LINC01140	rs7553864	1	87147675	Intron	T/C	0.14/0.19	1.88 (1.37, 2.6)	1.15×10^{-4}
MUL1	rs12566937	1	20506181	Intron	G/T	0.13/0.17	1.91 (1.37, 2.67)	1.33×10^{-4}
LYZL1, C10orf126	rs1889714	10	29099710	upstream, downstream	A/G	0.12/0.09	0.43 (0.28, 0.66)	1.47×10^{-4}
AMY1C, LOC101928476, LOC100129138	rs12026039	1	104028469	downstream, upstream	G/A	0.51/0.47	0.61 (0.47, 0.79)	1.74×10^{-4}
DMRT1, KANK1	rs912062	9	841152	upstream, downstream	C/A	0.17/0.22	1.76 (1.31, 2.37)	1.82×10^{-4}
GLI2, LINC01101	rs11688682	2	120590036	Upstream	C/G	0.08/0.04	2.62 (1.57, 4.37)	2.30×10^{-4}

SNP, single-nucleotide polymorphism; Chr, chromosome; OR, odds ratio; CI, confidence interval.

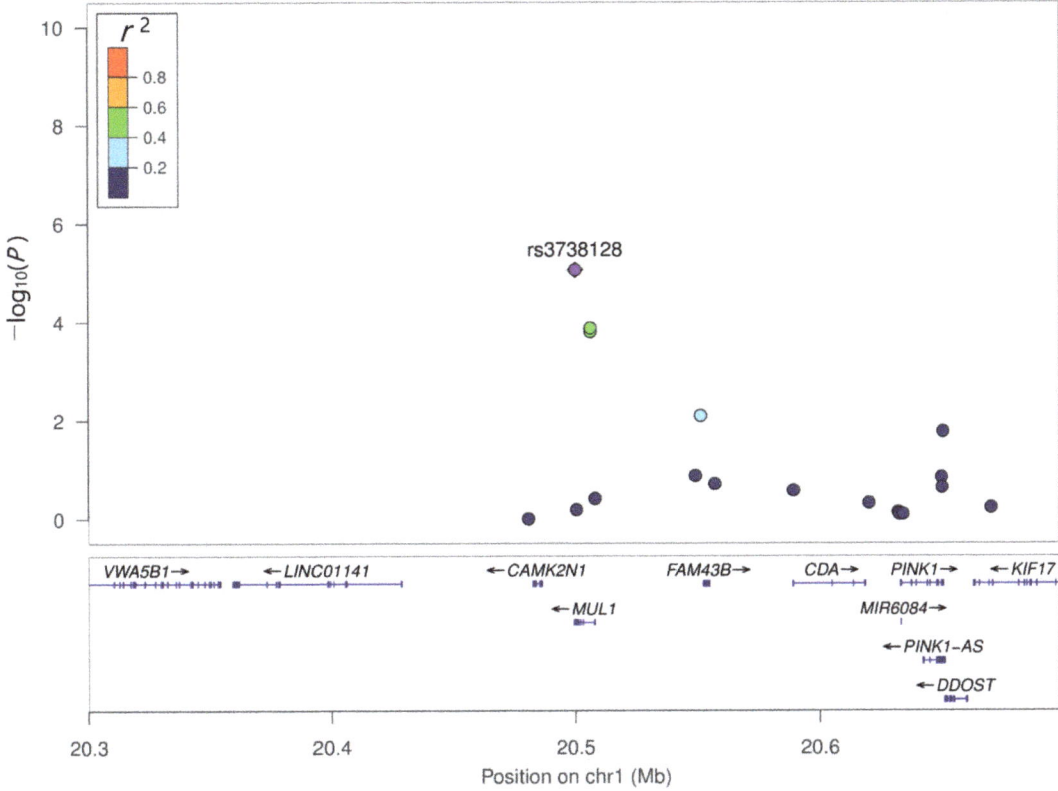

Figure 3. Regional association plot of the genetic variants of *MUL1*. *MUL1* SNP rs12566937 showed moderate linkage disequilibrium with *MUL1* SNP rs3738128.

4. Discussion

In this study, we identified genetic variants that were significantly associated with dementia in patients with PD with a median disease duration of over 12 years. The *MUL1* SNP rs3738128 showed the most significant association with dementia in PD. *ZHX2* and *ERP29* also showed correlations with dementia in PD. The *SNCA* locus showed the most significant association with susceptibility to PD, consistent with the results of previous GWAS [7–12].

There were few studies investigating the role of *MUL1* in the development of dementia in PD, and one case-control study conducted in China showed that *MUL1* SNP rs529974 was correlated with the development of PD [30]. *MUL1* encodes mitochondrial ubiquitin ligase 1, a mitochondrial E3 protein ligase that regulates mitofusin. The mitochondria are involved in cellular energy production and cell survival, playing an important role in the neurodegenerative process in PD [31]. Mitochondrial genes such as *parkin*, *PINK1*, *DJ-1*, *LRRK2*, *ATP13A2*, and *VPS35* are associated with PD [32]. An experimental study showed that *MUL1* suppressed the mitochondrial phenotype in *PINK1/parkin* mutant dopaminergic neuron, and the knockdown of *MUL1* in *parkin* knockout mouse cortical neurons augmented mitochondrial damage [33]. Therefore, mutants with *MUL1* and *parkin* mutations are employed in the development of animal models of PD [34]. *MUL1* overexpression has been shown to reduce the degeneration of dopaminergic neurons and enhance motor activity in neurons of flies fed with rotenone [35]. *MUL1* dysfunction renders dopaminergic neurons susceptible to mitochondrial damage. The loss of *MUL1* function may be more prominent when other mitochondrial dysfunctions exist as well, as a result of genetic variants or environmental toxins. The lack of correlation of *MUL1* with PD in this study may be explained by the adjunctive role of *MUL1* in mitochondrial function.

Considering that the *MUL1* pathway regulates mitochondrial damage in both dopaminergic and cortical neurons [33,36], defects in the *MUL1* pathway might affect the cognitive decline in PD. However, little is known about the association between *MUL1* and cognitive decline in PD or other neurodegenerative diseases that cause dementia. Mitochondrial dysfunction induces energy deficiency, intracellular calcium imbalance, and oxidative stress, leading to synaptic dysfunction and neuronal cell loss [37]. This mechanism explains how mitochondrial dysfunction mediates cognitive decline in neurodegenerative diseases, such as Alzheimer's disease. Mitochondrial dysfunction is also prominent among patients with PD [38]. When *MUL1* is downregulated, cortical neurons, as well as dopaminergic neurons, might become more susceptible to damage due to mitochondrial dysfunction, leading to the progression of cortical neuronal loss, synaptic dysfunction, and cognitive decline. In addition, recent studies have revealed that amyloid-beta and p-tau interact with mitochondrial proteins, resulting in increased mitochondrial fragmentation and reduced mitochondrial fusion in Alzheimer's disease [39]. Similarly, pathogenic alpha-synuclein and amyloid-beta found in the brains of patients with PDD [40,41] might interact with *MUL1*, leading to mitochondrial dysfunction. The significant association of *MUL1* with dementia in PD suggests the biological plausibility of the involvement of mitochondrial dysfunction in the development of dementia in PD. Further studies are needed to elucidate the exact pathogenic mechanisms underlying the involvement of *MUL1* in the development of dementia in PD.

Other genetic variants associated with dementia in PD were located in the loci of *ZHX2* and *ERP29*. Few clinical studies have investigated the role of *ZHX2* and *ERP29* in PD or dementia. The gene *ZHX2* encodes zinc-finger and homeodomain protein 2 (ZHX2) that regulates transcription and neuronal differentiation [42]. Genetic variants of *ZHX2* were found in two affected members of familial corticobasal degeneration, mutations of which were predicted to impair protein function [43]. Both corticobasal degeneration and PD are neurodegenerative diseases characterized by damage to cortical neurons and cognitive decline. Since *ZHX2* is also associated with cortical neuronogenesis [42], it may be associated with the progression to dementia. *ERP29* gene encodes a 29 kDa endoplasmic reticulum protein (ERp29), which is ubiquitously expressed in cells and regulates protein transport between the endoplasmic reticulum and Golgi apparatus [44]. ERp29 is involved in protein misfolding and mistrafficking [44,45], which are potent pathogenic features of PD and Alzheimer's disease [46]. Given that endoplasmic reticulum stress is related to Lewy body dementia [47], it is possible that ERp29 mutation also induces cortical neuronal damage and is linked to the progression of dementia in patients with PD.

In our study, the *SNCA* SNP rs11931074 was most significantly associated with susceptibility to PD, which is consistent with previous results [7,9–12,48]. Mutations in the *SNCA* gene were first found in familial PD with autosomal dominant inheritance [49,50], and several SNPs across the *SNCA* locus were also linked to the increased risk for sporadic PD in multiple GWAS [7–12]. The *SNCA* gene encodes alpha-synuclein, which is the main component of Lewy bodies, the pathologic hallmark of PD. Interestingly, *SNCA* SNP rs11931074, which showed the most significant association with PD in this study also has a distinct relationship with PD based on race [48]. The presence of *SNCA* SNP rs11931074 increases the risk of PD, as demonstrated by the allele model, homozygote model, and recessive model developed for the Asian population, while the association was found to be true only in an allele model developed for the Caucasian population. These results support the quality of PD samples used in this study and might emphasize the role of *SNCA* SNP rs11931074 in the development of PD in the Asian population.

We found that *SPHK1* and *FYN* SNPs were associated with PD. *SPHK1* gene encodes sphingosine kinase 1 protein, which phosphorylates sphingosine into sphingosine-1-phosphate (S1P). S1P synthesized by *SPHK1* exerts mitogenic and anti-apoptotic effects in an autocrine or paracrine manner [51]. The expression of sphingosine kinase 1 was downregulated in experimental models of PD, and inhibition of sphingosine kinase 1 decreases cell viability and enhances the production of reactive oxygen species [52]. *FYN* gene encodes the Fyc protein, which is a tyrosine phosphotransferase enzyme belonging to the Src family of nonreceptor tyrosine kinases. Fyc has been suggested to regulate alpha-synuclein phosphorylation, oxidative stress-induced dopaminergic neuronal death, and enhancement of neuroinflammation [53]. Therefore, both sphingosine kinase 1 protein and Fyc were suggested as potential therapeutic targets for PD [51,53], and our data support the protective effects of *SPHK1* and *FYN* in PD.

The strength of this study is that we used clinical diagnosis of dementia based on the long-term follow-up of patients with PD. The prevalence of dementia in patients with PD is 17% at 5 years after diagnosis and 46–75% at 10 years after diagnosis [6,54]. Therefore, including PD patients with a short follow-up duration would misclassify them as having PD without dementia. A previous GWAS investigating the cognitive decline in PD included patients whose median follow-up duration was 4 years [55], and another GWAS assessed cognition using cross-sectional MMSE scores or MoCA scores [14].

This study has a few limitations. First, the sample size was relatively small, which may explain why genetic variants associated with dementia in patients with PD did not remain statistically significant after stringent Bonferroni correction. Second, the biological functions of the genetic variants were not validated. However, the experimental studies on *SPHK1*, *FYN*, *MUL1*, *ZHX2*, and *ERP29* genes, as discussed above, might support the biological plausibility of the involvement of these genes in PD. Therefore, future functional studies are required to confirm our results.

5. Conclusions

This microarray genomic study identified the new loci of *MUL1* associated with dementia in PD, suggesting an essential role of mitochondrial dysfunction in the development of this nonmotor symptom of PD.

Supplementary Materials: The following are available online at https://www.mdpi.com/article/10.3390/genes12121975/s1, Figure S1: Quantile–quantile plot for Parkinson's disease, Figure S2: Quantile–quantile plot for dementia in Parkinson's disease, Table S1: Characteristics of the markers used in the microarray, Table S2: Staged verification of the markers, Table S3: Additional selection according to selection priority.

Author Contributions: Conceptualization, S.J. and S.J.C.; methodology, S.J., K.W.P., H.-S.R., and S.J.C.; formal analysis, S.J., Y.S.H., and S.H.L.; investigation, S.J., K.W.P., Y.S.H., S.H.L., and H.-S.R.; writing—original draft preparation, S.J. and S.J.C.; writing—review and editing, S.J., H.-S.R., and S.J.C.; funding acquisition, S.J.C. All authors have read and agreed to the published version of the manuscript.

Funding: This study was supported by the Korea Healthcare Technology R&D Project, Ministry of Health and Welfare, Republic of Korea (grant number: HI19C0256).

Institutional Review Board Statement: The study was conducted according to the guidelines of the Declaration of Helsinki, and approved by the Institutional Review Board of Asan Medical Center (2019–0533, 2019-04-22).

Informed Consent Statement: Informed consent was obtained from all subjects involved in the study.

Data Availability Statement: The data presented in this study are available on request from the corresponding author. The data are not publicly available due to the privacy of the study population.

Acknowledgments: We appreciate the unlimited academic support from Yoon Kim, Chairman of Samyang Holdings Corporation, Seoul, Korea.

Conflicts of Interest: The authors declare no conflict of interest.

References

1. GBD 2015 Neurological Disorders Collaborator Group. Global, regional, and national burden of neurological disorders during 1990–2015: A systematic analysis for the Global Burden of Disease Study 2015. *Lancet Neurol.* **2017**, *16*, 877–897. [CrossRef]
2. Hughes, A.J.; Daniel, S.E.; Kilford, L.; Lees, A.J. Accuracy of clinical diagnosis of idiopathic Parkinson's disease: A clinico-pathological study of 100 cases. *J. Neurol. Neurosurg. Psychiatry* **1992**, *55*, 181–184. [CrossRef] [PubMed]
3. Schapira, A.H.V.; Chaudhuri, K.R.; Jenner, P. Non-motor features of Parkinson disease. *Nat. Rev. Neurosci.* **2017**, *18*, 435–450. [CrossRef] [PubMed]
4. Pedersen, K.F.; Larsen, J.P.; Tysnes, O.B.; Alves, G. Prognosis of mild cognitive impairment in early Parkinson disease: The Norwegian ParkWest study. *JAMA Neurol.* **2013**, *70*, 580–586. [CrossRef] [PubMed]
5. Aarsland, D.; Kurz, M.W. The epidemiology of dementia associated with Parkinson disease. *J. Neurol. Sci.* **2010**, *289*, 18–22. [CrossRef]
6. Williams-Gray, C.H.; Mason, S.L.; Evans, J.R.; Foltynie, T.; Brayne, C.; Robbins, T.W.; Barker, R.A. The CamPaIGN study of Parkinson's disease: 10-year outlook in an incident population-based cohort. *J. Neurol. Neurosurg. Psychiatry* **2013**, *84*, 1258–1264. [CrossRef]
7. Nalls, M.A.; Blauwendraat, C.; Vallerga, C.L.; Heilbron, K.; Bandres-Ciga, S.; Chang, D.; Tan, M.; Kia, D.A.; Noyce, A.J.; Xue, A.; et al. Identification of novel risk loci, causal insights, and heritable risk for Parkinson's disease: A meta-analysis of genome-wide association studies. *Lancet Neurol.* **2019**, *18*, 1091–1102. [CrossRef]
8. Foo, J.N.; Chew, E.G.Y.; Chung, S.J.; Peng, R.; Blauwendraat, C.; Nalls, M.A.; Mok, K.Y.; Satake, W.; Toda, T.; Chao, Y.; et al. Identification of risk loci for Parkinson disease in Asians and comparison of risk between Asians and Europeans: A Genome-Wide Association Study. *JAMA Neurol.* **2020**, *77*, 746–754. [CrossRef]
9. Satake, W.; Nakabayashi, Y.; Mizuta, I.; Hirota, Y.; Ito, C.; Kubo, M.; Kawaguchi, T.; Tsunoda, T.; Watanabe, M.; Takeda, A.; et al. Genome-wide association study identifies common variants at four loci as genetic risk factors for Parkinson's disease. *Nat. Genet.* **2009**, *41*, 1303–1307. [CrossRef]
10. Simón-Sánchez, J.; Schulte, C.; Bras, J.M.; Sharma, M.; Gibbs, J.R.; Berg, D.; Paisan-Ruiz, C.; Lichtner, P.; Scholz, S.W.; Hernandez, D.G.; et al. Genome-wide association study reveals genetic risk underlying Parkinson's disease. *Nat. Genet.* **2009**, *41*, 1308–1312. [CrossRef]
11. Hamza, T.H.; Zabetian, C.P.; Tenesa, A.; Laederach, A.; Montimurro, J.; Yearout, D.; Kay, D.M.; Doheny, K.F.; Paschall, J.; Pugh, E.; et al. Common genetic variation in the HLA region is associated with late-onset sporadic Parkinson's disease. *Nat. Genet.* **2010**, *42*, 781–785. [CrossRef] [PubMed]
12. Nalls, M.A.; Pankratz, N.; Lill, C.M.; Do, C.B.; Hernandez, D.G.; Saad, M.; DeStefano, A.L.; Kara, E.; Bras, J.; Sharma, M.; et al. Large-scale meta-analysis of genome-wide association data identifies six new risk loci for Parkinson's disease. *Nat. Genet.* **2014**, *46*, 989–993. [CrossRef] [PubMed]
13. Chang, D.; Nalls, M.A.; Hallgrímsdóttir, I.B.; Hunkapiller, J.; van der Brug, M.; Cai, F.; Kerchner, G.A.; Ayalon, G.; International Parkinson's Disease Genomics Consortium; 23andMe Research Team; et al. A meta-analysis of genome-wide association studies identifies 17 new Parkinson's disease risk loci. *Nat. Genet.* **2017**, *49*, 1511–1516. [CrossRef] [PubMed]
14. Park, K.W.; Jo, S.; Kim, M.S.; Jeon, S.R.; Ryu, H.S.; Kim, J.; Park, Y.M.; Koh, S.B.; Lee, J.H.; Chung, S.J. Genomic association study for cognitive impairment in Parkinson's disease. *Front. Neurol.* **2020**, *11*, 579268. [CrossRef]
15. Guerreiro, R.; Ross, O.A.; Kun-Rodrigues, C.; Hernandez, D.G.; Orme, T.; Eicher, J.D.; Shepherd, C.E.; Parkkinen, L.; Darwent, L.; Heckman, M.G.; et al. Investigating the genetic architecture of dementia with Lewy bodies: A two-stage genome-wide association study. *Lancet Neurol.* **2018**, *17*, 64–74. [CrossRef]
16. Rongve, A.; Witoelar, A.; Ruiz, A.; Athanasiu, L.; Abdelnour, C.; Clarimon, J.; Heilmann-Heimbach, S.; Hernández, I.; Moreno-Grau, S.; de Rojas, I.; et al. GBA and APOE ε4 associate with sporadic dementia with Lewy bodies in European genome wide association study. *Sci. Rep.* **2019**, *9*, 7013. [CrossRef]
17. Chia, R.; Sabir, M.S.; Bandres-Ciga, S.; Saez-Atienzar, S.; Reynolds, R.H.; Gustavsson, E.; Walton, R.L.; Ahmed, S.; Viollet, C.; Ding, J.; et al. Genome sequencing analysis identifies new loci associated with Lewy body dementia and provides insights into its genetic architecture. *Nat. Genet.* **2021**, *53*, 294–303. [CrossRef]
18. Blauwendraat, C.; Reed, X.; Krohn, L.; Heilbron, K.; Bandres-Ciga, S.; Tan, M.; Gibbs, J.R.; Hernandez, D.G.; Kumaran, R.; Langston, R.; et al. Genetic modifiers of risk and age at onset in GBA associated Parkinson's disease and Lewy body dementia. *Brain* **2020**, *143*, 234–248. [CrossRef] [PubMed]
19. Jellinger, K.A. Dementia with Lewy bodies and Parkinson's disease-dementia: Current concepts and controversies. *J. Neural Transm.* **2018**, *125*, 615–650. [CrossRef]
20. Dubois, B.; Burn, D.; Goetz, C.; Aarsland, D.; Brown, R.G.; Broe, G.A.; Dickson, D.; Duyckaerts, C.; Cummings, J.; Gauthier, S.; et al. Diagnostic procedures for Parkinson's disease dementia: Recommendations from the movement disorder society task force. *Mov. Disord.* **2007**, *22*, 2314–2324. [CrossRef]
21. Ryu, H.S.; Park, K.W.; Choi, N.; Kim, J.; Park, Y.M.; Jo, S.; Kim, M.J.; Kim, Y.J.; Kim, J.; Kim, K.; et al. Genomic analysis identifies new loci associated with motor complications in Parkinson's disease. *Front. Neurol.* **2020**, *11*, 570. [CrossRef] [PubMed]
22. Orme, T.; Guerreiro, R.; Bras, J. The genetics of dementia with Lewy bodies: Current understanding and future directions. *Curr. Neurol. Neurosci. Rep.* **2018**, *18*, 67. [CrossRef] [PubMed]

23. Lambert, J.C.; Heath, S.; Even, G.; Campion, D.; Sleegers, K.; Hiltunen, M.; Combarros, O.; Zelenika, D.; Bullido, M.J.; Tavernier, B.; et al. Genome-wide association study identifies variants at CLU and CR1 associated with Alzheimer's disease. *Nat. Genet.* **2009**, *41*, 1094–1099. [CrossRef] [PubMed]
24. Seshadri, S.; Fitzpatrick, A.L.; Ikram, M.A.; DeStefano, A.L.; Gudnason, V.; Boada, M.; Bis, J.C.; Smith, A.V.; Carassquillo, M.M.; Lambert, J.C.; et al. Genome-wide analysis of genetic loci associated with Alzheimer disease. *JAMA* **2010**, *303*, 1832–1840. [CrossRef] [PubMed]
25. Lambert, J.C.; Ibrahim-Verbaas, C.A.; Harold, D.; Naj, A.C.; Sims, R.; Bellenguez, C.; DeStefano, A.L.; Bis, J.C.; Beecham, G.W.; Grenier-Boley, B.; et al. Meta-analysis of 74,046 individuals identifies 11 new susceptibility loci for Alzheimer's disease. *Nat. Genet.* **2013**, *45*, 1452–1458. [CrossRef] [PubMed]
26. Jansen, I.E.; Savage, J.E.; Watanabe, K.; Bryois, J.; Williams, D.M.; Steinberg, S.; Sealock, J.; Karlsson, I.K.; Hägg, S.; Athanasiu, L.; et al. Genome-wide meta-analysis identifies new loci and functional pathways influencing Alzheimer's disease risk. *Nat. Genet.* **2019**, *51*, 404–413. [CrossRef] [PubMed]
27. Villegas-Llerena, C.; Phillips, A.; Garcia-Reitboeck, P.; Hardy, J.; Pocock, J.M. Microglial genes regulating neuroinflammation in the progression of Alzheimer's disease. *Curr. Opin. Neurobiol.* **2016**, *36*, 74–81. [CrossRef]
28. Dendrou, C.A.; Petersen, J.; Rossjohn, J.; Fugger, L. HLA variation and disease. *Nat. Rev. Immunol.* **2018**, *18*, 325–339. [CrossRef]
29. Cingolani, P.; Platts, A.; Wang, L.L.; Coon, M.; Nguyen, T.; Wang, L.; Land, S.J.; Lu, X.; Ruden, D.M. A program for annotating and predicting the effects of single nucleotide polymorphisms, SnpEff: SNPs in the genome of Drosophila melanogaster strain w^{1118}; iso-2; iso-3. *Fly* **2012**, *6*, 80–92. [CrossRef]
30. Taximaimaiti, R.; Li, H. MUL1 gene polymorphisms and Parkinson's disease risk. *Acta Neurol. Scand.* **2019**, *139*, 483–487. [CrossRef] [PubMed]
31. Malpartida, A.B.; Williamson, M.; Narendra, D.P.; Wade-Martins, R.; Ryan, B.J. Mitochondrial dysfunction and mitophagy in Parkinson's disease: From mechanism to therapy. *Trends Biochem. Sci.* **2021**, *46*, 329–343. [CrossRef] [PubMed]
32. Grünewald, A.; Kumar, K.R.; Sue, C.M. New insights into the complex role of mitochondria in Parkinson's disease. *Prog. Neurobiol.* **2019**, *177*, 73–93. [CrossRef] [PubMed]
33. Yun, J.; Puri, R.; Yang, H.; Lizzio, M.A.; Wu, C.; Sheng, Z.H.; Guo, M. MUL1 acts in parallel to the PINK1/parkin pathway in regulating mitofusin and compensates for loss of PINK1/parkin. *eLife* **2014**, *3*, e01958. [CrossRef] [PubMed]
34. Doktór, B.; Damulewicz, M.; Pyza, E. Effects of MUL1 and PARKIN on the circadian clock, brain and behaviour in *Drosophila* Parkinson's disease models. *BMC Neurosci.* **2019**, *20*, 24. [CrossRef]
35. Doktór, B.; Damulewicz, M.; Pyza, E. Overexpression of mitochondrial ligases reverses rotenone-induced effects in a *Drosophila* model of Parkinson's disease. *Front. Neurosci.* **2019**, *13*, 94. [CrossRef] [PubMed]
36. Puri, R.; Cheng, X.T.; Lin, M.Y.; Huang, N.; Sheng, Z.H. Mul1 restrains Parkin-mediated mitophagy in mature neurons by maintaining ER-mitochondrial contacts. *Nat. Commun.* **2019**, *10*, 3645. [CrossRef]
37. Bhatti, J.S.; Bhatti, G.K.; Reddy, P.H. Mitochondrial dysfunction and oxidative stress in metabolic disorders—A step towards mitochondria based therapeutic strategies. *Biochim. Biophys. Acta Mol. Basis Dis.* **2017**, *1863*, 1066–1077. [CrossRef]
38. Reeve, A.K.; Grady, J.P.; Cosgrave, E.M.; Bennison, E.; Chen, C.; Hepplewhite, P.D.; Morris, C.M. Mitochondrial dysfunction within the synapses of substantia nigra neurons in Parkinson's disease. *NPJ Parkinson's Dis.* **2018**, *4*, 9. [CrossRef] [PubMed]
39. Reddy, P.H.; Oliver, D.M. Amyloid beta and phosphorylated tau-induced defective autophagy and mitophagy in Alzheimer's disease. *Cells* **2019**, *8*, 488. [CrossRef] [PubMed]
40. Tsuboi, Y.; Uchikado, H.; Dickson, D.W. Neuropathology of Parkinson's disease dementia and dementia with Lewy bodies with reference to striatal pathology. *Parkinsonism Relat. Disord.* **2007**, *13*, S221–S224. [CrossRef]
41. Smith, C.; Malek, N.; Grosset, K.; Cullen, B.; Gentleman, S.; Grosset, D.G. Neuropathology of dementia in patients with Parkinson's disease: A systematic review of autopsy studies. *J. Neurol. Neurosurg. Psychiatry* **2019**, *90*, 1234–1243. [CrossRef] [PubMed]
42. Wu, C.; Qiu, R.; Wang, J.; Zhang, H.; Murai, K.; Lu, Q. ZHX2 Interacts with Ephrin-B and regulates neural progenitor maintenance in the developing cerebral cortex. *J. Neurosci.* **2009**, *29*, 7404–7412. [CrossRef]
43. Fekete, R.; Bainbridge, M.; Baizabal-Carvallo, J.F.; Rivera, A.; Miller, B.; Du, P.; Kholodovych, V.; Powell, S.; Ondo, W. Exome sequencing in familial corticobasal degeneration. *Parkinsonism Relat. Disord.* **2013**, *19*, 1049–1052. [CrossRef]
44. Brecker, M.; Khakhina, S.; Schubert, T.J.; Thompson, Z.; Rubenstein, R.C. The probable, possible, and novel functions of ERp29. *Front. Physiol.* **2020**, *11*, 574339. [CrossRef] [PubMed]
45. Bhandary, B.; Marahatta, A.; Kim, H.R.; Chae, H.J. An involvement of oxidative stress in endoplasmic reticulum stress and its associated diseases. *Int. J. Mol. Sci.* **2012**, *14*, 434–456. [CrossRef]
46. Forloni, G.; Terreni, L.; Bertani, I.; Fogliarino, S.; Invernizzi, R.; Assini, A.; Ribizzi, G.; Negro, A.; Calabrese, E.; Volonté, M.A.; et al. Protein misfolding in Alzheimer's and Parkinson's disease: Genetics and molecular mechanisms. *Neurobiol. Aging* **2002**, *23*, 957–976. [CrossRef]
47. Baek, J.H.; Whitfield, D.; Howlett, D.; Francis, P.; Bereczki, E.; Ballard, C.; Hortobágyi, T.; Attems, J.; Aarsland, D. Unfolded protein response is activated in Lewy body dementias. *Neuropathol. Appl. Neurobiol.* **2016**, *42*, 352–365. [CrossRef] [PubMed]
48. Du, B.; Xue, Q.; Liang, C.; Fan, C.; Liang, M.; Zhang, Y.; Bi, X.; Hou, L. Association between alpha-synuclein (SNCA) rs11931074 variability and susceptibility to Parkinson's disease: An updated meta-analysis of 41,811 patients. *Neurol. Sci.* **2020**, *41*, 271–280. [CrossRef] [PubMed]

49. Singleton, A.B.; Farrer, M.; Johnson, J.; Singleton, A.; Hague, S.; Kachergus, J.; Hulihan, M.; Peuralinna, T.; Dutra, A.; Nussbaum, R.; et al. α-Synuclein locus triplication causes Parkinson's disease. *Science* **2003**, *302*, 841. [CrossRef] [PubMed]
50. Polymeropoulos, M.H.; Lavedan, C.; Leroy, E.; Ide, S.E.; Dehejia, A.; Dutra, A.; Pike, B.; Root, H.; Rubenstein, J.; Boyer, R.; et al. Mutation in the α-synuclein gene identified in families with Parkinson's disease. *Science* **1997**, *276*, 2045–2047. [CrossRef]
51. Motyl, J.; Strosznajder, J.B. Sphingosine kinase 1/sphingosine-1-phosphate receptors dependent signalling in neurodegenerative diseases. The promising target for neuroprotection in Parkinson's disease. *Pharmacol. Rep.* **2018**, *70*, 1010–1014. [CrossRef] [PubMed]
52. Pyszko, J.; Strosznajder, J.B. Sphingosine kinase 1 and sphingosine-1-phosphate in oxidative stress evoked by 1-methyl-4-phenylpyridinium (MPP+) in human dopaminergic neuronal cells. *Mol. Neurobiol.* **2014**, *50*, 38–48. [CrossRef] [PubMed]
53. Angelopoulou, E.; Paudel, Y.N.; Julian, T.; Shaikh, M.F.; Piperi, C. Pivotal role of Fyn Kinase in Parkinson's disease and levodopa-induced dyskinesia: A novel therapeutic target? *Mol. Neurobiol.* **2021**, *58*, 1372–1391. [CrossRef] [PubMed]
54. Aarsland, D.; Batzu, L.; Halliday, G.M.; Geurtsen, G.J.; Ballard, C.; Chaudhuri, K.R.; Weintraub, D. Parkinson disease-associated cognitive impairment. *Nat. Rev. Dis. Prim.* **2021**, *7*, 47. [CrossRef] [PubMed]
55. Tan, M.M.X.; Lawton, M.A.; Jabbari, E.; Reynolds, R.H.; Iwaki, H.; Blauwendraat, C.; Kanavou, S.; Pollard, M.I.; Hubbard, L.; Malek, N.; et al. Genome-wide association studies of cognitive and motor progression in Parkinson's disease. *Mov. Disord.* **2021**, *36*, 424–433. [CrossRef] [PubMed]

Review

Mapping the Diverse and Inclusive Future of Parkinson's Disease Genetics and Its Widespread Impact

Inas Elsayed [1,2,†], Alejandro Martinez-Carrasco [3,†], Mario Cornejo-Olivas [4,5] and Sara Bandres-Ciga [6,*]

1. Faculty of Pharmacy, University of Gezira, Wad Medani P.O. Box 20, Sudan; inaselsayed25@gmail.com
2. International Parkinson Disease Genomics Consortium (IPDGC)-Africa, University of Gezira, Wad Medani P.O. Box 20, Sudan
3. Queen Square Institute of Neurology, University College London (UCL), London WC1E 6BT, UK; alejandro.carrasco.20@ucl.ac.uk
4. Neurogenetics Research Center, Instituto Nacional de Ciencias Neurológicas, Lima 15003, Peru; mario.cornejo.o@incngen.org.pe
5. Center for Global Health, Universidad Peruana Cayetano Heredia, Lima 15003, Peru
6. Molecular Genetics Section, Laboratory of Neurogenetics, National Institute on Aging, NIH, Bethesda, MD 20892, USA
* Correspondence: sara.bandresciga@nih.gov
† These authors contributed equally to this work.

Abstract: Over the last decades, genetics has been the engine that has pushed us along on our voyage to understand the etiology of Parkinson's disease (PD). Although a large number of risk loci and causative mutations for PD have been identified, it is clear that much more needs to be done to solve the missing heritability mystery. Despite remarkable efforts, as a field, we have failed in terms of diversity and inclusivity. The vast majority of genetic studies in PD have focused on individuals of European ancestry, leading to a gap of knowledge on the existing genetic differences across populations and PD as a whole. As we move forward, shedding light on the genetic architecture contributing to PD in non-European populations is essential, and will provide novel insight into the generalized genetic map of the disease. In this review, we discuss how better representation of understudied ancestral groups in PD genetics research requires addressing and resolving all the challenges that hinder the inclusion of these populations. We further provide an overview of PD genetics in the clinics, covering the current challenges and limitations of genetic testing and counseling. Finally, we describe the impact of worldwide collaborative initiatives in the field, shaping the future of the new era of PD genetics as we advance in our understanding of the genetic architecture of PD.

Keywords: Parkinson's disease; genetics; diversity; post-GWAS era; genetic testing; genetics counselling

1. Introduction

Parkinson's disease (PD) is a complex neurodegenerative disorder whose prevalence is predicted to increase drastically, being more pronounced in older age people and with variations among sex and ancestry groups [1]. PD is clinically manifested as resting tremor, gait impairment, bradykinesia (slowness of movements), rigidity, and postural instability.

The heritability of PD driven by common genetic variation is estimated to be ~22% and only approximately one third of it has been uncovered with the largest genetic study in the European ancestry population to date [2]. PD is a complex genetic disease, and such heritable genetic variation has a different magnitude of effect, frequency, deleteriousness, and penetrance, so that we can differentiate between rare or common variants, pathogenic deleterious mutations or variants that slightly increase the risk for developing PD, and incomplete versus complete penetrance [3]. The vast majority of patients with PD are diagnosed as sporadic without a clear genetic cause but probably due to the interplay between genetic and environmental risk factors. However, up to 15% of PD patients have a positive PD family history and 5–10% respond to Mendelian inheritance [4]. The last twenty years have witnessed the discovery of recessively and dominantly inherited genes responsible for rare monogenic forms of PD [5]. Well-known, highly penetrant variants causing familial or early onset PD are found within *SNCA, VPS35, PARKN, DJ-1*, and *PINK1* genes. In addition, risk variants with incomplete penetrance have been reported within *GBA* and *LRRK2* [6], as well as 90 risk variants increasing the susceptibility of PD in Europeans and Asian populations [2,7]. Despite the progress made in understanding the genetic basis of PD, current available genetic testing is mostly used on familial and early onset PD cases requiring appropriate genetic counseling.

The discovery of genes and loci associated with PD allows us to redefine the genetic map of the disease, gaining knowledge of potential mechanisms contributing to PD. Understanding PD etiology gives us valuable insights to develop disease-modifying therapies that may stop or slow the progression of the disease. Genome wide association studies (GWAS) have been a powerful tool to better understand how genetics contribute to the risk, progression, and onset of PD [2,8,9]. However, a drawback of all the progress made in understanding PD genetics is that the vast majority of studies have focused on individuals of European ancestry, leading to a gap of knowledge on the likely existing genetic differences among populations.

In this review, we aim to outline the need, benefit, and challenges of exploring the genetic basis of PD across underrepresented populations. We provide an overview of the current state of the field and its applicability in clinical practice, as well as highlight the role of worldwide initiatives in shaping the future of the new era of PD genetics as we advance in our understanding of the genetic architecture of PD.

2. Parkinson's Disease Genetics in Underrepresented Populations: Need for Inclusivity

The recent decade has witnessed unprecedented growth in PD genomics research. This has produced a substantial improvement in therapeutic development. Despite efforts, the effective translation of PD research outcomes into health care optimization has failed at generalizability due to the limited ethnic diversity of studies, since most PD genetic studies have been conducted on populations of European descent and, lately, Asian ancestry populations [2,7]. Such paucity in the representation of ethnically diverse populations, including Africans, South Americans, and indigenous populations, can culminate in a serious disparity in the quality of health care delivered to PD patients [10,11].

The inclusion of ethnic diversity in PD genetics research is essential to improve PD health care in many aspects. First, the under-studying of non-European populations can lead to the underestimation of genetic risk factors specific to that population, which can serve as valuable markers for early disease detection and risk quantification. In addition, including diverse populations can help validate or refute previously identified risk loci in European populations and highlight potential variability in genetic variants' contributions to PD risk across different populations. An example of these differences can be found in the largest GWAS undertaken in the Asian population, in which associations between the PD phenotype and *GBA* or *MAPT* variants were not found [7].

Moreover, the ethnic diversity in PD research is crucial to improve our understanding of the disease's biology and pathogenesis. Addressing the variability in the genetic architecture of PD research across populations can help us capture a broader range of genetic and environmental factors implicated in disease development and progression, and tailor the ideal preventive measures and therapeutic interventions accordingly. For instance, these include rare genetic variation with key implications in disease pathogenesis, which could be better highlighted in certain populations while overlooked or completely missed in others because of naturally occurring, population-based variations in allele frequencies [12].

By addressing the variability in genetic architecture and environmental conditions, diversifying PD genetic research can also help us comprehend the interplay between the contributing common variants and environmental factors. This is important since many diagnostic/risk assessment algorithms like polygenic scores are based on the dosage of common variants' contributions to disease's risk/pathogenesis identified in European populations, which are not necessarily applicable to non-European populations [5,13]. Generally, translating genetic information derived from studies based on European ancestry populations to other ethnicities with a distinct genetic background and environmental exposure might produce limited gains in the future, or potentially even worse clinical outcomes. This underscores the importance of diversifying PD genetic research, which represents a current global priority.

A better representation of understudied populations in PD genetics research requires addressing, and resolving, all the challenges that hinder the inclusion of these populations (Figure 1). Several factors were found to be responsible for the underrepresentation of non-European populations in PD research. The major recognized challenge limiting the access of populations living in low and middle-income countries (LMICs) to PD research is the lack of funding and infrastructure. Hence, allocating funds to support genetics studies in PD in these countries can help to improve the accessibility of these populations to research [14]. Besides the financial and logistic limitations, the availability of a trained scientific workforce is another major challenge in LMICs. Fortunately, establishing training programs targeting the scientific workforce in under-developed countries is currently more feasible with the aid of virtual tools and technologies [15]. To address this to some degree, the Global Parkinson's Genetics Program (GP2, https://www.gp2.org/, accessed on 20 September 2021) [16] has recently established a virtual center of excellence with resources and expertise to serve the training needs across these populations.

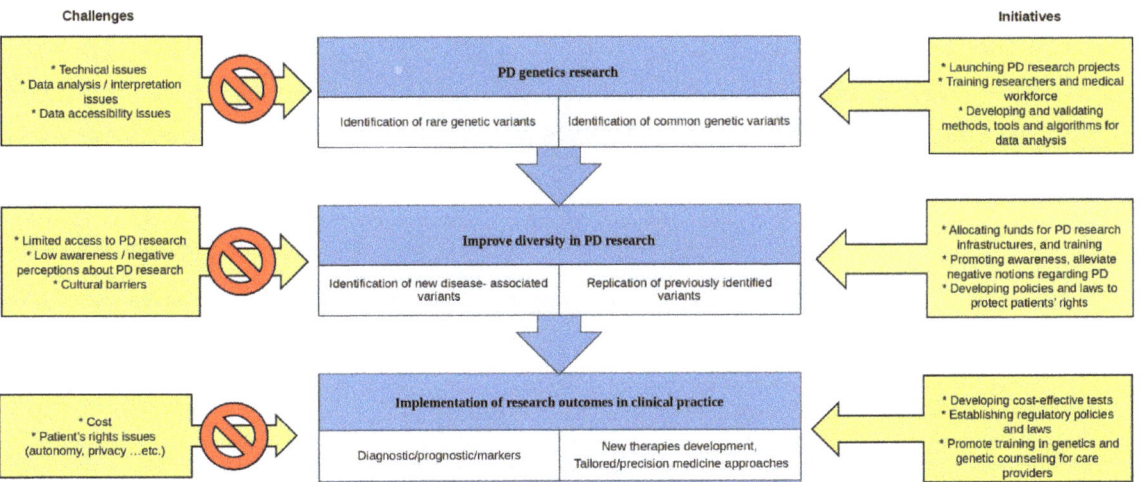

Figure 1. The Parkinson's disease genetics path: From research to clinics. Parkinson's disease (PD).

In addition, the lack of motives to participate in genetics research either due to low awareness about research benefits or negative perceptions about research or towards research procedures, especially invasive procedures like blood sampling, can also significantly limit the inclusion of certain populations in PD genetics research [17–19]. This combined with potential restricting cultural and/or religious barriers were found to limit the participation of understudied populations both in LMICs as well as minorities living in high-income countries [19]. Organizing educational programs in targeted communities to improve populations' awareness about PD genetics research and reduce cultural stigma, combined with developing policies and regulations to protect participants' confidentiality and safety is essential to guarantee the better engagement of targeted populations in genetic studies [20].

When thinking about genetic research in non-European populations, there are some limitations to take into account. One of them is the variation in linkage disequilibrium patterns and haplotype structure between ethnically diverse populations that can complicate GWAS imputation while using genotyping panels designed for European populations in other populations [13,21]. Furthermore, the heterogeneity of certain populations, particularly African and Latino populations, where complex ancestry admixture exists, represents another major challenge [13].

Fortunately, in the last few years, several national and international endeavors have been launched to improve population diversity in PD research. One of the prominent international initiatives supporting ethnic diversity in PD genetic research is The Global Parkinson's Genetics Program (GP2) from the Aligning Science Across Parkinson's (ASAP) initiative [16]. Aiming to enhance PD genetics research and population diversity to generate comprehensive, reproducible, and accessible data, GP2 has devoted significant resources to establish research infrastructure and train researchers in PD research around the globe. An example of its commitment is the GP2 Black and African American Connections to Parkinson's disease study (BLAAC PD) launched in 2021. This project targets one of the most underrepresented in neurodegenerative disease studies, which are the African American and Black American populations [19]. Similarly, The International Parkinson Disease Genomics Consortium-Africa (IPDGC-Africa) and Latin American Research Consortium on the Genetics of PD (LARGE-PD) have established PD research programs targeting underrepresented populations in Africa and Latin America respectively [22–24]. Besides enhancing the representation of understudied populations, these initiatives aim to improve PD research facilities, train the local workforce, and engage the communities through promoting research-supporting concepts and alleviating negative notions [16,22]. Such endeavors are expected to improve diversity in PD genetics research and warrant equity in medical services provided for PD patients around the globe.

3. Parkinson's Disease Genetics in the Clinic: Interpretation of Genetic Testing and Genetic Counseling. Challenges and Limitations

Genetic testing is mostly defined as DNA-based testing performed within a medical context for health care purposes with the intention to counsel individuals or families on the risk of diseases or implications to health and life decision-making. Depending on the specific purpose of the genetic test it can be diagnostic or predictive [25]. Genetic testing is often expensive, time-consuming, and not necessarily accessible in some countries. Diagnostic genetic testing, when positive, not only stops expensive diagnostic tests, but also has therapy implications and allows appropriate counseling on severe life decisions. On the other hand, a negative report might reorient a differential diagnosis or lead to future reassessment and further investigation [25].

Genetic testing and appropriate counselling for complex disorders like PD has been rapidly evolving, however there are many aspects to be considered when requiring a genetic test in PD. The increasing knowledge in the PD genetics field and the advent of technology have revolutionized genetic testing and counseling over the past decades. The technology used in genetic testing has rapidly evolved from single-gene approaches to next-generation sequencing, including exome and genome sequencing. As a result, genomic data for diagnostic purposes has been generated at a large scale and in an unprecedented manner, often requiring high capacities and resources for a clinician to interpret it [26]. The increasing application of genetic testing in clinical practice has been related to the decline of genotyping and sequencing costs. However, downstream requirements for genomic interpretation still limits its broader use in complex disorders like PD [27].

Since the identification of missense variants in the *SNCA* gene in 1997 [28], genetic mutations in about twenty genes have been described (Table 1), with at least six of them showing consistent evidence for causality [29]. Many genetic variants in *SNCA*, *VPS35*, *PRKN*, *LRRK2*, *PINK1*, and *DJ1* among other genes, have been consistently linked to monogenic PD forms representing approximately 5% of all PD cases [30,31]. However, incomplete penetrance, often seen in *LRRK2* and *GBA* variants, implies limited use for establishing individual risk in clinical practice. Additionally, a relevant situation to mention relates to these two genes harboring one or few founder mutations that are particularly frequent in certain populations.

Table 1. List of genes reported to be linked with Parkinson disease.

Gene	Year of Discovery	Reported Variants	Frequency	Inheritance	Confidence as a PD Gene
SNCA *	1997, 2003	Missense or multiplication	Very rare	Dominant	Very high
PRKN *	1998	Missense or loss of function	Rare	Recessive	Very high
UCHL1	1998	Missense	Unclear	Dominant	Low
PARK7 *	2003	Missense	Very rare	Recessive	Very high
LRRK2 *	2004	Missense	Common	Dominant	Very high
PINK1 *	2004	Missense or loss of function	Rare	Recessive	Very high
POLG	2004	Missense or loss of function	Rare	Dominant	High
HTRA2	2005	Missense	Unclear	Dominant	Low
ATP13A2 *	2006	Missense or loss of function	Very rare	Recessive	Very high
FBXO7 *	2008	Missense	Very rare	Recessive	Very high
GIGYF2	2008	Missense	Unclear	Dominant	Low
GBA *	2009	Missense or loss of function	Common	Dominant (incomplete penetrance)	Very high
PLA2G6 *	2009	Missense or loss of function	Rare	Recessive	Very high
EIF4G1	2011	Missense	Unclear	Dominant	Low
VPS35 *	2011	Missense	Very rare	Dominant	Very high
DNAJC6	2012	Missense or loss of function	Very rare	Recessive	High
SYNJ1	2013	Missense or loss of function	Very rare	Recessive	High
DNAJC13	2014	Missense	Unclear	Dominant	Low
TMEM230	2016	Missense	Unclear	Dominant	Low
VPS13C	2016	Missense or loss of function	Rare	Recessive	High
LRP10	2018	Missense or loss of function	Unclear	Dominant	Low
NUS1	2018	Missense	Unclear	Recessive	Low

* Gene of PD clinical significance adapted from Blauwendraat et al., 2019. Parkinson's Disease (PD).

LRRK2 and SNCA mutations are often screened in the presence of family history and suspicion of monogenic autosomal dominantly inherited PD. Affected PD patients carrying LRRK2 mutations have been reported worldwide with higher frequencies among Ashkenazi Jewish and Tunisian Barber populations [32], and lower frequencies among East Asians and Latinos with high Amerindian ancestry [33,34]. LRRK2-G2019S and ROC (Ras of complex) domain variants (R1441G/C/H) are among the most common variants associated with PD. Despite the fact that motor symptoms and responses to levodopa do not differ from idiopathic PD, some studies suggest a lower frequency of non-motor symptoms and mild cognitive impairment [35]. Age-dependent penetrance has been consistently demonstrated, with higher rates within LRRK2-G2019S carriers [36]. Genetic testing among putative LRRK2 carriers can be useful from the patient and research perspective, as clinical trials of LRRK2 kinase inhibitors have started showing promising results as the first personalized therapies for monogenic LRRK2 patients [30,37]. Missense and copy number variants (duplications and triplications) in the SNCA gene have been linked to monogenic autosomal inherited PD [38]. Clinical phenotypes of SNCA carriers are quite variable but often severe, with some SNCA mutations and rearrangements being related to a higher frequency of cognitive impairment, psychosis, and depression [39,40]. Thus, genetic diagnostic panels that include LRRK2 and SNCA should be considered when affected individuals with autosomal dominant familial PD are seen in clinics.

Biallelic rare variants within the PRKN, PINK1 and DJ-1 genes are consistently associated with early onset recessive PD. These three genes encode proteins sharing a common pathway, mitochondrial quality control and regulation [41]. Main clinical features related to variants within these genes are mostly consistent with early onset disease with slower progression, excellent Dopa response, frequent dystonia, dyskinesia, and uncommon cognitive decline [42]. Genetic testing for *PRKN*, *PINK1*, and *DJ-1* in familial forms of PD with recessive patterns and in early onset PD cases might be considered on diagnostic and therapy algorithms. Early onset PRKN PD cases usually have a prolonged and consistent response to low doses of levodopa, however they tend to develop levodopa-induced dyskinesias as well as compulsive disorders with the use of dopamine agonists [43]. Other treatment options including DBS have demonstrated positive results in selected cases [44].

Genetic testing and counseling for PD common risk variants is hard to interpret for individual cases in clinical practice. GWAS studies have nominated potential susceptibility factors in *LRRK2* and *SNCA* linked to sporadic PD [3]. *GBA*, the coding gene for glucocerebrosidase, is the most common genetic factor for developing PD and an important risk factor for other synucleinopathies for which multiple clinical trials are ongoing. Heterozygous variants in *GBA* are present in up to 15–20% of PD patients in certain populations and bear a higher risk of non-motor features, such as cognitive decline and dementia [45,46]. The consistent association of *GBA* with PD risk contrasts with the large number of *GBA* carriers who do not develop PD given the low penetrance of this gene [47,48].

On the other hand, there is controversial evidence suggesting risk conferred by heterozygous *PRKN* and *PINK1* variants in PD etiology. Large-scale studies systematically interrogating *PINK1* variants failed to confirm its role as risk factor for PD [49]. Given the lack of replication and controversial findings across research studies, genetic testing seeking this specific heterozygous variation is not recommended.

Regular genetic testing for sporadic late-onset PD is not currently recommended as standard clinical practice. PD is considered a complex disorder with the coexistence of a genetic predisposition together with variable environmental exposure [50]. Sporadic PD is the most common form of neurodegenerative Parkinsonism, representing the vast majority of cases seen in regular clinical practice. Despite the tremendous advances in understanding the genetic architecture of PD, it is still challenging to explore individual genetic risk in sporadic forms. Novel multi-OMIC approaches are being investigated to predict PD risk, including polygenic risk stratification and multimodal data integration.

Genetic testing for PD should be performed within appropriate genetic counseling approaches depending on the individual clinical profile. Genetic testing might be highly valuable in the presence of a positive family history, early onset PD, or specific high-risk ancestry like Ashkenazi Jewish [51]. While there is strong evidence for a potential use of genetic testing for monogenic PD, not only for diagnostic purposes, but also for precision medicine decisions, there are still significant limitations including the existence of variants with variable penetrance, variants of uncertain significance, and the presence of other susceptibility genetic factors [52]. Given the complexity of PD, it is strongly recommended to discuss the benefits and limitations of genetic testing during pre-test counseling sessions, including the risk of privacy loss and discrimination [53]. Since direct-to-consumer testing for common variants of *LRRK2* and *GBA* genes is currently available, care providers must be trained in genetic counseling to address consultants' concerns. Comprehensive genetic education and training of clinicians and patients together with efforts to promote universal access to genetic services are needed to massively translate PD genetic testing into clinical practice across the globe [54].

4. The New Era of Parkinson's Disease Genetics: Increasing Knowledge about Disease Etiology

The new era of PD genetics holds promise. Genetic studies conducted in underrepresented populations have started to emerge [55–57] as well as consortium setups [58] (GP2, https://www.gp2.org/, accessed on 20 September 2021) with programs strongly focused on increasing diversity in PD research so that the applications of genetic discoveries can be extrapolated to the entire population. As a result, an important analytical approach in our field will be the implementation of trans-ethnic GWAS meta-analysis, such as GWAMA [57] and MANTRA [59], that will allow us to combine genetic information from different ancestries to further delineate the etiology of this complex disease. GWAS data from non-European samples is considerably increasing, ensuring higher statistical power to improve fine-mapping strategies by leveraging the LD structure from different populations [59]. A considerable improvement in the fine-mapping resolution when studying data across highly ancestral heterogeneous samples has been shown as opposed to fine-mapping based on European ancestry only data [60].

When it comes to further exploring nominated loci from GWAS, the altered molecular pathways contributing to the phenotype of interest may be diverse [61]. Additionally, among the nominated loci, it is usually challenging to detect the causal variant underneath the peak, often masked by other non-causal alleles falling within the same haplotype block as a result of the underlying LD structure. In this context, the tuning of genotyping approaches as well as the development and implementation of novel bioinformatics tools are of paramount importance. On the one hand, some novel genotyping platforms, such as the recently created Neuro Booster array (unpublished manuscript), are focused on a wide SNP coverage, including more than 1.8 million variants (as compared to the roughly 400,000 variants from previous arrays [62,63]), and a custom content of approximately 95K neurological disease related variants. On the other hand, a wide range of publicly available and useful data science approaches allows us to interpret GWAS outcomes and further dissect potential loci. Fine-mapping methods represent a means to come up with the likely causal variant from a specific locus for a given phenotype and to determine the functional implications of such loci [64]. In very few situations, the causal variant will be within the locus of interest, affecting the protein conformation if the mutation is coding. Most likely, a causal genetic variant can be found within a non-modifiable or regulatory region, resulting in the dysregulation of the gene product of interest. Colocalization methods allow us to explore whether GWAS studies share a common genetic causal variant with tissue level and cell-state-specific expression quantitative trait loci (eQTL datasets), allowing us to link GWAS single nucleotide polymorphisms (SNPs) to the regulation of gene expression [64]. Moreover, functional fine-mapping methods give us insight into the putative epigenetic signatures of GWAS nominated loci, such as DNA methylation or histone modification of regulatory elements, as well as the formation of chromatin loops [65].

The identification of culprit variants affecting PD risk may be possible by the implementation of state of the art high-throughput long read sequencing technologies. Causal variants do not necessarily have to be SNPs, but can also be more complex genomic variation, such as repeat expansions or structural variants which are easily overlooked in short-read sequencing, and/or technologically challenging to genotype due to repetitive sequences or high GC content. PD studies looking at non-SNV variation are starting to emerge [66].

Undertaking integration of different level data (i.e., clinical data, genetics, transcriptomics, proteomics, and metabolomics) can be challenging and costly. Fortunately, it is worth highlighting that some frameworks that facilitate the process for post-GWAS analyses are available [67]. These platforms include large integrated biological datasets, making the automatization of concrete and parallel analyses possible, easing reproducibility and transparency. As we move forward, standardization and harmonization of datasets, as well as automating data processing is key. An example of this is GenoML (https://genoml.com/, accessed on 20 September 2021) which enables automatic machine learning in genetic studies and has been widely applied in the PD genetics field [68,69].

Overall, post-GWAS analyses are focused on approaches to prioritize molecular pathways and promising targets for biomarkers and drug development. By discovering and validating potential findings in independent cohorts, we can nominate pathways to be assessed in cell lines and animal models or build up networks. Moreover, novel datasets for PD genetics research are currently being made public resources to the research community. The Foundational Data Initiative for Parkinson's Disease (FOUNDIN-PD) [70] is an international, collaborative, and multi-year project, aiming to produce a multi-layered molecular dataset in a large cohort of 95 induced pluripotent stem cell (iPSC) lines at multiple time points during differentiation to dopaminergic (DA) neurons (https://www.foundinpd.org/#Foundinpd, accessed on 20 September 2021).

5. Future Perspectives

Over the last 20 years, in many ways, genetics has been the engine that has pushed us along on our voyage to gain knowledge about PD etiology. As we move forward, shedding light on the genetic architecture contributing to PD in non-European populations is essential and will provide novel insights regarding the generalised genetic map of the disease. This is a major commitment and a significant step forward for our field in an effort to understand how the basis of disease varies across populations.

We envisage that we will continue increasing the number of known genetic risk loci disease-causing mutations for the complex and variable manifestations of PD. Our field will keep investigating risk loci to saturation, genetic modifiers of disease, and genetically defined disease subtypes. In terms of genetic players underlying PD etiology, we anticipate that our field will expand our understanding of structural and repeat variability involved in disease through the application of long-read sequencing, which so far has been relatively difficult to explore using traditional genome sequencing methods.

However, in the future, our field will not just strive to improve our understanding of the role genetics plays in PD on a global scale, but to also make that understanding actionable. Worldwide initiatives will be key to the creation of publicly available resources for the scientific research community [16]. Multimodal data integration will facilitate translation of genetic maps to mechanisms and will improve our ability to develop more accurate models of disease prediction and prognosis.

It is not enough to just make data available to the wider PD research community, we must train the next generation of scientists. Training individuals with exceptional drive and talent will be key to success. The future of PD genetics ultimately aims to inform biology, improve disease modeling, promote target prioritization, inform trial design and efficiency, and develop therapeutic strategies matching patients to specific treatments.

Author Contributions: Authors have contributed equally to writing and revising the manuscript. All authors have read and agreed to the published version of the manuscript.

Funding: This research was supported, in part, by the Intramural Research Program of the National Institutes of Health (National Institute on Aging, National Institute of Neurological Disorders and Stroke: project numbers 1ZIA-NS003154, Z01-AG000949-02 and Z01-ES10198).

Institutional Review Board Statement: Not applicable.

Informed Consent Statement: Not applicable.

Data Availability Statement: Not applicable.

Acknowledgments: This work was carried out with the support and guidance of 'GP2 Trainee Network' which is part of the Global Parkinson's Disease Genetics Program and funded by the Aligning Science Across Parkinson's (ASAP) initiative.

Conflicts of Interest: The authors declare no conflict of interest.

References

1. Williams, C.B.; Bedenne, L. Management of Colorectal Polyps: Is All the Effort Worthwhile? *J. Gastroenterol. Hepatol.* **1990**, *5*, 144–165. [CrossRef]
2. Nalls, M.A.; Blauwendraat, C.; Vallerga, C.L.; Heilbron, K.; Bandres-Ciga, S.; Chang, D.; Tan, M.; Kia, D.A.; Noyce, A.J.; Xue, A.; et al. Identification of Novel Risk Loci, Causal Insights, and Heritable Risk for Parkinson's Disease: A Meta-Analysis of Genome-Wide Association Studies. *Lancet Neurol.* **2019**, *18*, 1091–1102. [CrossRef]
3. Blauwendraat, C.; Nalls, M.A.; Singleton, A.B. The Genetic Architecture of Parkinson's Disease. *Lancet Neurol.* **2020**, *19*, 170–178. [CrossRef]
4. Deng, H.; Wang, P.; Jankovic, J. The Genetics of Parkinson Disease. *Ageing Res. Rev.* **2018**, *42*, 72–85. [CrossRef]
5. Klein, C.; Westenberger, A. Genetics of Parkinson's Disease. *Cold Spring Harb. Perspect. Med.* **2012**, *2*, a008888. [CrossRef]
6. Hall, A.; Bandres-Ciga, S.; Diez-Fairen, M.; Quinn, J.P.; Billingsley, K.J. Genetic Risk Profiling in Parkinson's Disease and Utilizing Genetics to Gain Insight into Disease-Related Biological Pathways. *Int. J. Mol. Sci.* **2020**, *21*, 7332. [CrossRef]

7. Foo, J.N.; Chew, E.G.Y.; Chung, S.J.; Peng, R.; Blauwendraat, C.; Nalls, M.A.; Mok, K.Y.; Satake, W.; Toda, T.; Chao, Y.; et al. Identification of Risk Loci for Parkinson Disease in Asians and Comparison of Risk between Asians and Europeans: A Genome-Wide Association Study. *JAMA Neurol.* **2020**, *77*, 746–754. [CrossRef]
8. Iwaki, H.; Blauwendraat, C.; Leonard, H.L.; Kim, J.J.; Liu, G.; Maple-Grødem, J.; Corvol, J.-C.; Pihlstrøm, L.; van Nimwegen, M.; Hutten, S.J.; et al. Genomewide Association Study of Parkinson's Disease Clinical Biomarkers in 12 Longitudinal Patients' Cohorts. *Mov. Disord.* **2019**, *34*, 1839–1850. [CrossRef]
9. Blauwendraat, C.; Heilbron, K.; Vallerga, C.L.; Bandres-Ciga, S.; von Coelln, R.; Pihlstrøm, L.; Simón-Sánchez, J.; Schulte, C.; Sharma, M.; Krohn, L.; et al. Parkinson's Disease Age at Onset Genome-Wide Association Study: Defining Heritability, Genetic Loci, and α-Synuclein Mechanisms. *Mov. Disord.* **2019**, *34*, 866–875. [CrossRef]
10. Popejoy, A.B.; Fullerton, S.M. Genomics Is Failing on Diversity. *Nature* **2016**, *538*, 161–164. [CrossRef]
11. Sirugo, G.; Williams, S.M.; Tishkoff, S.A. The Missing Diversity in Human Genetic Studies. *Cell* **2019**, *177*, 26–31. [CrossRef]
12. Cai, M.; Liu, Z.; Li, W.; Wang, Y.; Xie, A. Association between rs823128 Polymorphism and the Risk of Parkinson's Disease: A Meta-Analysis. *Neurosci. Lett.* **2018**, *665*, 110–116. [CrossRef]
13. Ibanez, L.; Dube, U.; Saef, B.; Budde, J.; Black, K.; Medvedeva, A.; Del-Aguila, J.L.; Davis, A.A.; Perlmutter, J.S.; Harari, O.; et al. Parkinson Disease Polygenic Risk Score Is Associated with Parkinson Disease Status and Age at Onset but Not with α-Synuclein Cerebrospinal Fluid Levels. *BMC Neurol.* **2017**, *17*, 198. [CrossRef]
14. Tekola-Ayele, F.; Rotimi, C.N. Translational Genomics in Low- and Middle-Income Countries: Opportunities and Challenges. *Public Health Genomics* **2015**, *18*, 242–247. [CrossRef]
15. Sarfo, F.S.; Adamu, S.; Awuah, D.; Ovbiagele, B. Tele-Neurology in Sub-Saharan Africa: A Systematic Review of the Literature. *J. Neurol. Sci.* **2017**, *380*, 196–199. [CrossRef]
16. Global Parkinson's Genetics Program GP2: The Global Parkinson's Genetics Program. *Mov. Disord.* **2021**, *36*, 842–851. [CrossRef]
17. Rizig, M.; Okubadejo, N.; Salama, M.; Thomas, O.; Akpalu, A.; Gouider, R.; IPDGC Africa. The International Parkinson Disease Genomics Consortium Africa. *Lancet Neurol.* **2021**, *20*, 335. [CrossRef]
18. Erves, J.C.; Mayo-Gamble, T.L.; Malin-Fair, A.; Boyer, A.; Joosten, Y.; Vaughn, Y.C.; Sherden, L.; Luther, P.; Miller, S.; Wilkins, C.H. Needs, Priorities, and Recommendations for Engaging Underrepresented Populations in Clinical Research: A Community Perspective. *J. Community Health* **2017**, *42*, 472–480. [CrossRef]
19. Sanderson, S.C.; Diefenbach, M.A.; Zinberg, R.; Horowitz, C.R.; Smirnoff, M.; Zweig, M.; Streicher, S.; Jabs, E.W.; Richardson, L.D. Willingness to Participate in Genomics Research and Desire for Personal Results among Underrepresented Minority Patients: A Structured Interview Study. *J. Community Genet.* **2013**, *4*, 469–482. [CrossRef]
20. Nuytemans, K.; Manrique, C.P.; Uhlenberg, A.; Scott, W.K.; Cuccaro, M.L.; Luca, C.C.; Singer, C.; Vance, J.M. Motivations for Participation in Parkinson Disease Genetic Research Among Hispanics versus Non-Hispanics. *Front. Genet.* **2019**, *10*, 658. [CrossRef]
21. Schneider, M.G.; Swearingen, C.J.; Shulman, L.M.; Ye, J.; Baumgarten, M.; Tilley, B.C. Minority Enrollment in Parkinson's Disease Clinical Trials. *Parkinsonism Relat. Disord.* **2009**, *15*, 258–262. [CrossRef] [PubMed]
22. Schurz, H.; Müller, S.J.; van Helden, P.D.; Tromp, G.; Hoal, E.G.; Kinnear, C.J.; Möller, M. Evaluating the Accuracy of Imputation Methods in a Five-Way Admixed Population. *Front. Genet.* **2019**, *10*, 34. [CrossRef]
23. International Parkinson Disease Genomics Consortium (IPDGC) Ten Years of the International Parkinson Disease Genomics Consortium: Progress and Next Steps. *J. Parkinson's Dis.* **2020**, *10*, 19–30. [CrossRef]
24. Zabetian, C.P.; Mata, I.F. Latin American Research Consortium on the Genetics of PD (LARGE-PD) LARGE-PD: Examining the Genetics of Parkinson's Disease in Latin America. *Mov. Disord.* **2017**, *32*, 1330–1331. [CrossRef]
25. Sequeiros, J.; Paneque, M.; Guimarães, B.; Rantanen, E.; Javaher, P.; Nippert, I.; Schmidtke, J.; Kääriäainen, H.; Kristoffersson, U.; Cassiman, J.-J. The Wide Variation of Definitions of Genetic Testing in International Recommendations, Guidelines and Reports. *J. Community Genet.* **2012**, *3*, 113–124. [CrossRef] [PubMed]
26. Horton, R.H.; Lucassen, A.M. Recent Developments in Genetic/genomic Medicine. *Clin. Sci.* **2019**, *133*, 697–708. [CrossRef]
27. Payne, K.; Gavan, S.P.; Wright, S.J.; Thompson, A.J. Cost-Effectiveness Analyses of Genetic and Genomic Diagnostic Tests. *Nat. Rev. Genet.* **2018**, *19*, 235–246. [CrossRef]
28. Polymeropoulos, M.H.; Lavedan, C.; Leroy, E.; Ide, S.E.; Dehejia, A.; Dutra, A.; Pike, B.; Root, H.; Rubenstein, J.; Boyer, R.; et al. Mutation in the α-Synuclein Gene Identified in Families with Parkinson's Disease. *Science* **1997**, *276*, 2045–2047. [CrossRef]
29. Bandres-Ciga, S.; Diez-Fairen, M.; Kim, J.J.; Singleton, A.B. Genetics of Parkinson's Disease: An Introspection of Its Journey towards Precision Medicine. *Neurobiol. Dis.* **2020**, *137*, 104782. [CrossRef]
30. Tolosa, E.; Vila, M.; Klein, C.; Rascol, O. LRRK2 in Parkinson Disease: Challenges of Clinical Trials. *Nat. Rev. Neurol.* **2020**, *16*, 97–107. [CrossRef] [PubMed]
31. Pickrell, A.M.; Youle, R.J. The Roles of PINK1, Parkin, and Mitochondrial Fidelity in Parkinson's Disease. *Neuron* **2015**, *85*, 257–273. [CrossRef]
32. Ozelius, L.J.; Senthil, G.; Saunders-Pullman, R.; Ohmann, E.; Deligtisch, A.; Tagliati, M.; Hunt, A.L.; Klein, C.; Henick, B.; Hailpern, S.M.; et al. LRRK2 G2019S as a Cause of Parkinson's Disease in Ashkenazi Jews. *N. Engl. J. Med.* **2006**, *354*, 424–425. [CrossRef] [PubMed]
33. Shu, L.; Zhang, Y.; Sun, Q.; Pan, H.; Tang, B. A Comprehensive Analysis of Population Differences in Variant Distribution in Parkinson's Disease. *Front. Aging Neurosci.* **2019**, *11*, 13. [CrossRef]

34. Cornejo-Olivas, M.; Torres, L.; Velit-Salazar, M.R.; Inca-Martinez, M.; Mazzetti, P.; Cosentino, C.; Micheli, F.; Perandones, C.; Dieguez, E.; Raggio, V.; et al. Variable Frequency of Variants in the Latin American Research Consortium on the Genetics of Parkinson's Disease (LARGE-PD), a Case of Ancestry. *NPJ Parkinson's Dis.* **2017**, *3*, 19. [CrossRef] [PubMed]
35. Kestenbaum, M.; Alcalay, R.N. Clinical Features of LRRK2 Carriers with Parkinson's Disease. *Adv. Neurobiol.* **2017**, *14*, 31–48.
36. Trinh, J.; Guella, I.; Farrer, M.J. Disease Penetrance of Late-Onset Parkinsonism: A Meta-Analysis. *JAMA Neurol.* **2014**, *71*, 1535–1539. [CrossRef] [PubMed]
37. Rosborough, K.; Patel, N.; Kalia, L.V. α-Synuclein and Parkinsonism: Updates and Future Perspectives. *Curr. Neurol. Neurosci. Rep.* **2017**, *17*, 31. [CrossRef]
38. Lunati, A.; Lesage, S.; Brice, A. The Genetic Landscape of Parkinson's Disease. *Rev. Neurol.* **2018**, *174*, 628–643. [CrossRef]
39. Kasten, M.; Klein, C. The Many Faces of α-Synuclein Mutations. *Mov. Disord.* **2013**, *28*, 697–701. [CrossRef]
40. Fuchs, J.; Nilsson, C.; Kachergus, J.; Munz, M.; Larsson, E.-M.; Schüle, B.; Langston, J.W.; Middleton, F.A.; Ross, O.A.; Hulihan, M.; et al. Phenotypic Variation in a Large Swedish Pedigree due to SNCA Duplication and Triplication. *Neurology* **2007**, *68*, 916–922. [CrossRef]
41. Ryan, B.J.; Hoek, S.; Fon, E.A.; Wade-Martins, R. Mitochondrial Dysfunction and Mitophagy in Parkinson's: From Familial to Sporadic Disease. *Trends Biochem. Sci.* **2015**, *40*, 200–210. [CrossRef] [PubMed]
42. Kasten, M.; Hartmann, C.; Hampf, J.; Schaake, S.; Westenberger, A.; Vollstedt, E.-J.; Balck, A.; Domingo, A.; Vulinovic, F.; Dulovic, M.; et al. Genotype-Phenotype Relations for the Parkinson's Disease Genes Parkin, PINK1, DJ1: MDSGene Systematic Review. *Mov. Disord.* **2018**, *33*, 730–741. [CrossRef] [PubMed]
43. Niemann, N.; Jankovic, J. Juvenile Parkinsonism: Differential Diagnosis, Genetics, and Treatment. *Parkinsonism Relat. Disord.* **2019**, *67*, 74–89. [CrossRef] [PubMed]
44. Ligaard, J.; Sannæs, J.; Pihlstrøm, L. Deep Brain Stimulation and Genetic Variability in Parkinson's Disease: A Review of the Literature. *NPJ Parkinson's Dis.* **2019**, *5*, 18. [CrossRef]
45. Riboldi, G.M.; Di Fonzo, A.B. Gaucher Disease, and Parkinson's Disease: From Genetic to Clinic to New Therapeutic Approaches. *Cells* **2019**, *8*, 364. [CrossRef]
46. Cilia, R.; Tunesi, S.; Marotta, G.; Cereda, E.; Siri, C.; Tesei, S.; Zecchinelli, A.L.; Canesi, M.; Mariani, C.B.; Meucci, N.; et al. Survival and Dementia in GBA-Associated Parkinson's Disease: The Mutation Matters. *Ann. Neurol.* **2016**, *80*, 662–673. [CrossRef] [PubMed]
47. Do, J.; McKinney, C.; Sharma, P.; Sidransky, E. Glucocerebrosidase and Its Relevance to Parkinson Disease. *Mol. Neurodegener.* **2019**, *14*, 36. [CrossRef]
48. Hruska, K.S.; LaMarca, M.E.; Scott, C.R.; Sidransky, E. Gaucher Disease: Mutation and Polymorphism Spectrum in the Glucocerebrosidase Gene (GBA). *Hum. Mutat.* **2008**, *29*, 567–583. [CrossRef]
49. Krohn, L.; Grenn, F.P.; Makarious, M.B.; Kim, J.J.; Bandres-Ciga, S.; Roosen, D.A.; Gan-Or, Z.; Nalls, M.A.; Singleton, A.B.; Blauwendraat, C.; et al. Comprehensive Assessment of PINK1 Variants in Parkinson's Disease. *Neurobiol. Aging* **2020**, *91*, 168.e1–168.e5. [CrossRef]
50. Balestrino, R.; Schapira, A.H.V. Parkinson Disease. *Eur. J. Neurol.* **2020**, *27*, 27–42. [CrossRef]
51. Payne, K.; Walls, B.; Wojcieszek, J. Approach to Assessment of Parkinson Disease with Emphasis on Genetic Testing. *Med. Clin. N. Am.* **2019**, *103*, 1055–1075. [CrossRef]
52. Goldman, J.S. Predictive Genetic Counseling for Neurodegenerative Diseases: Past, Present, and Future. *Cold Spring Harb. Perspect. Med.* **2019**, *23*, a036525. [CrossRef] [PubMed]
53. Crook, A.; Jacobs, C.; Newton-John, T.; O'Shea, R.; McEwen, A. Genetic Counseling and Testing Practices for Late-Onset Neurodegenerative Disease: A Systematic Review. *J. Neurol.* **2021**. [CrossRef] [PubMed]
54. Alcalay, R.N.; Kehoe, C.; Shorr, E.; Battista, R.; Hall, A.; Simuni, T.; Marder, K.; Wills, A.-M.; Naito, A.; Beck, J.C.; et al. Genetic Testing for Parkinson Disease: Current Practice, Knowledge, and Attitudes among US and Canadian Movement Disorders Specialists. *Genet. Med.* **2020**, *22*, 574–580. [CrossRef]
55. Rajan, R.; Divya, K.P.; Kandadai, R.M.; Yadav, R.; Satagopam, V.P.; Madhusoodanan, U.K.; Agarwal, P.; Kumar, N.; Ferreira, T.; Kumar, H.; et al. Genetic Architecture of Parkinson's Disease in the Indian Population: Harnessing Genetic Diversity to Address Critical Gaps in Parkinson's Disease Research. *Front. Neurol.* **2020**, *11*, 524. [CrossRef] [PubMed]
56. Loesch, D.P.; Horimoto, A.R.V.R.; Heilbron, K.; Sarihan, E.I.; Inca-Martinez, M.; Mason, E.; Cornejo-Olivas, M.; Torres, L.; Mazzetti, P.; Cosentino, C.; et al. Characterizing the Genetic Architecture of Parkinson's Disease in Latinos. *Ann. Neurol.* **2021**, *90*, 353–365. [CrossRef]
57. Tipton, P.W.; Jaramillo-Koupermann, G.; Soto-Beasley, A.I.; Walton, R.L.; Soler-Rangel, S.; Romero-Osorio, Ó.; Díaz, C.; Moreno-López, C.L.; Ross, O.A.; Wszolek, Z.K.; et al. Genetic Characterization of Parkinson's Disease Patients in Ecuador and Colombia. *Parkinsonism Relat. Disord.* **2020**, *75*, 27–29. [CrossRef] [PubMed]
58. Williams, U.; Bandmann, O.; Walker, R. Parkinson's Disease in Sub-Saharan Africa: A Review of Epidemiology, Genetics and Access to Care. *J. Mov. Disord.* **2018**, *11*, 53–64. [CrossRef]
59. Morris, A.P. Transethnic Meta-Analysis of Genomewide Association Studies. *Genet. Epidemiol.* **2011**, *35*, 809–822. [CrossRef]
60. Asimit, J.L.; Hatzikotoulas, K.; McCarthy, M.; Morris, A.P.; Zeggini, E. Trans-Ethnic Study Design Approaches for Fine-Mapping. *Eur. J. Hum. Genet.* **2016**, *24*, 1330–1336. [CrossRef]

61. Bandres-Ciga, S.; Saez-Atienzar, S.; Kim, J.J.; Makarious, M.B.; Faghri, F.; Diez-Fairen, M.; Iwaki, H.; Leonard, H.; Botia, J.; Ryten, M.; et al. Large-Scale Pathway Specific Polygenic Risk and Transcriptomic Community Network Analysis Identifies Novel Functional Pathways in Parkinson Disease. *Acta Neuropathol.* **2020**, *140*, 341–358. [CrossRef] [PubMed]
62. Nalls, M.A.; Bras, J.; Hernandez, D.G.; Keller, M.F.; Majounie, E.; Renton, A.E.; Saad, M.; Jansen, I.; Guerreiro, R.; Lubbe, S.; et al. NeuroX, a Fast and Efficient Genotyping Platform for Investigation of Neurodegenerative Diseases. *Neurobiol. Aging* **2015**, *36*, 1605.e7–1605.e12. [CrossRef] [PubMed]
63. Blauwendraat, C.; Faghri, F.; Pihlstrom, L.; Geiger, J.T.; Elbaz, A.; Lesage, S.; Corvol, J.-C.; May, P.; Nicolas, A.; Abramzon, Y.; et al. NeuroChip, an Updated Version of the NeuroX Genotyping Platform to Rapidly Screen for Variants Associated with Neurological Diseases. *Neurobiol. Aging* **2017**, *57*, 247.e9–247.e13. [CrossRef]
64. Benner, C.; Spencer, C.C.A.; Havulinna, A.S.; Salomaa, V.; Ripatti, S.; Pirinen, M. FINEMAP: Efficient Variable Selection Using Summary Data from Genome-Wide Association Studies. *Bioinformatics* **2016**, *32*, 1493–1501. [CrossRef] [PubMed]
65. Kichaev, G.; Yang, W.-Y.; Lindstrom, S.; Hormozdiari, F.; Eskin, E.; Price, A.L.; Kraft, P.; Pasaniuc, B. Integrating Functional Data to Prioritize Causal Variants in Statistical Fine-Mapping Studies. *PLoS Genet.* **2014**, *10*, e1004722. [CrossRef] [PubMed]
66. Bustos, B.I.; Billingsley, K.; Blauwendraat, C.; Raphael Gibbs, J.; Gan-Or, Z.; Krainc, D.; Singleton, A.B.; Lubbe, S.J. For the International Parkinson's Disease Genomics Consortium (IPDGC). Genome-Wide Contribution of Common Short-Tandem Repeats to Parkinson's Disease Genetic Risk. *medRxiv* **2021**. [CrossRef]
67. Watanabe, K.; Taskesen, E.; van Bochoven, A.; Posthuma, D. Functional Mapping and Annotation of Genetic Associations with FUMA. *Nat. Commun.* **2017**, *8*, 1826. [CrossRef]
68. Faghri, F.; Hashemi, S.H.; Leonard, H.; Scholz, S.W.; Campbell, R.H.; Nalls, M.A.; Singleton, A.B. Predicting Onset, Progression, and Clinical Subtypes of Parkinson Disease Using Machine Learning. *bioRxiv* **2018**, 338913. [CrossRef]
69. Makarious, M.B.; Leonard, H.L.; Vitale, D.; Iwaki, H.; Sargent, L.; Dadu, A.; Violich, I.; Hutchins, E.; Saffo, D.; Bandres-Ciga, S.; et al. Multi-Modality Machine Learning Predicting Parkinson's Disease. *bioRxiv* **2021**. [CrossRef]
70. Bressan, E.; Reed, X.; Bansal, V.; Hutchins, E.; Cobb, M.M.; Webb, M.G.; Alsop, E.; Grenn, F.P.; Illarionova, A.; Savytska, N.; et al. The Foundational Data Initiative for Parkinson's Disease (FOUNDIN-PD): Enabling Efficient Translation from Genetic Maps to Mechanism. *bioRxiv* **2021**. [CrossRef]

Article

Comparative Transcriptome Analysis in Monocyte-Derived Macrophages of Asymptomatic *GBA* Mutation Carriers and Patients with GBA-Associated Parkinson's Disease

Tatiana Usenko [1,2,*], Anastasia Bezrukova [1], Katerina Basharova [1], Alexandra Panteleeva [1,2], Mikhail Nikolaev [1,2], Alena Kopytova [1], Irina Miliukhina [1,2,3], Anton Emelyanov [1,2], Ekaterina Zakharova [4] and Sofya Pchelina [1,2]

1. Petersburg Nuclear Physics Institute Named by B.P. Konstantinov of National Research Centre «Kurchatov Institute», Gatchina 188300, Russia; bezrukova_ai@pnpi.nrcki.ru (A.B.); kbasharova@yandex.ru (K.B.); panteleeva_aa@pnpi.nrcki.ru (A.P.); nikolaev_ma@pnpi.nrcki.ru (M.N.); kopytova_ae@pnpi.nrcki.ru (A.K.); milyukhinaiv@yandex.ru (I.M.); emelyanov_ak@pnpi.nrcki.ru (A.E.); pchelina_sn@pnpi.nrcki.ru (S.P.)
2. Pavlov First Saint-Petersburg State Medical University, Saint-Petersburg 197022, Russia
3. Institute of the Human Brain of RAS, Saint-Petersburg 197376, Russia
4. Research Center for Medical Genetics, Moscow 115522, Russia; doctor.zakharova@gmail.com
* Correspondence: usenko_us@pnpi.nrcki.ru; Tel.: +81137146093

Abstract: Mutations of the *GBA* gene, encoding for lysosomal enzyme glucocerebrosidase (GCase), are the greatest genetic risk factor for Parkinson's disease (PD) with frequency between 5% and 20% across the world. N370S and L444P are the two most common mutations in the *GBA* gene. PD carriers of severe mutation L444P in the *GBA* gene is characterized by the earlier age at onset compared to N370S. Not every carrier of *GBA* mutations develop PD during one's lifetime. In the current study we aimed to find common gene expression signatures in PD associated with mutation in the *GBA* gene (GBA-PD) using RNA-seq. We compared transcriptome of monocyte-derived macrophages of 5 patients with GBA-PD (4 L444P/N, 1 N370S/N) and 4 asymptomatic *GBA* mutation carriers (GBA-carriers) (3 L444P/N, 1 N370S/N) and 4 controls. We also conducted comparative transcriptome analysis for L444P/N only GBA-PD patients and GBA-carriers. Revealed deregulated genes in GBA-PD independently of *GBA* mutations (L444P or N370S) were involved in immune response, neuronal function. We found upregulated pathway associated with zinc metabolism in L444P/N GBA-PD patients. The potential important role of *DUSP1* in the pathogenesis of GBA-PD was suggested.

Keywords: Parkinson's disease; *GBA*; macrophages; RNA-seq; transcriptome

1. Introduction

Parkinson's disease (PD) is a neurodegenerative disorder that is characterized by the accumulation of abnormal protein aggregates of alpha-synuclein in the brain [1,2]. Several genetic factors have been associated with an increased risk of PD development. Mutations in the *GBA* gene are the highest genetic risk factors for PD with an increase of PD risk (of seven to eight times) and with a frequency of 5% to 20% in all populations [3,4]. The *GBA* gene, encoding the lysosomal enzyme glucocerebrosidase (GCase), is the key enzyme in ceramide metabolism and catalyzes the hydrolysis of glucosylceramide to glucose and ceramide. GCase is expressed in most tissues, especially in the brain, endocrine issue, liver, spleen, skin (https://www.proteinatlas.org, accessed on 21 September 2021). *GBA* mutations resulted in the most common lysosomal storage disorder (LSD), Gaucher disease (GD), characterized with lysosphingolipid accumulation, presumably in blood macrophages. Generally, the two most common mutations in the *GBA* gene N370S (c.1226A > G) and L444P (c.1448 T > C) account for 60–70% of the mutant alleles amongst others [4,5]. PD carriers of the severe L444P mutation in the *GBA* gene are characterized by an earlier age at onset and rapid progression [6] compared to N370S and other mild mutations. The molecular mechanisms of an association between *GBA* mutations and PD are unclear [7]. We, and others, have previously demonstrated that mutations in the *GBA* gene lead to a decrease of GCase activity and an increase of blood lysosphingolipid concentration, even in heterozygous carriers of *GBA* mutations [8–11]. However, not all carriers of *GBA* mutations develop PD. GCase dysfunction does not seem to be enough to launch the pathogenic mechanism of PD among *GBA* mutation carriers.

Transcriptome analysis using next-generation sequencing (RNA-seq) is a powerful method to analyze the genome transcriptomic profile with high-resolution. Although variations in the transcriptome are tissue specific, the blood and brain demonstrated significant gene expression similarities [12,13]. It is worth noting that RNA-seq revealed the difference between transcriptomic profiles in the peripheral blood of symptomatic and asymptomatic G2019S *LRRK2* mutation carriers and identified common differentially expression genes functionally involved in the pathways and related with LRRK2-PD pathogenesis, such as Akt signaling, glucose metabolism, or immunity [14,15]. Monocyte-derived macrophages represent one of the most promising models for investigating molecular mechanisms of GCase dysfunction, as this cell type is vulnerable for disturbances in ceramide metabolism [16,17]. In particular, we and others demonstrated high potential of peripheral blood monocyte-derived macrophages to reflect individual sensitivity for drugs influencing GCase activity [18,19]. Here, we first generated the transcriptomic profiles for GBA-PD patients, asymptomatic *GBA* mutation carriers (GBA carriers), and controls in monocyte-derived macrophages, in order to investigate what variations in monocyte-derived macrophage transcriptomes can be attributed to the presence of *GBA* mutation and what can be viewed as a trigger of PD in *GBA* mutation carriers. Our results will be useful to others looking for potential triggers of PD among *GBA* mutation carriers, and provides future directions for PD preclinical research.

2. Materials and Methods

This project was approved by the Pavlov First Saint-Petersburg State Medical University. A formal written consent form was provided to all included subjects to read and sign prior to the study.

2.1. Subjects

Five patients with GBA-PD, four GBA carriers, and four neurologically healthy individuals were enrolled for the current study. Demographic data of the studied groups are summarized in Table 1. Controls had no history of parkinsonism. GBA-PD patients were diagnosed at two neurological clinic centers in St. Petersburg, Russia: Pavlov First Saint-Petersburg State Medical University and the Institute of the Human Brain of RAS. A standard neurologic clinical examination was performed for all participants and the diagnosis of PD was based on previously published criteria [20]. GBA-PD patients were recruited by genotyping of N370S, L444P mutations in the *GBA* gene among PD patients, as previously described [3]. GBA carriers were collected from first-degree relatives of GD patients at the Research Centre for Medical Genetics where *GBA* mutations were confirmed by target sequencing of all exons in the *GBA* gene.

Table 1. Demographic characteristics of the compared groups.

Groups	Age at Exam, Mean ± SD, Years	Age at Onset, Mean ± SD, Years	Gender (Male:Female)	Mutations in the *GBA* Gene
GBA-PD, N = 5	53.5 ± 8.73	49.0 ± 10.89	3:2	4 L444P/N 1 N370S/N
GBA-carriers, N = 4	54.9 ± 8.9	-	2:2	3 L444P/N 1 N370S/N
Controls, N = 4	54.4 ± 9.5	-	2:2	-

GBA-PD—Parkinson's disease associated with mutations in the *GBA* gene; GBA-carriers—asymptomatic *GBA* mutation carriers; SD—standard deviation.

2.2. Differentiation of Human Monocytes to Macrophages

Peripheral blood mononuclear cells (PBMCs) were isolated from 24 mL of peripheral blood from participants, by density gradient centrifugation (Ficoll–Paque PLUS, GE Healthcare, Chicago, IL, USA). PBMCs were differentiated by the macrophage colony-stimulating factor (M-CSF) (10 ng/ml) (Sigma-Aldrich, Burlington, MA, USA) in RPMI 1640 medium (Gibco, Waltham, MA, USA) supplemented with 10% FCS (Gibco, Waltham, MA, USA) with harvesting after 5 days. Phenotypical maturation of monocyte-derived macrophages was confirmed by light microscopy and flow cytometry with specific antibodies to CD14+ and CD68+ (eBioscience, San Diego, CA, USA), as described earlier [18,21].

2.3. RNA Isolation and RNA Sequencing (RNA-Seq)

RNA was isolated from monocyte-derived macrophages and amplified following the user manual of the SMART-Seq™ v4 Ultra™ Low Input RNA Kit for sequencing. Sequencing libraries were generated using the NEBNext® Ultra™ DNA Library Prep Kit for Illumina® (NEB, Ipswich, MA, USA), following the manufacturer's recommendations. The RNA molecules that contained polyA were then sequenced on the Illumina HiSeq1500 platform.

2.4. Quality Control

Quality control for each sample was performed by FastQC (v0.11.9) [22] and RSeQC (v4.0.0)) [23]. In this step, clean data (clean reads) were obtained by removing low-quality reads, reads containing adapters, and reads containing ploy-N from raw data. The removal adapter was conducted with Cutadapt [24]. All downstream analyses were based on clean data.

2.5. Reads Mapping to Reference Genome

Human reference genome assembly GRCh38 (hg38) and gene model annotation files were downloaded from the Gencode website (https://www.gencodegenes.org/human/ (accessed on 9 September 2021)) directly (release 37). HISAT2 (v2.2.1) [25] was used with default parameters to build the index of the reference genome and mapping reads to the genome.

2.6. Quantification of Gene Expression Level

Counting sequencing reads mapping to each gene after the alignment step was performed using the HTSeq-count function from the HTSeq framework (v.0.6.1) [26].

2.7. Analysis of Gene Differential Expression

Gene differential expression analyses of three groups were performed using the DESeq2 package (v.1.30.1) [27] in R (v.4.0.3). DESeq2 provides statistical routines for determining differential expression in digital gene expression data using a model based on negative binomial distribution. The resulting p-values were adjusted using Benjamini and Hochberg's approach for controlling the false discovery rate (FDR). Detected differential expression of genes was considered statistically significant at FDR ≤ 0.05 and a fold change (FC) threshold >1.5. The differentially expressed genes were visualized in a volcano plot built by using ggplot (v.3.3.3) in R (v4.0.3).

2.8. GO Enrichment Analysis of Differentially Expressed Genes

Gene Ontology (GO) enrichment analysis of differentially expressed genes was performed using GO resource (http://geneontology.org (accessed on 9 September 2021)) and was carried out using the apps ClueGO v. 2.5.7 [28] and CluePedia v. 1.5.3 [29] for Cytoscape v. 3.6.1. GO terms with a corrected p-value of less than 0.05. Term groups were selected by ClueGO based on the number of common genes/terms (>50%). Term clusters were selected based on common genes. A network of selected metabolic processes and DEGs was built using CluePedia v. 1.5.7.

3. Results

3.1. RNA-Seq Experiments

A whole-transcriptome analysis of monocyte-derived macrophages obtained from four patients with L444P/N GBA-PD, three L444P/N GBA carriers, and controls without any *GBA* mutations ($N = 4$) was performed. Transcriptome analysis of monocyte-derived macrophages was also conducted for all GBA-PD patients (L444P/N, $N = 4$, N370S/N, $N = 1$), and GBA carriers (L444P/N, $N = 3$, N370S/N, $N = 1$). Using the Illumina HiSeq 1500 sequencer, we generated 10–14M raw reads, trimming from the 13 samples, with a read length of 50 bp. After strict quality control, more than 20G clean bases were retained. Overall, 21,980 genes were identified in each of the 13 samples. Post-trimming and mapping results for all groups are provided in Table S1. Between 85.20% and 95.97% of the clean reads was aligned to the reference genome. Raw data were subjected to differential expression testing with DESeq2.

3.2. Changes in the Transcriptome Attributed to the Presence of GBA L444P/N Mutation

First, we conducted comparative transcriptome analysis of GBA-PD patients baring L444P/N mutation and controls, which revealed 32 DEGs, and asymptomatic carriers of the *GBA* L444P/N mutation and controls, which revealed 18 DEGs (Tables S2 and S3, Figure 1A,B). Moreover, 36 DEGs were revealed between L444P/N GBA-PD patients and L444P/N GBA carriers (Table S4, Figure 1C). The top list of revealed DEGs in L444P/N GBA-PD patients compared to controls included the genes, *JUNB*, *NR4A2*, and *EGR1*, which played roles in neurogenesis. GBA-PD was characterized by downregulated expression of those genes. GO term enrichment analysis was conducted for all determined DEGs. We considered "metabolic process" terms with a *p*-value (Bonferroni corrected) <0.05 and all types of GO terms to gene connections. Significant terms are presented in Table 2 and networks are performed (Figure 2A–C). Pathways from GO databases enriched by DEGs that were found when comparing GBA-PD patients to the controls were associated with cytokine secretion (cellular response to chemokine (GO:1990869) and immune response (monocyte chemotaxis (GO:0002548), neutrophil chemotaxis (GO:0030593), and myeloid leukocyte migration (GO: 0097529)) (Table 2, Figure 2A). Altered biological GO pathways in L444P/N GBA-PD patients compared to L444P/N GBA carriers were the pathways related to cellular response to cadmium ion (GO:0071276), cellular response to zinc ion (GO:0071294), cellular zinc ion homeostasis (GO:0006882), detoxification of copper ion (GO:0010273), cellular response to copper ion (GO:0071280) (Table 2, Figure 2B). The 13 genes deregulated in L444P/N GBA carriers compared to the controls were involved in the enriched pathways related to immune response (system development (GO:0048731), immune system development (GO:0002520), myeloid leukocytes differentiation (GO:0002573)), and regulation negation axon extension involved in regeneration (GO:0048692) and axon extension involved in regeneration (GO:0048677) (Table 2, Figure 2C). The Venn diagram demonstrated one upregulated DEG, *KIAA0319*, which was upregulated in both L444P/N GBA-PD patients and L444P/N GBA carriers compared to the controls and two DEGs, *DUSP1* and *ARL4C*, which were downregulated in L444P/N GBA-PD patients compared to both L444P/N GBA carriers and controls (Figure 3A). The comparison between the list of DEGs from the GO analysis and the list of DEGs obtained by the Venn diagram revealed five genes (*IL31RA*, *ACOD1*, *OSCAR*, *MT1M*, *TBX3*) downregulated in L444P/N GBA carriers compared to L444P/N GBA-PD and controls, and one downregulated gene (*DUSP1*) in L444P/N GBA-PD patients compared to L444P/N GBA carriers and controls (Figure 3A), and one upregulated gene (*KIAA0319*) in L444P/N GBA-PD patients and L444P/N GBA carriers compared to controls (Figure 3B).

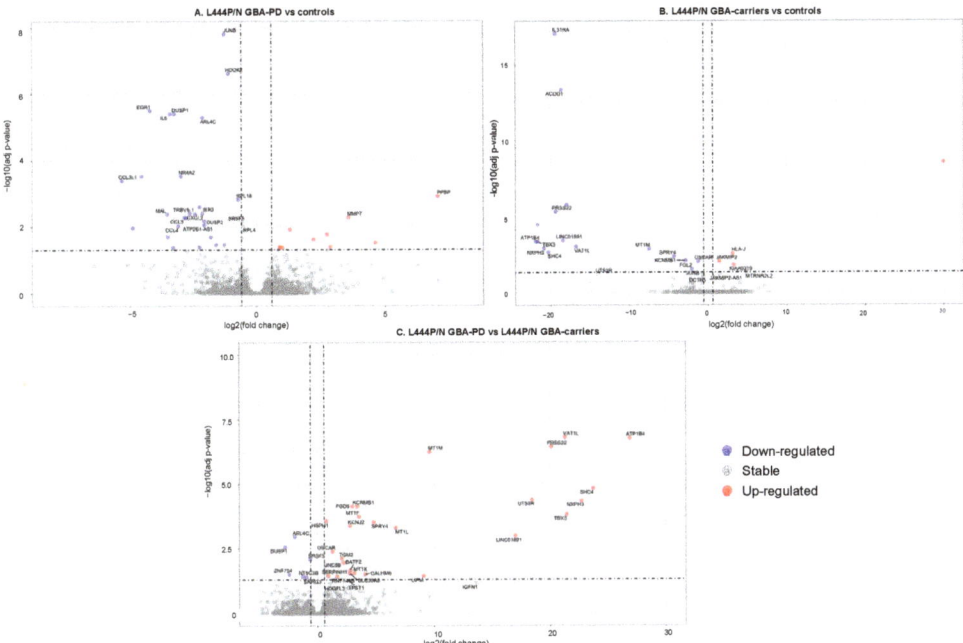

Figure 1. Volcano plot for DEGs between the studied groups (FDR < 0.05 and |FC| > 1.5); the upregulated genes are represented by red dots and the downregulated genes are represented by blue dots. (**A**). L444P/N GBA-PD patients and controls, (**B**). L444P/N GBA carriers and controls, (**C**). L444P/N GBA-PD patients, and L444P/N GBA carriers. (GBA-PD—Parkinson's disease associated with mutations in the *GBA* gene; GBA carriers—asymptomatic *GBA* mutation carriers; DEGs—differentially expressed genes).

Table 2. Functional clusters selected according to the results of the GO analysis between L444P/N GBA-PD patients, L444P/N GBA carriers, and controls.

(GO ID) GO Terms	P adjusted	DEGs
L444P/N GBA-PD vs. Controls		
(GO:0097529) myeloid leukocyte migration	6.93×10^{-9}	CCL3, CCL3L1, CCL4, CXCL2, CXCL5, DUSP1, IL6, PPBP
(GO:0002548) monocyte chemotaxis	6.93×10^{-9}	CCL3, CCL3L1, CCL4, DUSP1, IL6
(GO:1990869) cellular response to chemokine	6.93×10^{-9}	CCL3, CCL3L1, CCL4, CXCL2, CXCL5, DUSP1, PPBP
(GO:0030593) neutrophil chemotaxis	6.93×10^{-9}	CCL3, CCL3L1, CCL4, CXCL2, CXCL5, PPBP
L444P/N GBA-PD vs. L444P/N GBA carriers		
(GO:0006882) cellular zinc ion homeostasis	8.09×10^{-9}	MT1F, MT1L, MT1M, MT1X, SLC39A8
(GO:0010273) detoxification of copper ion	9.81×10^{-7}	MT1F, MT1L, MT1M, MT1X
(GO:0071276) cellular response to cadmium ion	9.81×10^{-7}	MT1F, MT1L, MT1M, MT1X
(GO:0071280) cellular response to copper ion	9.81×10^{-7}	MT1F, MT1L, MT1M, MT1X
(GO:0071294) cellular response to zinc ion	9.81×10^{-7}	MT1F, MT1L, MT1M, MT1X
L444P/N GBA carriers vs. controls		
(GO:0048731) system development	0.001035	ACOD1, IL31RA, KIAA0319, OSCAR, TBX3
(GO:0002520) immune system development	0.001035	ACOD1, IL31RA, OSCAR
(GO:0002573) myeloid leukocyte differentiation	0.001011	IL31RA, OSCAR
(GO:0048692) negative regulation of axon extension involved in regeneration	0.000615	KIAA0319
(GO:0048677) axon extension involved in regeneration	0.000615	KIAA0319

GBA-PD—Parkinson's disease associated with mutations in the *GBA* gene; GBA carriers—asymptomatic *GBA* mutation carriers; DEGs—differentially expressed genes; GO—gene ontology.

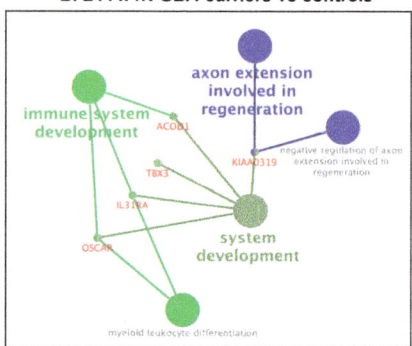

Figure 2. Networks of selected metabolic processes and DEGs in (**A**). L444P/N GBA-PD vs. controls; (**B**). L444P/N GBA carriers vs. controls; (**C**). L444P/N GBA-PD vs. L444P/N GBA carriers (obtained using CluePedia v. 1.5.7 + ClueGo v.2.5.7). (GBA-PD—Parkinson's disease associated with mutations in the *GBA* gene; GBA carriers—asymptomatic *GBA* mutation carriers; DEGs—differentially expressed genes).

3.3. Differentially Expressed Genes and Enriched Pathways in GBA-PD Patients (L444P/N +N370S/N) and GBA Carriers (L444P/N +N370S/N) Compared to Controls

Comparative transcriptome analysis of monocyte-derived macrophages revealed 23 DEGs between GBA-PD patients and GBA carriers, 28 DEGs between GBA-PD patients and controls. Moreover, eight DEGs were found between GBA carriers compared to controls (Figure 4A,B, Tables S5–S7.) The top list also revealed DEGs in GBA-PD patients compared to controls, including the genes, *JUNB*, *NR4A2*, *EGR1*. Significant terms of GO analysis between GBA-PD patients, GBA carriers, and controls are presented in Table S8, and networks are presented in Figure 5A,B. A total of 25 genes were enriched in 17 GO pathways. The altered biological pathways in GBA-PD patients compared to the controls were directly related to the functioning of the immune system, immune response, cytokine metabolism, and the immune response ((GO:0019221) cytokine-mediated signaling pathway, (GO:0006935) chemotaxis, (GO:0002548) monocyte chemotaxis, (GO:1990869) cellular response to chemokine), and apoptosis ((GO:0010941) regulation of cell death, (GO:0010942) positive regulation of cell death). The main of the alerted GO pathways in GBA carriers compared to the controls was the pathway associated with cytokine metabolism ((GO:0071345) cellular response to cytokine stimulus). The Venn diagram demonstrated two genes *HOOK2*, *JUNB* downregulated in GBA-PD patients and GBA carriers compared to controls that can be attributed to the presence of *GBA* mutations. The *HOOK2*, *JUNB* genes were also involved in the enriched pathway (response to cytokines (GO:0034097)) identified by GO analysis in GBA-PD patients compared to controls (Figure 6A).

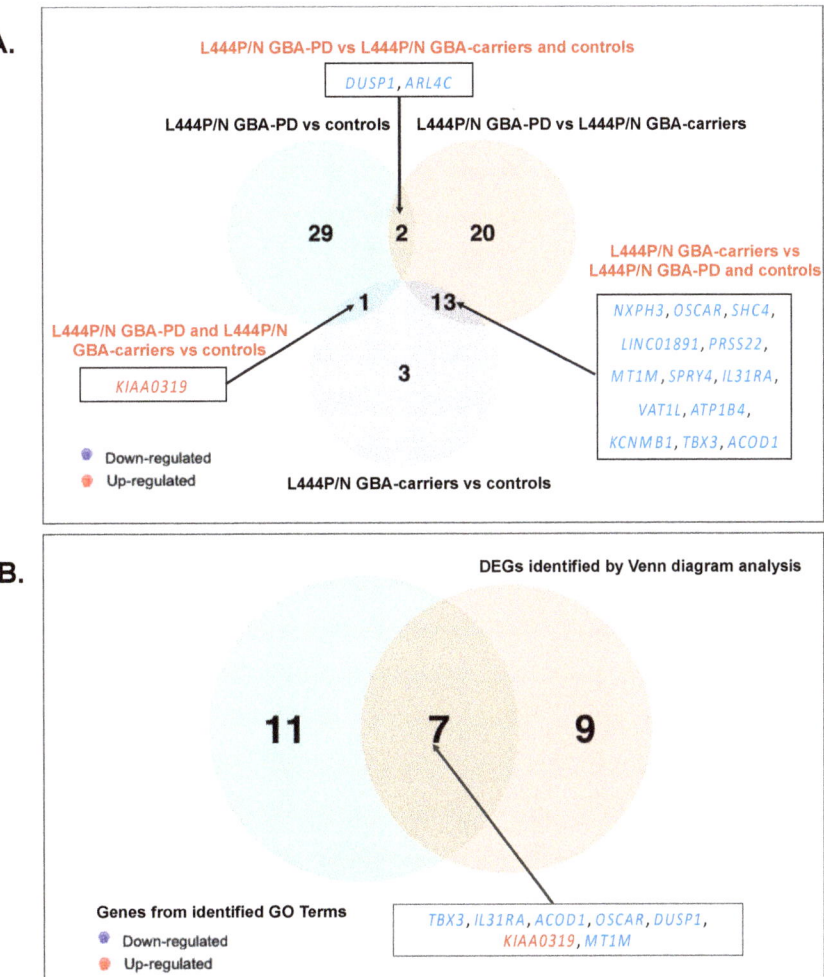

Figure 3. Venn diagram of (**A**). DEGs in monocyte-derived macrophages of L444P/N GBA-PD patients to controls compared to L444P/N GBA-PD patients to L444P/N GBA carriers, and compared to L444P/N GBA carriers and controls. B. DEGs determined be the Venn diagram in (**B**) and DEGs determined by GO analysis for L444P/N GBA-PD patients, L444P/N GBA carriers, controls. All data are presented as the number of genes with a p-value < 0.05 and |FC| more than 1.5. Three Venn diagrams were developed using the library VennDiagram (v.1.6.20) in R (v.4.0.3). (GBA-PD—Parkinson's disease associated with mutations in the *GBA* gene; GBA carriers—asymptomatic *GBA* mutation carriers; DEGs—differentially expressed genes; GO—gene ontology).

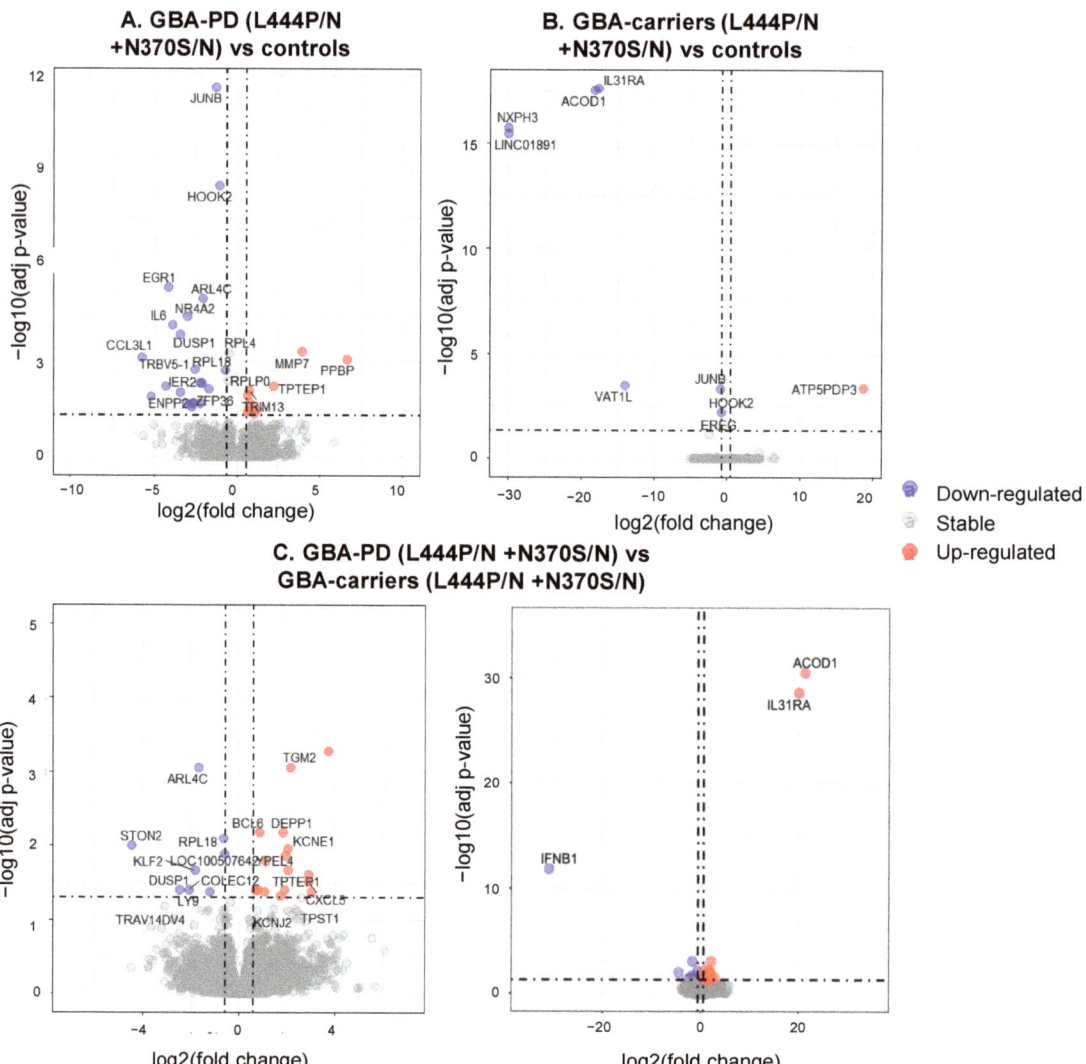

Figure 4. Volcano plot for DEGs between the studied groups (FDR < 0.05 and |FC| > 1.5); the upregulated genes are represented by red dots and the downregulated genes are represented by blue dots. (**A**). GBA-PD patients and controls; (**B**). GBA carriers and controls; (**C**). GBA-PD patients and GBA carriers. (GBA-PD—Parkinson's disease associated with mutations in the *GBA* gene; GBA carriers—asymptomatic *GBA* mutation carriers; DEGs—differentially expressed genes).

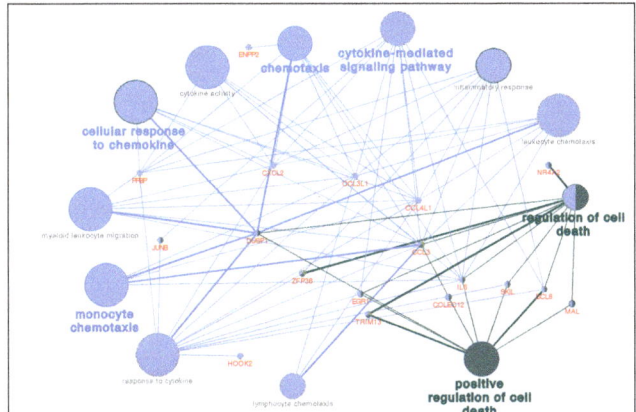
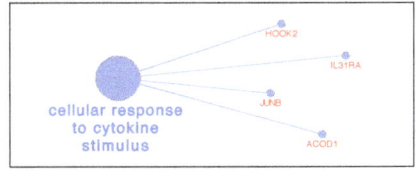
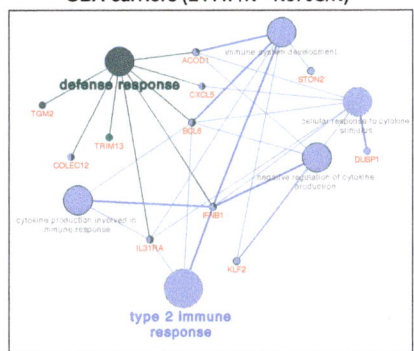

Figure 5. Networks of selected metabolic processes and DEGs in (**A**). GBA-PD (L444P/N +N370S/N) vs. controls; (**B**). GBA carriers (L444P/N +N370S/N) vs. controls; (**C**). GBA-PD (L444P/N +N370S/N) vs. GBA carriers (L444P/N +N370S/N) (obtained using CluePedia v. 1.5.7 + ClueGo v.2.5.7). (GBA-PD—Parkinson's disease associated with mutations in the *GBA* gene; GBA carriers—asymptomatic *GBA* mutation carriers; DEGs—differentially expressed genes).

3.4. Differential Expression of Genes and Pathways in GBA-PD Patients (L444P/N +N370S/N) and GBA Carriers (L444P/N +N370S/N)

Differential expression analysis of monocyte-derived macrophages resulted in 23 DEGs in GBA-PD patients compared to GBA carriers (Figure 4C). GO analysis showed the main altered pathways that are related to immune response ((GO:0042092) type 2 immune response, (GO:0006952) defense response) (Table S8, Figure 5C). The Venn diagram allowed us to reveal seven overlapping DEGs (*DUSP1, ALR4C, RPL16, TPTEP1, COLEC12, TRIM13, BCL6*) among GBA-PD patients when compared with GBA carriers and controls and two genes (*ACOD1, IL31RA*) between GBA carriers compared with GBA-PD patients and controls (Figure 6B). The comparison between the list of DEGs from GO analysis and list of DEGs obtained by the Venn diagram revealed two genes (*IL31RA, ACOD1*) downregulated in GBA carriers, compared to GBA-PD and controls, and four deregulated genes (two (*DUSP1, COLEC12*) downregulated and two (*TRIM13, BCL6*) upregulated) in GBA-PD patients, compared to GBA carriers and controls (Figure 6B).

3.5. Searching the Overlapping DEGs between our and Publicly Available Dataset

To identify the similarities between lists of DEGs from our and previously published studies we used Venn diagram. We revealed overlapping DEGs between list of DEGs from our analysis of GBA-PD (L444P/N +N370S/N) and list of DEGs of G2019S LRRK2-PD from study of Infante and colleagues: two genes encoding monocyte attracting chemokines, such as *CCL3L1* gene, in GBA-PD, G2019S LRRK2-PD and PD in comparison with controls, and the *CCL3* gene, when comparing GBA-PD to controls and G2019 LRRK2-PD to PD and also, *JUNB* gene when comparing GBA-PD, GBA-carriers, G2019 LRRK2-PD and G2019 LRRK2-carriers [15]. Dataset of Infante and colleagues' study is available for download following this link: https://ars.els-cdn.com/content/image/1-s2.0-S0197458015005382-mmc1.doc [15] (accessed on 9 September 2021).

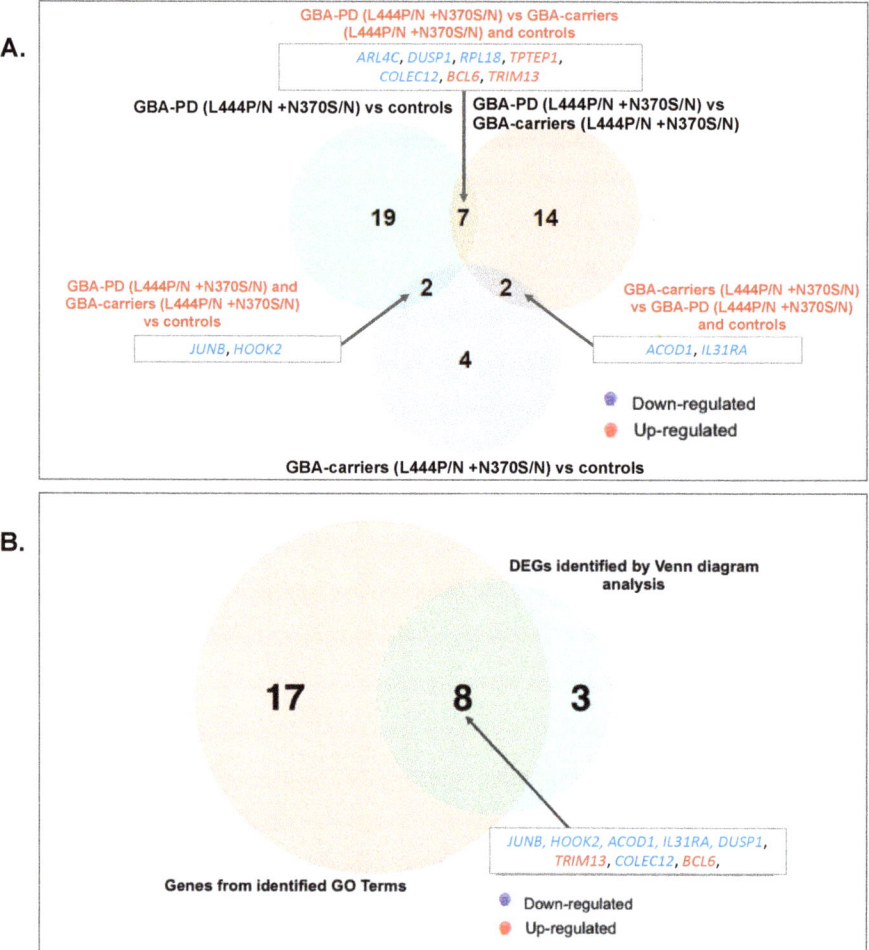

Figure 6. Venn diagram of (**A**). DEGs in monocyte-derived macrophages of GBA-PD (L444P/N +N370S/N) patients to controls compared to GBA-PD (L444P/N +N370S/N) patients to GBA carriers (L444P/N +N370S/N), and compared to GBA carriers (L444P/N +N370S/N) and controls; B. DEGs determined be Venn diagram in (**B**) and DEGs determined by GO analysis for GBA-PD (L444P/N +N370S/N) patients, GBA carriers (L444P/N +N370S/N), controls. All data are presented as the number of genes with a p-value < 0.05 and |FC| more than 1.5. Three Venn diagrams were developed using the library VennDiagram (v.1.6.20) in R (v.4.0.3). (GBA-PD—Parkinson's disease associated with mutations in the *GBA* gene; GBA carriers—asymptomatic *GBA* mutation carriers; DEGs—differentially expressed genes; GO—gene ontology).

4. Discussion

This is the first whole-transcriptome analysis of monocyte-derived macrophages in GBA-PD patients, GBA carriers, and controls. We intended to cover molecular pathways involved in GBA-PD pathogenesis and study the differences in the transcriptome between *GBA* mutation carriers with and without PD. To date, the last review of genome-wide transcriptomic studies in sporadic PD identified a total of 96 studies during the period between 2004 and 2017: 12 meta-analyses, 21 re-analyses of exiting data, and 63 original studies carried out by means of different genome-wide technologies [30]. Several studies analyzed transcriptomic profiles in the blood, brain tissue, and dopaminergic neurons in autosomal dominant PD associated with mutations in the *LRRK2* gene (LRRK2-PD) (OMIM no.609007) [13,31–33], with only one research study conducted with RNA-seq technology [14]. In fact, presently, only one research study has examined the transcriptomic profile in GBA-PD [34] despite the (obvious) actual problems of incomplete penetrance of *GBA* mutations.

Here, we compared the gene expression profile in monocyte-derived macrophages between L444P/N *GBA* mutation carriers, discordant for clinical manifestation of PD and controls. This mutation is more severe compared to N370S, and characterized by an earlier age of PD onset, as well as motor, psychiatric, cognitive, and olfactory symptoms [6]. It also results in more pronounced alpha-synuclein accumulation in in vitro and in vivo models of PD [35]. According to our previous data, GCase enzyme activity decreases more strongly and the plasma level of oligomeric alpha-synuclein is higher in L444P/N GBA-PD patients compared to N370S/N GBA-PD patients [10]. We revealed 32 DEGs between L444P/N GBA-PD and the controls, 36 between L444P/N GBA-PD and L444P/N GBA carriers and 18 between L444P/N GBA carriers and controls. First, we focused on searching for molecular biomarkers involved in PD pathogenesis among L444P/N *GBA* mutation carriers. We revealed two potential biomarkers for PD in L444P/N *GBA* mutation carriers (downregulation of the *DUSP1* and *ARL4C* gene expression). The *DUSP1* gene encodes the mitogen-activated protein kinase 1 (MKP-1) phosphatase that participates in regulation of apoptosis, endoplasmic reticulum (ER) stress, cell cycle, and autophagy, with the cellular process playing a pivotal role in PD [36]. MKP-1 belongs to the class I classical cysteine-based protein phosphatases (DUSP family) that have the dual ability to dephosphorylate phospho-serine/threonine and phospho-tyrosine residues [37,38]. MKP-1 is expressed during embryonic development in the midbrain, including dopaminergic neurons, as well as in adulthood in substantia nigra (SN) and can act as a neuroprotective agent. *ARL4C*, known also as *ARL7*, participates in cholesterol transport between the perinuclear compartment and the plasma membrane for ABCA1-associated removal and, thus, may be integral to the LXR-dependent efflux pathway [39]. Dysregulation of cholesterol metabolism has been implicated in PD [40].

Next, we aimed to find similarities in symptomatic and asymptomatic L444P/N *GBA* mutation carriers that can be attributed to the presence of L444P/N *GBA* mutations. Moreover, all L444P/N *GBA* mutation carriers were characterized by an increased *KIAA0319* expression level. The *KIAA0319* gene was involved in the pathway associated with the axon extension, involved in regeneration (GO:0048677). The genetic variants of *KIAA0319* were found to be associated with dyslexia [41,42].

Additionally, the transcriptomic analysis was conducted for both L444P and N370S *GBA* mutations. We revealed 28 DEGs between GBA-PD and controls, 23 between GBA-PD and GBA carriers, and 8 between GBA carriers and controls. We suggested that four genes, *DUSP1*, *COLEC12*, *TRIM13*, *BCL6*, deregulated in GBA-PD patients, might be potential candidates for PD biomarkers among *GBA* mutations carriers. Downregulated expression of *DUSP1* and *COLEC12* genes and upregulated expression of the *TRIM13* and *BCL6* genes were found in GBA-PD patients compared to both GBA carriers and controls. *DUSP1* and *TRIM13* are involved in initiation of autophagy and in the ubiquitin-proteasome pathway of protein degradation during ER stress that may play a critical role in alpha-synuclein degradation. It has been shown that repression of endogenous *TRIM13* inhibits autophagy induced by ER stress [43]. Family DUSPs have many substrates and modulate diverse neural functions, such as neurogenesis, differentiation, and apoptosis. DUSP1 critically contributes to the resolution of acute inflammatory responses of macrophages and mediates protective glucocorticoids effects, which potently inhibit pro-inflammatory responses, and are widely used for the treatment of inflammatory diseases [44]. We revealed decreased expression level of the *DUSP1* gene in GBA-PD patients compared to GBA carriers and controls. Thus, a decreased expression level of the DUSP1 gene may lead to impairment of macrophage's inflammatory response and, therefore, contribute increasing inflammation levels. TRIM13 is a negative regulator of MDA5-mediated type I interferon (IFN) production and may impact RIG-I-mediated type I IFN production. Proper regulation of the type I IFN response contributes to maintaining immune homeostasis [45]. Since macrophages are vital to immune response, dysregulation of the TRIM13 gene may lead to disturbance of immune homeostasis and levels of cytokines, which act as important mediators of the immune system. The *COLEC12* gene, known also as *SCARA4*, *SRCLI*, *SRCLII*, *CL-P1*, is implicated in innate immune responses [46], and is associated with lipid metabolism and phagosome formation. In particularly, *COLEC12* protein functions as a receptor for the detection, uptake, and degradation of oxidized modified low-density lipoproteins by vascular endothelial cells [47]. *BCL6* is a critical marker in cell apoptosis and contributed to the inflammation activation of macrophages [48]. Previous studies on mouse and human macrophages showed that COLEC12 is a novel receptor involved in myelin uptake by phagocytes and may play a role in active multiple sclerosis, which is a chronic, inflammatory, neurodegenerative disease [49]. Considering its role in the uptake of myelin, COLEC12 likely plays an important role in the pathophysiology of neurodegenerative disease, but as an uptake of myelin leads to both demyelination and central nervous system repair, depending on whether it concerns intact myelin or myelin debris, COLEC12-mediated myelin uptake can be beneficial or detrimental. BCL6 is a critical marker in cell apoptosis and contributes to the inflammation activation of macrophages. BCL6 overexpression was found to inhibit the increase in reactive oxygen species ROS. Mitochondrial functions lead to exacerbation of ROS generation and susceptibility to oxidative stress involved in PD pathogenesis [50].

Next, we found similarities in symptomatic and asymptomatic *GBA* mutation carriers that consisted of the decreased *JUNB* and *HOOK2* gene expression in both GBA-PD patients and GBA carriers compared to controls. *HOOK2* encodes the Hook2 protein that belongs to a family of cytoplasmic linker proteins. Hook2 is implicated in the formation of aggresomes, vesicle trafficking, and fusion, particularly in degradation of neuronal tau aggregates in Alzheimer's disease (AD) [51–53].

Comparing the transcriptomes between the GBA-PD independent of the type of mutation (L444P, N370S) as well as in L444P/N GBA-PD revealed three genes (*JUNB*, *EGR1*, *NR4A2*) encoding transcriptional regulators involved in the maintenance of dopaminergic neuron function, neuronal differentiation, and neurogenesis from the top of the DEGs list. It is worth noting that a previous transcriptomic analysis conducted for the blood and brain for sporadic PD revealed an alteration in the pathways that include dopamine metabolism, mitochondrial function, oxidative stress, protein degradation, neuroinflammation, vesicular transport, and synaptic transmission [30]. Our data support the statement that neurodegenerative mechanisms could be detectable from a peripheral tissue. *JUNB*, *EGR1*, *NR4A2* belong to immediate-early genes (IEGs) and encode the transcription factors, JunB, Egr-1, NR4A2, respectively [54–57]. These factors are activated in respond to a variety of cellular stimuli and control specific neuronal functions, including neuronal activity. Both JunB and Egr-1 are key mediators of apoptosis and the inflammatory response [58,59]. It is interesting to note that *JUNB* overexpression protections against cell death of nigral neurons [60]. Furthermore, JunB modulates expression of canonical markers of alternative activation in macrophages [61]. The latest study demonstrated that a large share of EGR1 target regions in macrophages are enhancers associated to the inflammatory response [59]. Egr1 inhibits pro-inflammatory gene expression in macrophages [59]. Egr-1 activation promotes neuroinflammation and dopaminergic neurodegeneration in an experimental model of PD [62]. *NR4A2* (Nurr1) is critical in the development and maintenance of the dopaminergic neurons. It coordinates several key proteins, including tyrosine hydroxylase (TH), dopamine transporter (DAT), and vesicular monoamine transporter (SCL18A2/VMAT) [63]. Previous studies demonstrated an association between *NR4A2* polymorphisms with PD [64–66] and showed that sporadic PD patients is characterized by decreased *NR4A2* gene expression in PBMCs [67]. Nurr1 also appears to restrain inflammatory processes by polarizing macrophages to the M2 type [68]. Thus, the role of these genes in neuroimmune interaction could not be excluded as monocytes; macrophages may migrate across the blood–brain barrier and induce the neuroinflammatory processes in the brain and, therefore, contribute to brain pathology, such as neurodegeneration [69].

According to the Human Protein Atlas (https://www.proteinatlas.org (accessed on 9 September 2021)), the top of DEGs in GBA-PD patients compared to controls, *JUNB*, *EGR1*, *NR4A2*, and potential biomarkers of GBA-PD (*DUSP1*, *COLEC12*, *TRIM13*, *BCL6*, *ARL4C*) express not only in the blood, but in brain tissues.

GO enrichment analysis revealed several altered pathways in GBA-PD patients independent of the type of mutation in the *GBA* genes (L444P, N370S) generally related to the immune system. Growing evidence suggests that neuroinflammation may contribute to the development of Parkinson's disease and elevated levels of inflammation-related mediators in the brain and cerebrospinal fluid. Many studies focused on peripheral inflammatory processes have found a significant association between immune markers and disease severity. We should note that, previously, we (and others) demonstrated elevation of proinflammatory cytokine secretion in plasma of GBA-PD patients compared to sporadic PD patients and controls [70,71].

Presently, only one paper performed transcriptomic analysis for PD patients baring *GBA* mutations. The study was fulfilled on iPSC-derived dopamine neurons from three GBA-PD patients with the N370S *GBA* variant [34]. Single-cell profiling demonstrated disease relevant pathways, even in the carriers of the same mutation. Thus, in one initially diagnosed as a patient with PD, the patient's cellular profile prompted a clinical reassessment, leading to a revised diagnosis of progressive supranuclear palsy (PSP). Nevertheless, on iPSC-derived dopamine neurons from two other patients with N370S GBA-PD, the authors found 60 deregulated genes that included downregulated genes implicated in neuronal function, and upregulated genes involved in zinc ion transport [33]. Similar to Lang and colleagues, we also found upregulation of genes, *MT1F*, *MT1L*, *MT1M*, *MT1X*, and *SLC39A8*, involved in the zinc metabolism pathway in GBA-PD patients, compared to GBA carriers. Alterations of zinc homeostasis have long been implicated in PD. Zn2+, besides its role in multiple cellular functions, also acts as a synaptic transmitter in the brain. Recent meta-analysis studies, though, point to lower zinc levels in serum and plasma and CSF of PD patients compared to healthy controls. The association between deregulated levels of circulating zinc and PD has been explained by its antioxidant role since this trace element is essential for a variety of enzymes and proteins (superoxide dismutase oxidative, metallothioneins, and interleukins) involved in oxidative stress [72]. Moreover, dysregulated zinc homeostasis zinc plays a critical role in the innate immune system, especially for maintaining the function of macrophages due to participation in impairment phagocytosis and an abnormal inflammatory response [73]. The following ingenuity pathway analysis conducted by Lang and colleagues showed that, among 60 deregulated genes in GBA-PD, eight (*PRKCB*, *RTN1*, *ATP1A3*, *TSPAN7*, *NTM*, *L1CAM*, *BDNF*, *SLC2A1*) are regulated with histone deacetylase 4 (HDAC4). In our study, both GBA-PD and L444P/N GBA-PD patients demonstrated decreased expression of the *DUSP1* gene involved in ER stress, implicated previously in PD pathogenesis, particularly GBA-PD [74]. Notably, Lang and colleagues found downregulation of another gene from the same MKP family—the *DUSP4* gene encoding MKP-2 that is closely related with *DUSP1*/MPK-1 [34,38]. There are currently few studies assessing the role of the DUSP genes in PD. However, one study reported decreased *DUSP1* mRNA expression in the brain tissue in an idiopathic PD patient [37]. *DUSP1* overexpression protects dopaminergic neurons against neurotoxicity induced with 6-hydroxydopamine in vitro [75]. Strategies aimed at increasing the expression of DUSP1 have been discussed as potential therapeutic approaches for PD [37]. Taken together, our results highlight the potential important role of the DUSP family in the pathogenesis of GBA-PD. To summarize our results with the study by Lang and colleagues, we could make a conclusion about the involvement of the downregulation of genes related to neuronal functions and upregulation of pathways related to immune response and zinc ion homeostasis in GBA-PD pathogenesis.

It is interesting to note that, oppositely, regarding the number of DEGs attributed to a presence of *GBA* mutations revealed in our preset study, RNA-seq conducted in *LRRK2* G2019S mutation carriers suggested that G2019S mutation in the *LRRK2* gene markedly altered blood transcriptome in comparison with sporadic PD [14]. Infante and colleagues found 174 genes with significant differential expression in the blood between LRRK2-PD patients with G2019S mutation and asymptomatic carriers and 1139 DEGs between asymptomatic carriers of G2019S *LRRK2* mutation and controls [14]. These data allow us to suggest that the *GBA* mutation had less influence on the transcriptome profile in comparison with *LRRK2* mutations. We compared our gene set with the gene set presented in the study by Infante and colleagues, and revealed overlap genes, encoding monocyte-attracting chemokines, such as *CCL3L1* gene, when comparing GBA-PD, G2019S LRRK2-PD, and PD to controls, and the *CCL3* gene, when comparing GBA-PD to controls and G2019 LRRK2-PD to G2019S LRRK2-carriers. That observation supports the hypothesis involving the role of immune response in PD pathogenesis [76]. It is also important to mention that the difference between the amount of the revealed differentially expressed genes in our study compared to the study by Infante and colleagues can be explained due to the fact that their study conducted whole-blood transcriptomic analysis. We could not exclude the possibility that such discrepancies are attributed to monotype cell populations used in the present study for transcriptomic analysis.

The current study has some limitations. The small size of the studied groups may influence the outcome of differential expression analysis for genes with small differences in expression levels, eliminating nonspecific gene expression differences. Moreover, the influence of L-DOPA treatment in GBA-PD patients on the gene expression level cannot be ruled out. In addition, we could not exclude PD manifestation among GBA carriers later in their lives, as only 10% of carriers of mutations in the *GBA* gene develop PD at the age of 60, 16% at the age of 70, and 19% at the age of 80 [77]. It is interesting to note that Lang et al. demonstrated that the genome profile in sporadic PD could—in some cases—be similar to GBA-PD, suggesting that findings from GBA-PD could be extrapolated to a subset of sporadic PD patients [34]. A further limitation of our study is the absence of PD patients without *GBA* mutations.

5. Conclusions

In conclusion, this study provides new insights into the global transcriptome in GBA-PD and asymptomatic *GBA* mutation carriers. Potential involvement of genes of neuronal functions, inflammation, and zinc metabolism in the pathogenesis of GBA-PD was shown. Alteration expression of *DUSP1* may be considered a potential biomarker of PD among *GBA* mutations carriers. This knowledge could assist in answering the fundamental question about potential triggers, which is important for future studies devoted toward determining the pathogenesis of PD among *GBA* mutations carriers.

Supplementary Materials: The following are available online at https://www.mdpi.com/article/10.3390/genes12101545/s1, Table S1: Post-trimming and mapping results to reference genome for all groups, Table S2: Differentially expressed genes from L444P/N GBA-PD patients compared to controls in monocyte-derived macrophages, Table S3: Differentially expressed genes from L444P/N GBA-carriers compared to controls in monocyte-derived macrophages, Table S4: Differentially expressed genes from L444P/N GBA-PD patients compared to L444P/N GBA-carriers in monocyte-derived macrophages, Table S5: Differentially expressed genes from GBA-PD (L444P/N+N370S/N) patients compared to controls in monocyte-derived macrophages, Table S6: Differentially expressed genes from GBA-carriers (L444P/N+N370S/N) compared to controls in monocyte-derived macrophages, Table S7: Differentially expressed genes from GBA-PD (L444P/N+N370S/N) patients compared to GBA-carriers (L444P/N+N370S/N) in monocyte-derived macrophages, Table S8: Functional clusters selected according to the results of GO analysis between GBA-PD (L444P/N+N370S/N) patients, GBA-carriers (L444P/N+N370S/N) and controls.

Author Contributions: T.U.: writing—original draft, formal analysis, visualization, investigation, A.B.: formal analysis, visualization, investigation, K.B.: formal analysis, visualization, investigation, A.P.: investigation, M.N.: investigation, A.K.; investigation, I.M.: investigation, A.E.: investigation, E.Z.: investigation, S.P.: conceptualization, resources, funding acquisition, supervision, writing—review and editing. All authors have read and agreed to the published version of the manuscript.

Funding: The study was supported by the Russian Science Foundation grant No. 19-15-00315.

Institutional Review Board Statement: The study was conducted according to the guidelines of the Declaration of Helsinki, and approved by Local Ethics Committee of Pavlov First Saint-Petersburg State Medical University (Approval Code: 244, Approval Date: 25 January 2021).

Informed Consent Statement: Written informed consent has been obtained from the patients to publish this paper.

Data Availability Statement: The data discussed in this publication have been deposited in NCBI's Gene Expression Omnibus (Edgar et al., 2002) and are accessible through GEO Series accession number GSE184956 (https://www.ncbi.nlm.nih.gov/geo/query/acc.cgi?acc=GSE184956 (accessed on 9 September 2021)) and at ArrayExpress database at EMBL-EBI (www.ebi.ac.uk/arrayexpress (accessed on 9 September 2021)) under accession number E-MTAB-11029 or https://www.ebi.ac.uk/arrayexpress/arrays/E-MTAB-11029 (accessed on 9 September 2021) for array design "E-MTAB-11029".

Acknowledgments: The authors are grateful to all patients whose participation in the study made this analysis possible.

Conflicts of Interest: The authors declare no conflict of interest.

References

1. Dickson, D.W.; Braak, H.; Duda, J.E.; Duyckaerts, C.; Gasser, T.; Halliday, G.M.; Hardy, J.; Leverenz, J.B.; Del Tredici, K.; Wszolek, Z.K.; et al. Neuropathological assessment of Parkinson's disease: Refining the diagnostic criteria. *Lancet Neurol.* **2009**, *8*, 1150–1157. [CrossRef]
2. Xu, L.; Pu, J. Alpha-Synuclein in Parkinson's Disease: From Pathogenetic Dysfunction to Potential Clinical Application. *Parkinson's Dis.* **2016**, *2016*, 1720621. [CrossRef] [PubMed]
3. Emelyanov, A.K.; Usenko, T.S.; Tesson, C.; Senkevich, K.A.; Nikolaev, M.A.; Miliukhina, I.V.; Kopytova, A.E.; Timofeeva, A.A.; Yakimovsky, A.F.; Lesage, S.; et al. Mutation analysis of Parkinson's disease genes in a Russian data set. *Neurobiol. Aging* **2018**, *71*, 267-e7. [CrossRef] [PubMed]
4. Sidransky, E.; Nalls, M.A.; Aasly, J.O.; Aharon-Peretz, J.; Annesi, G.; Barbosa, E.R.; Bar-Shira, A.; Berg, D.; Bras, J.; Brice, A.; et al. Multicenter Analysis of Glucocerebrosidase Mutations in Parkinson's Disease. *N. Engl. J. Med.* **2009**, *361*, 1651–1661. [CrossRef]
5. Lesage, S.; Anheim, M.; Condroyer, C.; Pollak, P.; Durif, F.; Dupuits, C.; Viallet, F.; Lohmann, E.; Corvol, J.-C.; Honoré, A.; et al. Large-scale screening of the Gaucher's disease-related glucocerebrosidase gene in Europeans with Parkinson's disease. *Hum. Mol. Genet.* **2011**, *20*, 202–210. [CrossRef]
6. Malek, N.; Weil, R.S.; Bresner, C.; Lawton, M.A.; Grosset, K.A.; Tan, M.; Bajaj, N.; Barker, R.A.; Burn, D.J.; Foltynie, T.; et al. Features of GBA-associated Parkinson's disease at presentation in the UK Tracking Parkinson's study. *J. Neurol. Neurosurg. Psychiatry* **2018**, *89*, 702–709. [CrossRef]
7. Blumenreich, S.; Barav, O.B.; Jenkins, B.J.; Futerman, A.H. Lysosomal Storage Disorders Shed Light on Lysosomal Dysfunction in Parkinson's Disease. *Int. J. Mol. Sci.* **2020**, *21*, 4966. [CrossRef]
8. Alcalay, R.N.; Levy, O.A.; Waters, C.C.; Fahn, S.; Ford, B.; Kuo, S.H.; Mazzoni, P.; Pauciulo, M.W.; Nichols, W.C.; Gan-Or, Z.; et al. Glucocerebrosidase activity in Parkinson's disease with and without *GBA* mutations. *Brain* **2015**, *138*, 2648–2658. [CrossRef]
9. Guedes, L.C.; Chan, R.B.; Gomes, M.A.; Conceição, V.A.; Machado, R.B.; Soares, T.; Xu, Y.; Gaspar, P.; Carriço, J.A.; Alcalay, R.N.; et al. Serum lipid alterations in GBA-associated Parkinson's disease. *Parkinsonism Relat. Disord.* **2017**, *44*, 58–65. [CrossRef]
10. Pchelina, S.; Emelyanov, A.; Baydakova, G.; Andoskin, P.; Senkevich, K.; Nikolaev, M.; Miliukhina, I.; Yakimovskii, A.; Timofeeva, A.; Fedotova, E.; et al. Oligomeric α-synuclein and glucocerebrosidase activity levels in GBA-associated Parkinson's disease. *Neurosci. Lett.* **2017**, *636*, 70–76. [CrossRef]
11. Pchelina, S.; Baydakova, G.; Nikolaev, M.; Senkevich, K.; Emelyanov, A.; Kopytova, A.; Miliukhina, I.; Yakimovskii, A.; Timofeeva, A.; Berkovich, O.; et al. Blood lysosphingolipids accumulation in patients with parkinson's disease with glucocerebrosidase 1 mutations. *Mov. Disord.* **2018**, *33*, 1325–1330. [CrossRef]
12. Cooper-Knock, J.; Kirby, J.; Ferraiuolo, L.; Heath, P.R.; Rattray, M.; Shaw, P.J. Gene expression profiling in human neurodegenerative disease. *Nat. Rev. Neurol.* **2012**, *8*, 518–530. [CrossRef]
13. Mutez, E.; Nkiliza, A.; Belarbi, K.; de Broucker, A.; Vanbesien-Mailliot, C.; Bleuse, S.; Duflot, A.; Comptdaer, T.; Semaille, P.; Blervaque, R.; et al. Involvement of the immune system, endocytosis and EIF2 signaling in both genetically determined and sporadic forms of Parkinson's disease. *Neurobiol. Dis.* **2014**, *63*, 165–170. [CrossRef]
14. Infante, J.; Prieto, C.; Sierra, M.; Sánchez-Juan, P.; González-Aramburu, I.; Sánchez-Quintana, C.; Berciano, J.; Combarros, O.; Sainz, J. Identification of candidate genes for Parkinson's disease through blood transcriptome analysis in LRRK2-G2019S carriers, idiopathic cases, and controls. *Neurobiol. Aging* **2015**, *36*, 1105–1109. [CrossRef]

15. Infante, J.; Prieto, C.; Sierra, M.; Sánchez-Juan, P.; González-Aramburu, I.; Sánchez-Quintana, C.; Berciano, J.; Combarros, O.; Sainz, J. Comparative blood transcriptome analysis in idiopathic and LRRK2 G2019S-associated Parkinson's disease. *Neurobiol. Aging* **2016**, *38*, 214.e1–214.e5. [CrossRef]
16. Aflaki, E.; Stubblefield, B.K.; Maniwang, E.; Lopez, G.; Moaven, N.; Goldin, E.; Marugan, J.; Patnaik, S.; Dutra, A.; Southall, N.; et al. Macrophage models of Gaucher disease for evaluating disease pathogenesis and candidate drugs. *Sci. Transl. Med.* **2014**, *6*, 240ra73. [CrossRef]
17. Pandey, M.K.; Grabowski, G.A. Immunological cells and functions in Gaucher disease. *Crit. Rev. Oncog.* **2013**, *18*, 197–220. [CrossRef]
18. Kopytova, A.E.; Rychkov, G.N.; Nikolaev, M.A.; Baydakova, G.V.; Cheblokov, A.A.; Senkevich, K.A.; Bogdanova, D.A.; Bolshakova, O.I.; Miliukhina, I.V.; Bezrukikh, V.A.; et al. Ambroxol increases glucocerebrosidase (GCase) activity and restores GCase translocation in primary patient-derived macrophages in Gaucher disease and Parkinsonism. *Parkinsonism Relat. Disord.* **2021**, *84*, 112–121. [CrossRef]
19. Welsh, N.J.; Gewinner, C.A.; Mistry, K.; Koglin, M.; Cooke, J.; Butler, M.; Powney, B.; Roberts, M.; Staddon, J.M.; Schapira, A.H. V Functional assessment of glucocerebrosidase modulator efficacy in primary patient-derived macrophages is essential for drug development and patient stratification. *Haematologica* **2020**, *105*, e206–e209. [CrossRef]
20. Postuma, R.B.; Berg, D.; Stern, M.; Poewe, W.; Olanow, C.W.; Oertel, W.; Obeso, J.; Marek, K.; Litvan, I.; Lang, A.E.; et al. MDS clinical diagnostic criteria for Parkinson's disease. *Mov. Disord.* **2015**, *30*, 1591–1601. [CrossRef]
21. Nikolaev, M.A.; Kopytova, A.E.; Baidakova, G.V.; Emel'yanov, A.K.; Salogub, G.N.; Senkevich, K.A.; Usenko, T.S.; Gorchakova, M.V.; Koval'chuk, Y.P.; Berkovich, O.A.; et al. Human Peripheral Blood Macrophages as a Model for Studying Glucocerebrosidase Dysfunction. *Cell Tissue Biol.* **2019**, *13*, 100–106. [CrossRef]
22. Andrews, S. FastQC: A Quality Control Tool for High Throughput Sequence Data. Available online: http://www.bioinformatics.babraham.ac.uk/projects/fastqc (accessed on 9 September 2021).
23. Wang, L.; Wang, S.; Li, W. RSeQC: Quality control of RNA-seq experiments. *Bioinformatics* **2012**, *28*, 2184–2185. [CrossRef] [PubMed]
24. Martin, M. Cutadapt removes adapter sequences from high-throughput sequencing reads. *EMBnet. J.* **2011**, *17*, 10–12. [CrossRef]
25. Kim, D.; Langmead, B.; Salzberg, S.L. HISAT: A fast spliced aligner with low memory requirements. *Nat. Methods* **2015**, *12*, 357–360. [CrossRef] [PubMed]
26. Anders, S.; Pyl, P.T.; Huber, W. HTSeq—A Python framework to work with high-throughput sequencing data. *Bioinformatics* **2015**, *31*, 166–169. [CrossRef]
27. Love, M.I.; Huber, W.; Anders, S. Moderated estimation of fold change and dispersion for RNA-seq data with DESeq2. *Genome Biol.* **2014**, *15*, 550. [CrossRef]
28. Bindea, G.; Mlecnik, B.; Hackl, H.; Charoentong, P.; Tosolini, M.; Kirilovsky, A.; Fridman, W.-H.; Pagès, F.; Trajanoski, Z.; Galon, J. ClueGO: A Cytoscape plug-in to decipher functionally grouped gene ontology and pathway annotation networks. *Bioinformatics* **2009**, *25*, 1091–1093. [CrossRef]
29. Bindea, G.; Galon, J.; Mlecnik, B. CluePedia Cytoscape plugin: Pathway insights using integrated experimental and in silico data. *Bioinformatics* **2013**, *29*, 661–663. [CrossRef]
30. Borrageiro, G.; Haylett, W.; Seedat, S.; Kuivaniemi, H.; Bardien, S. A review of genome-wide transcriptomics studies in Parkinson's disease. *Eur. J. Neurosci.* **2018**, *47*, 1–16. [CrossRef]
31. Botta-Orfila, T.; Sànchez-Pla, A.; Fernández, M.; Carmona, F.; Ezquerra, M.; Tolosa, E. Brain transcriptomic profiling in idiopathic and LRRK2-associated Parkinson's disease. *Brain Res.* **2012**, *1466*, 152–157. [CrossRef]
32. Pallos, J.; Jeng, S.; McWeeney, S.; Martin, I. Dopamine neuron-specific LRRK2 G2019S effects on gene expression revealed by translatome profiling. *Neurobiol. Dis.* **2021**, *155*, 105390. [CrossRef]
33. Reinhardt, P.; Schmid, B.; Burbulla, L.F.; Schöndorf, D.C.; Wagner, L.; Glatza, M.; Höing, S.; Hargus, G.; Heck, S.A.; Dhingra, A.; et al. Genetic correction of a LRRK2 mutation in human iPSCs links parkinsonian neurodegeneration to ERK-dependent changes in gene expression. *Cell Stem Cell* **2013**, *12*, 354–367. [CrossRef]
34. Lang, C.; Campbell, K.R.; Ryan, B.J.; Carling, P.; Attar, M.; Vowles, J.; Perestenko, O.V.; Bowden, R.; Baig, F.; Kasten, M.; et al. Single-Cell Sequencing of iPSC-Dopamine Neurons Reconstructs Disease Progression and Identifies HDAC4 as a Regulator of Parkinson Cell Phenotypes. *Cell Stem Cell* **2019**, *24*, 93–106.e6. [CrossRef] [PubMed]
35. Maor, G.; Rapaport, D.; Horowitz, M. The effect of mutant GBA1 on accumulation and aggregation of α-synuclein. *Hum. Mol. Genet.* **2019**, *28*, 1768–1781. [CrossRef] [PubMed]
36. Wang, J.; Zhou, J.-Y.; Kho, D.; Reiners, J.J.J.; Wu, G.S. Role for DUSP1 (dual-specificity protein phosphatase 1) in the regulation of autophagy. *Autophagy* **2016**, *12*, 1791–1803. [CrossRef] [PubMed]
37. An, N.; Bassil, K.; Al Jowf, G.I.; Steinbusch, H.W.M.; Rothermel, M.; de Nijs, L.; Rutten, B.P.F. Dual-specificity phosphatases in mental and neurological disorders. *Prog. Neurobiol.* **2021**, *198*, 101906. [CrossRef] [PubMed]
38. Bhore, N.; Wang, B.-J.; Chen, Y.-W.; Liao, Y.-F. Critical Roles of Dual-Specificity Phosphatases in Neuronal Proteostasis and Neurological Diseases. *Int. J. Mol. Sci.* **2017**, *18*, 1963. [CrossRef] [PubMed]
39. Hong, C.; Walczak, R.; Dhamko, H.; Bradley, M.N.; Marathe, C.; Boyadjian, R.; Salazar, J.V.; Tontonoz, P. Constitutive activation of LXR in macrophages regulates metabolic and inflammatory gene expression: Identification of ARL7 as a direct target. *J. Lipid Res.* **2011**, *52*, 531–539. [CrossRef] [PubMed]

40. Courtney, R.; Landreth, G.E. LXR Regulation of Brain Cholesterol: From Development to Disease. *Trends Endocrinol. Metab.* **2016**, *27*, 404–414. [CrossRef]
41. Humphreys, P.; Kaufmann, W.E.; Galaburda, A.M. Developmental dyslexia in women: Neuropathological findings in three patients. *Ann. Neurol.* **1990**, *28*, 727–738. [CrossRef]
42. Galaburda, A.M.; LoTurco, J.; Ramus, F.; Fitch, R.H.; Rosen, G.D. From genes to behavior in developmental dyslexia. *Nat. Neurosci.* **2006**, *9*, 1213–1217. [CrossRef]
43. Tomar, D.; Singh, R.; Singh, A.K.; Pandya, C.D.; Singh, R. TRIM13 regulates ER stress induced autophagy and clonogenic ability of the cells. *Biochim. Biophys. Acta* **2012**, *1823*, 316–326. [CrossRef]
44. Hoppstädter, J.; Ammit, A.J. Role of Dual-Specificity Phosphatase 1 in Glucocorticoid-Driven Anti-inflammatory Responses. *Front. Immunol.* **2019**, *10*, 1446. [CrossRef] [PubMed]
45. Narayan, K.; Waggoner, L.; Pham, S.T.; Hendricks, G.L.; Waggoner, S.N.; Conlon, J.; Wang, J.P.; Fitzgerald, K.A.; Kang, J. TRIM13 is a negative regulator of MDA5-mediated type I interferon production. *J. Virol.* **2014**, *88*, 10748–10757. [CrossRef]
46. Zani, I.A.; Stephen, S.L.; Mughal, N.A.; Russell, D.; Homer-Vanniasinkam, S.; Wheatcroft, S.B.; Ponnambalam, S. Scavenger receptor structure and function in health and disease. *Cells* **2015**, *4*, 178–201. [CrossRef] [PubMed]
47. Selman, L.; Skjodt, K.; Nielsen, O.; Floridon, C.; Holmskov, U.; Hansen, S. Expression and tissue localization of collectin placenta 1 (CL-P1, SRCL) in human tissues. *Mol. Immunol.* **2008**, *45*, 3278–3288. [CrossRef] [PubMed]
48. Kutyavin, V.I.; Chawla, A. BCL6 regulates brown adipocyte dormancy to maintain thermogenic reserve and fitness. *Proc. Natl. Acad. Sci. USA* **2019**, *116*, 17071–17080. [CrossRef]
49. Bogie, J.F.J.; Mailleux, J.; Wouters, E.; Jorissen, W.; Grajchen, E.; Vanmol, J.; Wouters, K.; Hellings, N.; van Horssen, J.; Vanmierlo, T.; et al. Scavenger receptor collectin placenta 1 is a novel receptor involved in the uptake of myelin by phagocytes. *Sci. Rep.* **2017**, *7*, 44794. [CrossRef]
50. Dias, V.; Junn, E.; Mouradian, M.M. The role of oxidative stress in Parkinson's disease. *J. Parkinsons. Dis.* **2013**, *3*, 461–491. [CrossRef]
51. Herrmann, L.; Wiegmann, C.; Arsalan-Werner, A.; Hilbrich, I.; Jäger, C.; Flach, K.; Suttkus, A.; Lachmann, I.; Arendt, T.; Holzer, M. Hook proteins: Association with Alzheimer pathology and regulatory role of hook3 in amyloid beta generation. *PLoS ONE* **2015**, *10*, e0119423. [CrossRef]
52. Szebenyi, G.; Wigley, W.C.; Hall, B.; Didier, A.; Yu, M.; Thomas, P.; Krämer, H. Hook2 contributes to aggresome formation. *BMC Cell Biol.* **2007**, *8*, 19. [CrossRef] [PubMed]
53. Walenta, J.H.; Didier, A.J.; Liu, X.; Krämer, H. The Golgi-associated hook3 protein is a member of a novel family of microtubule-binding proteins. *J. Cell Biol.* **2001**, *152*, 923–934. [CrossRef] [PubMed]
54. Jakaria, M.; Haque, M.E.; Cho, D.-Y.; Azam, S.; Kim, I.-S.; Choi, D.-K. Molecular Insights into NR4A2(Nurr1): An Emerging Target for Neuroprotective Therapy Against Neuroinflammation and Neuronal Cell Death. *Mol. Neurobiol.* **2019**, *56*, 5799–5814. [CrossRef] [PubMed]
55. Qin, X.; Wang, Y.; Paudel, H.K. Early Growth Response 1 (Egr-1) Is a Transcriptional Activator of β-Secretase 1 (BACE-1) in the Brain. *J. Biol. Chem.* **2016**, *291*, 22276–22287. [CrossRef]
56. Yoshitomi, Y.; Ikeda, T.; Saito-Takatsuji, H.; Yonekura, H. Emerging Role of AP-1 Transcription Factor JunB in Angiogenesis and Vascular Development. *Int. J. Mol. Sci.* **2021**, *22*, 2804. [CrossRef] [PubMed]
57. Zukin, R.; Jover-Mengual, T.; Yokota, H.; Calderone, A.; Simionescu, M.; Lau, C.G. Molecular and Cellular Mechanisms of Ischemia-Induced Neuronal Death. *Stroke Pathophysiol. Diagn. Manag.* **2004**. [CrossRef]
58. Lee, J.K.H.; Pearson, J.D.; Maser, B.E.; Ingham, R.J. Cleavage of the JunB transcription factor by caspases generates a carboxyl-terminal fragment that inhibits activator protein-1 transcriptional activity. *J. Biol. Chem.* **2013**, *288*, 21482–21495. [CrossRef]
59. Trizzino, M.; Zucco, A.; Deliard, S.; Wang, F.; Barbieri, E.; Veglia, F.; Gabrilovich, D.; Gardini, A. EGR1 is a gatekeeper of inflammatory enhancers in human macrophages. *Sci. Adv.* **2021**, *7*, eaaz8836. [CrossRef]
60. Winter, C.; Weiss, C.; Martin-Villalba, A.; Zimmermann, M.; Schenkel, J. JunB and Bcl-2 overexpression results in protection against cell death of nigral neurons following axotomy. *Brain Res. Mol. Brain Res.* **2002**, *104*, 194–202. [CrossRef]
61. Fontana, M.F.; Baccarella, A.; Pancholi, N.; Pufall, M.A.; Herbert, D.R.; Kim, C.C. JUNB is a key transcriptional modulator of macrophage activation. *J. Immunol.* **2015**, *194*, 177–186. [CrossRef]
62. Yu, Q.; Huang, Q.; Du, X.; Xu, S.; Li, M.; Ma, S. Early activation of Egr-1 promotes neuroinflammation and dopaminergic neurodegeneration in an experimental model of Parkinson's disease. *Exp. Neurol.* **2018**, *302*, 145–154. [CrossRef] [PubMed]
63. Jankovic, J.; Chen, S.; Le, W.D. The role of Nurr1 in the development of dopaminergic neurons and Parkinson's disease. *Prog. Neurobiol.* **2005**, *77*, 128–138. [CrossRef]
64. Le, W.-D.; Xu, P.; Jankovic, J.; Jiang, H.; Appel, S.H.; Smith, R.G.; Vassilatis, D.K. Mutations in NR4A2 associated with familial Parkinson disease. *Nat. Genet.* **2003**, *33*, 85–89. [CrossRef] [PubMed]
65. Liu, H.; Liu, H.; Li, T.; Cui, J.; Fu, Y.; Ren, J.; Sun, X.; Jiang, P.; Yu, S.; Li, C. NR4A2 genetic variation and Parkinson's disease: Evidence from a systematic review and meta-analysis. *Neurosci. Lett.* **2017**, *650*, 25–32. [CrossRef] [PubMed]
66. Sleiman, P.M.A.; Healy, D.G.; Muqit, M.M.K.; Yang, Y.X.; Van Der Brug, M.; Holton, J.L.; Revesz, T.; Quinn, N.P.; Bhatia, K.; Diss, J.K.J.; et al. Characterisation of a novel NR4A2 mutation in Parkinson's disease brain. *Neurosci. Lett.* **2009**, *457*, 75–79. [CrossRef]

67. Ruiz-Sánchez, E.; Yescas, P.; Rodríguez-Violante, M.; Martínez-Rodríguez, N.; Díaz-López, J.N.; Ochoa, A.; Valdes-Rojas, S.S.; Magos-Rodríguez, D.; Rojas-Castañeda, J.C.; Cervantes-Arriaga, A.; et al. Association of polymorphisms and reduced expression levels of the NR4A2 gene with Parkinson's disease in a Mexican population. *J. Neurol. Sci.* **2017**, *379*, 58–63. [CrossRef]
68. Mahajan, S.; Saini, A.; Chandra, V.; Nanduri, R.; Kalra, R.; Bhagyaraj, E.; Khatri, N.; Gupta, P. Nuclear Receptor Nr4a2 Promotes Alternative Polarization of Macrophages and Confers Protection in Sepsis. *J. Biol. Chem.* **2015**, *290*, 18304–18314. [CrossRef]
69. Mammana, S.; Fagone, P.; Cavalli, E.; Basile, M.S.; Petralia, M.C.; Nicoletti, F.; Bramanti, P.; Mazzon, E. The Role of Macrophages in Neuroinflammatory and Neurodegenerative Pathways of Alzheimer's Disease, Amyotrophic Lateral Sclerosis, and Multiple Sclerosis: Pathogenetic Cellular Effectors and Potential Therapeutic Targets. *Int. J. Mol. Sci.* **2018**, *19*, 831. [CrossRef]
70. Brockmann, K.; Schulte, C.; Schneiderhan-Marra, N.; Apel, A.; Pont-Sunyer, C.; Vilas, D.; Ruiz-Martinez, J.; Langkamp, M.; Corvol, J.-C.; Cormier, F.; et al. Inflammatory profile discriminates clinical subtypes in LRRK2-associated Parkinson's disease. *Eur. J. Neurol.* **2017**, *24*, 427-e6. [CrossRef]
71. Miliukhina, I.V.; Usenko, T.S.; Senkevich, K.A.; Nikolaev, M.A.; Timofeeva, A.A.; Agapova, E.A.; Semenov, A.V.; Lubimova, N.E.; Totolyan, A.A.; Pchelina, S.N. Plasma Cytokines Profile in Patients with Parkinson's Disease Associated with Mutations in *GBA* Gene. *Bull. Exp. Biol. Med.* **2020**, *168*, 423–426. [CrossRef]
72. Sikora, J.; Ouagazzal, A.-M. Synaptic Zinc: An Emerging Player in Parkinson's Disease. *Int. J. Mol. Sci.* **2021**, *22*, 4724. [CrossRef]
73. Gao, H.; Dai, W.; Zhao, L.; Min, J.; Wang, F. The Role of Zinc and Zinc Homeostasis in Macrophage Function. *J. Immunol. Res.* **2018**, *2018*, 6872621. [CrossRef] [PubMed]
74. Fernandes, H.J.R.; Hartfield, E.M.; Christian, H.C.; Emmanoulidou, E.; Zheng, Y.; Booth, H.; Bogetofte, H.; Lang, C.; Ryan, B.J.; Sardi, S.P.; et al. ER Stress and Autophagic Perturbations Lead to Elevated Extracellular α-Synuclein in GBA-N370S Parkinson's iPSC-Derived Dopamine Neurons. *Stem Cell Rep.* **2016**, *6*, 342–356. [CrossRef]
75. Collins, L.M.; O'Keeffe, G.W.; Long-Smith, C.M.; Wyatt, S.L.; Sullivan, A.M.; Toulouse, A.; Nolan, Y.M. Mitogen-activated protein kinase phosphatase (MKP)-1 as a neuroprotective agent: Promotion of the morphological development of midbrain dopaminergic neurons. *Neuromol. Med.* **2013**, *15*, 435–446. [CrossRef] [PubMed]
76. Huang, Y.; Halliday, G.M. Aspects of innate immunity and Parkinson's disease. *Front. Pharmacol.* **2012**, *3*, 33. [CrossRef]
77. Balestrino, R.; Tunesi, S.; Tesei, S.; Lopiano, L.; Zecchinelli, A.L.; Goldwurm, S. Penetrance of Glucocerebrosidase (GBA) Mutations in Parkinson's Disease: A Kin Cohort Study. *Mov. Disord.* **2020**, *35*, 2111–2114. [CrossRef] [PubMed]

Article

C9orf72-G$_4$C$_2$ Intermediate Repeats and Parkinson's Disease; A Data-Driven Hypothesis

Hila Kobo [1,2,†], Orly Goldstein [1,†], Mali Gana-Weisz [1], Anat Bar-Shira [1], Tanya Gurevich [2,3], Avner Thaler [2,3,4], Anat Mirelman [2,3,4], Nir Giladi [2,3,4] and Avi Orr-Urtreger [1,2,*]

[1] The Genomic Research Laboratory for Neurodegeneration, Neurological Institute, Tel Aviv Sourasky Medical Center, Tel Aviv 64239, Israel; Hilakobo@tauex.tau.ac.il (H.K.); orlyg@tlvmc.gov.il (O.G.); maligw@tlvmc.gov.il (M.G.-W.); anatbn@tlvmc.gov.il (A.B.-S.)
[2] Sackler Faculty of Medicine, Sagol School of Neuroscience, Tel Aviv University, Tel Aviv 6997801, Israel; tanyag@tlvmc.gov.il (T.G.); avnert@tlvmc.gov.il (A.T.); anatmi@tlvmc.gov.il (A.M.); nirg@tlvmc.gov.il (N.G.)
[3] Movement Disorders Unit, Neurological Institute, Tel Aviv Sourasky Medical Center, Tel Aviv 64239, Israel
[4] Laboratory for Early Markers of Neurodegeneration, Neurological Institute, Tel Aviv Sourasky Medical Center, Tel Aviv 64239, Israel
* Correspondence: aviorr@tlvmc.gov.il; Tel.: +972-3-6974704
† The two authors contributed equally to the work.

Abstract: Pathogenic *C9orf72*-G$_4$C$_2$ repeat expansions are associated with ALS/FTD, but not with Parkinson's disease (PD); yet the possible link between intermediate repeat lengths and PD remains inconclusive. We aim to study the potential involvement of these repeats in PD. The number of *C9orf72*-repeats were determined by flanking and repeat-primed PCR assays, and the risk-haplotype was determined by SNP-array. Their association with PD was assessed in a stratified manner: in PD-patients-carriers of mutations in *LRRK2*, *GBA*, or *SMPD1* genes (n = 388), and in PD-non-carriers (NC, n = 718). Allelic distribution was significantly different only in PD-NC compared to 600 controls when looking both at the allele with higher repeat's size ($p = 0.034$) and at the combined number of repeats from both alleles ($p = 0.023$). Intermediate repeats (20–60 repeats) were associated with PD in PD-NC patients ($p = 0.041$; OR = 3.684 (CI 1.05–13.0)) but not in PD-carriers ($p = 0.684$). The *C9orf72* risk-haplotype, determined in a subgroup of 588 PDs and 126 controls, was observed in higher frequency in PD-NC (dominant model, OR = 1.71, CI 1.04–2.81, $p = 0.0356$). All 19 alleles within the risk-haplotype were associated with higher *C9orf72* RNA levels according to the GTEx database. Based on our data, we suggest a model in which intermediate repeats are a risk factor for PD in non-carriers, driven not only by the number of repeats but also by the variants' genotypes within the risk-haplotype. Further studies are needed to elucidate this possible role of *C9orf72* in PD pathogenesis.

Keywords: Parkinson's disease (PD); *C9orf72*; intermediate repeats; hexanucleotide expansions

Citation: Kobo, H.; Goldstein, O.; Gana-Weisz, M.; Bar-Shira, A.; Gurevich, T.; Thaler, A.; Mirelman, A.; Giladi, N.; Orr-Urtreger, A. *C9orf72*-G$_4$C$_2$ Intermediate Repeats and Parkinson's Disease; A Data-Driven Hypothesis. *Genes* **2021**, *12*, 1210. https://doi.org/10.3390/genes12081210

Academic Editor: Suzanne Lesage

Received: 6 July 2021
Accepted: 3 August 2021
Published: 5 August 2021

Publisher's Note: MDPI stays neutral with regard to jurisdictional claims in published maps and institutional affiliations.

Copyright: © 2021 by the authors. Licensee MDPI, Basel, Switzerland. This article is an open access article distributed under the terms and conditions of the Creative Commons Attribution (CC BY) license (https://creativecommons.org/licenses/by/4.0/).

1. Introduction

Parkinson's disease (PD) is a common neurodegenerative disorder, affecting about 2% of the elderly population worldwide [1]. Its complex genetic background has been revealed in the past decades, implicating many genes associated with the disease. A wide variety of genetic changes and mechanisms are involved in PD, including rare and common variants, recessive, dominant, and oligogenic inheritance, and epigenetics [2–8]. However, the full range of genetic changes in PD is still evolving.

G$_4$C$_2$ Hexanucleotide repeat expansions in *C9orf72* are strongly associated with amyotrophic lateral sclerosis (ALS) and frontotemporal dementia (FTD) [9,10], mostly in European and North American populations [11,12]. Although 30 repeats and over are considered pathogenic for ALS and FTD, most patients that are *C9orf72*-associated-ALS or -FTD, carry an expanded allele with hundreds, or even thousands, of repeats [12–14]. Interestingly, parkinsonism was observed in more than 40% of FTD and FTD/ALS patients with pathogenic *C9orf72* expansions [15]. This observation has led researchers to investigate the possible association of *C9orf72* expansions with PD.

While rare cases of PD patients with 30 to 60 repeat expansions or more in *C9orf72* were detected (<0.7%), there was no association with PD [16–20]. Few studies have suggested that intermediate-size repeat lengths in *C9orf72* may be a risk factor for PD; however, these studies suggested different repeat lengths for this association: ≥7 repeats in Han Chinese [21], and ≥20 repeats in Caucasians [17]. In a multi-center meta-analysis of mostly Caucasian and Asian populations, the pathogenicity expansions threshold was determined as >60 repeats, and the intermediate repeat size was set as 17–60 repeats [22], suggesting an effect on PD-risk for the cutoff of 17 repeats, and even as little as 10 repeats as a stand-alone or as a cutoff. In a recent comprehensive review by Bourinaris and Houlden [23], *C9orf72* intermediate repeat lengths were reported in several parkinsonism and movement disorders, including in Dopa-responsive PD, atypical parkinsonisms including PSP and MSA, essential-tremor plus parkinsonism, and spinocerebellar ataxia.

As previous studies have shown the pivotal role of genetically homogeneous populations, such as the Ashkenazi Jews (AJ), in understanding the genetic background of neurodegenerative diseases [24–27], we hereby determined *C9orf72* repeats' size in PD patients of Ashkenazi origin and examined their potential association with PD. We also studied the possible association of the shared risk-haplotype, which is observed in carriers of intermediate repeats, with PD-risk.

2. Methods

2.1. Population

This study included a cohort of consecutively recruited unrelated 1106 PD patients of full Ashkenazi Jewish origin (Table 1). Patients were recruited between the years 2005 and 2015. The diagnostic criteria, recruitment, and genotyping for *LRRK2*, *GBA*, and *SMPD1* mutations have been previously described [24,25,28]. Carrier patients (PD-carriers) were defined with one or more of the following mutations in *LRRK2* (p.G2019S), *SMPD1* (p. L302P), or any of the 10 *GBA* mutations (c.84insG, IVS2 + 1G > A, p.V394L, p.N370S, p.L444P, p.R496H, p.E326K, p.T369M, p.R44C, and 370Rec). Patients who did not carry any of these mutations were determined as non-carriers (PD-NC). The cohort of 600 ethnically matched control individuals used in this study has been previously described [26].

Table 1. Characteristics of the 1106 Parkinson's disease patients of Ashkenazi Jewish origin.

	Non-Carrier PD Patients [a]	PD Patients Carriers of *LRRK2*, *GBA*, or *SMPD1* Mutations
N	718	388 [c]
Women, N (%)	266 (37.0)	171 (44.1)
AAE, mean (SD), y	68.5 (10.2) [d]	65.4 (10.1)
AAO, mean (SD), y	61.4 (11.5)	58.4 (10.6)
Family history of PD [b], N (%)	144 (20)	120 (30.9)

Abbreviations: PD, Parkinson disease; N, number of individuals; AAE, age at enrollment; SD, standard deviation; y, years; AAO, age at disease onset. [a] Patients without the *LRRK2*, *GBA*, or *SMPD1* mutations (specified in the Methods section). [b] Patients with 1st or 2nd degree family members with PD. [c] Carrier patients included: 133 individuals with *LRRK2* mutation, 223 individuals with *GBA* mutations, 8 individuals with *SMPD1* mutation, 23 individuals with mutations in both *LRRK2* and *GBA*, and 1 patient with both *GBA* and *SMPD1* mutations. [d] Data regarding AAE was not available for one individual.

2.2. Determining the G_4C_2 Hexanucleotide Repeat Length in the C9orf72 Gene

To determine the number of repeats in *C9orf72*, flanking and repeat-primed PCR assays were performed as previously described [9,10], with some modifications [26]. This method detects all repeats expansion but can determine the number of repeats up to 55 repeats. Therefore, in all individuals that carried an allele with 30 repeats or more, an additional method was used to determine accurately the repeat number up to 145 repeats (Asuragen assay kit AmplideX® PCR/CE C9orf72 Kit; Asuragen Genetics; Austin, TX, USA).

2.3. Assembly of the Risk-Haplotype within the C9orf72 Locus

To determine the presence of the risk-haplotype in *C9orf72*-locus, we used the genotype data (from Vacic et al. [29], Affymetrixs SNP6.0) of 127 AJ controls (all part of our cohort of 600 controls) and 597 AJ-PD patients (594 are part of our cohort of 1106 PD patients). When determination of the presence of the risk-haplotype was impossible due to missing genotypes, these individuals were excluded (Control = 1, PD = 5). One PD was excluded due to low genotype rate, one was not tested for repeat size, and two PDs who carried >145 repeats were also excluded. In total, 126 controls and 588 PD patients were analyzed.

2.4. Statistical Analyses of C9orf72 G_4C_2 Hexanucleotide Repeats

All statistical analyses were performed using IBM SPSS statistics software v25 (IBM Corporation, New York, USA). Differences in continuous variables were tested using Mann–Whitney U test or *t*-test (2-tailed). To test the difference in *C9orf72* repeat lengths between patients and controls, both alleles repeat sizes (per individual) were included.

Categorical variables were compared using 2-sided χ^2-test, or Fisher's exact test when numbers were less than 5. Odds ratio (OR), with 95% confidence interval (CI), was applied to assess the association of *C9orf72* G_4C_2 repeat lengths with PD. This association was examined using the longest repeat size (per individual) as the independent variable. Association of the risk-haplotype with PD was examined using a dominant model. Logistic regression analysis was performed when using repeat units as a quantitative trait (the largest allele or the sum of both alleles).

3. Results

3.1. Allele Frequencies of C9orf72 G_4C_2 Hexanucleotide Repeats in Ashkenazi PD Patients

Allele frequencies of *C9orf72* repeats were determined in our cohort of 1106 Ashkenazi PD patients (Table 1), that was divided into two groups based on their genotypic status, either carriers of PD-associated mutations (see Methods section), or non-carrier patients (PD-NC). We ran a stratified analysis based on the carrier status in *LRRK2*, *GBA*, and *SMPD1*, as a high percentage of our PD cohort carry risk alleles in these 3 genes (35.1%, 388/1106), and these carrier-patients may mask the effect of the hexanucleotide repeat length on PD-risk in non-carrier patients. The most frequent alleles found were 2, 8 and 5 repeat units (66.2%, 12.1%, 8.9% in carrier, and 63.1%, 12.7%, 9.7% in non-carrier patients; Figure 1 and Supplement Table S1). These alleles were also shown as the most common alleles in our previously published data of Ashkenazi controls (66.4%, 11.0%, and 10.2%) [26]. No significant difference in allele distribution was observed between patients with mutations (in *LRRK2*, *GBA*, or *SMPD1* genes) and controls (Mann–Whitney U test p = 0.756; 4.23 ± 6.13, N = 776 and 4.09 ± 5.36, N = 1200, respectively). However, a significant difference in allele distribution was detected in PD-NC (Mann–Whitney U test p = 0.034; 4.71 ± 8.40, N = 1436). This was also significant for the total number of repeats (combining the numbers of repeats from both alleles; excluding individuals with expanded alleles of >145 repeats) in PD-NC (Mann–Whitney U test p = 0.023; 8.63 ± 5.57, N = 714, and 7.94 ± 5.01, N = 599) and was not significant in PD-carriers (Mann–Whitney U test p = 0.565; 8.10 ± 4.90, N = 387).

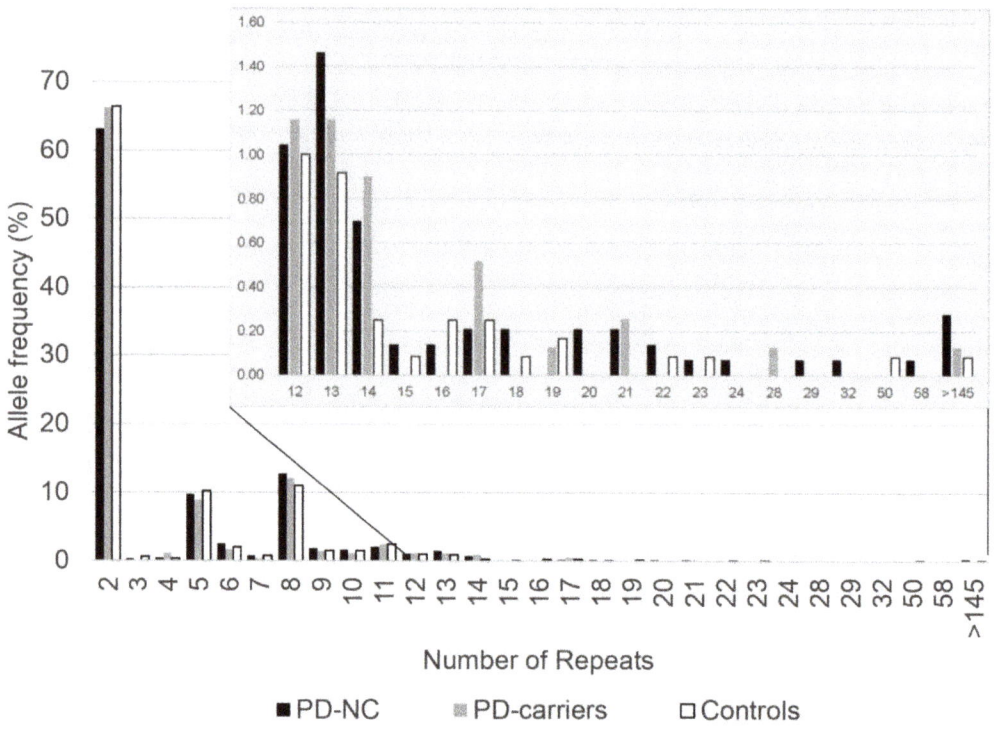

Figure 1. Graphic representation of *C9orf72* G_4C_2 hexanucleotide repeat allele frequencies in Ashkenazi Parkinson's disease patients and controls. The repeats' allele frequencies for each group are presented: PD patients without *LRRK2*, *GBA*, and *SMPD1* mutations (PD-non-carriers, black); PD patients carrying mutation in *LRRK2*, *GBA*, or *SMPD1* (PD-carriers, gray); and controls as previously published (Reference [26], white). Upper panel is a zoomed-in graph of 12 repeats and higher.

3.2. The Association of C9orf72 G_4C_2 Hexanucleotide Intermediate Repeat Lengths with PD in Ashkenazi Patients

We examined the association of the longest repeat allele in each individual with PD (in PD-carriers and PD-NC). First, we examined whether large expansion lengths (>145 repeats) are present within our cohort of PD patients and controls. We found one PD-carrier patient (carrying the *LRRK2* p.G2019S mutation, 1/388 = 0.3%), four PD-NC (4/718 = 0.6%), and one control (1/600 = 0.2%) that carried an expanded allele. No significant association was detected in any of the patients' groups compared to controls (Fisher's Exact Test $p = 1.000$ and $p = 0.384$, respectively). Of interest, one of these PD patients was later diagnosed with ALS, and two had dementia. For further analysis, we excluded these five carriers with *C9orf72* G_4C_2 repeats expansion (all with > 145 repeats). None of our patients or controls carried alleles between 60 and 145 repeats.

Next, we examined if intermediate *C9orf72* hexanucleotide repeats (20–60 repeats) were associated with PD in our cohort. Among the 1101 PD patients, 3 PD-carriers (0.8%, 3/387) and 13 PD-NC (1.8%, 13/714) carried an allele with intermediate repeat lengths, compared to 3 among the controls (0.5%, 3/599). An association with increased risk for PD was observed in PD-NC patients compared to controls (Fisher's exact test, $p = 0.041$; OR = 3.684, CI = 1.045–12.990; Table 2, Figure 1), but no association was detected when comparing PD-carrier patients to controls ($p = 0.684$, Table 2).

Table 2. The association 2–60 *C9orf72* hexanucleotide repeats with Parkinson's disease.

Cohort	Non-Carrier PD Patients [a]	PD Patients Carriers of *LRRK2*, *GBA*, or *SMPD1* Mutations	Controls [b]
2–19 repeats, N (%)	701 (98.2)	384 (99.2)	596 (99.5)
20–60 repeats, N (%)	13 (1.8)	3 (0.8)	3 (0.5)
Odds Ratio (95% CI)	3.684 (1.045–12.990)	1.552 (0.312–7.729)	
p-value [c]	0.041	0.684	

Abbreviations: PD, Parkinson's disease; N, Number of individuals; CI, confidence interval. The longest allele in each individual was recorded. [a] Patients without the *LRRK2*, *GBA*, or *SMPD1* mutations (specified in the Methods section). [b] The values in the control cohort were previously published [26]. [c] 2 × 2 Fisher's exact test (2-sided).

Measuring the effect of the risk associated with an increasing number of repeats (β, regression analysis for each one repeat unit) shows a significant effect in PD-NC when looking either at the largest allele or at the sum of alleles ($\beta = 0.032$, $p = 0.015$, OR = 1.032, CI = 1.006–1.059; $\beta = 0.025$, $p = 0.021$, OR = 1.025, CI = 1.004–1.047; respectively). No effect was observed in PD-carriers ($\beta = 0.009$, $p = 0.542$, OR = 1.01, CI = 0.979–1.041; $\beta = 0.007$, $p = 0.617$, OR = 1.007, CI = 0.981–1.033; respectively). No effect on age of motor symptoms onset was observed in PD-NC (*p* value = 0.978).

3.3. The C9orf72 Risk-Haplotype Is Associated with Higher RNA Expression Levels and with PD

We have previously shown that the *C9orf72* expansions in AJ shared a risk-haplotype that expands 107Kb (Goldstein et al. [26], hg19: chr9:27484575-27591569), encompasses the complete *C9orf72* gene, and includes 44 informative single nucleotide variants (SNVs), with significant association with higher number of repeats (over 8 repeats). We further examined here the effects of these 44 SNVs (within the risk haplotype) on the RNA expression levels (eQTL) and splicing (sQTL) of their adjacent genes, as reported by the Genotype-Tissue Expression (GTEx) project [30]. Of them, 11 SNVs had no QTLs, 4 had low effect size (absolute NES < 0.2), and one had no significant effect (*m*-value < 0.9). Nine other variants were also excluded due to high allele frequency in AJs (> 0.5 in gnomAD v3.1 database of non-neuro cases). The other 19 SNVs within this 107Kb risk-haplotype were all associated with higher *C9orf72* RNA expression levels compared to the expression levels of the non-risk alleles (Table 3, Figure 2). Cerebellum and Nucleus-accumbens were the tissues with the highest normalized effect size (NES). Moreover, as GTEx evaluates the effect of each SNV on the neighboring genes within a 2 Mb interval (1 Mb upstream and 1 Mb downstream), it is important to note that all 19 SNVs affected exclusively *C9orf72*-RNA levels and not any other genes within that 2 Mb window. These SNVs also affected splice variants (sQTL) in an exclusive manner, only for *C9orf72*, mainly in cerebellum (Table 3).

We, therefore, tested if carrying the risk-haplotype is associated with PD. Genotyping data from 127 AJ controls and 597 AJ-PD patients were assembled (see Methods), and the presence of the 19-SNVs-risk-haplotype was determined. Overall enrichment of the risk-haplotype was observed in PDs compared to controls: 167 out of 588 PDs carried one or two copies of the risk-haplotype (28.4%) compared to 24 out of 126 controls (19.0%). When stratifying based on mutation carrier status (PD-carriers and PD-NC), a significant association was detected in PD-NC: 28.6% of them carried one or two copies of the risk-haplotype (OR = 1.71, CI = 1.04–2.81, $p = 0.0356$), and tendency was shown in PD-carriers (28.0%, OR = 1.65, CI = 0.97–2.82, $p = 0.0656$).

Table 3. The effect of the 19 variations within the risk-haplotype on RNA expression levels and splice variants as reported by GTEx database.

Location, chr9 (hg38)	rs ID	Gene	Risk Allele	Ref > Alt [a]	Highest eQTL for C9orf72, NES (Tissue)	eQTL for C9orf72 in Cerebellum, NES	Highest sQTL for C9orf72, NES (Tissue)	GnomAD v3.1- AF for the Risk Allele (in Non-Neuro Cases)		
								AJ	European (Non-Finnish)	All Populations
27488094	rs17779457	MOB3B	G	T > G	**0.464** (N.A.)	0.341	−0.88 (Cer)	0.2097	0.2428	0.2490
27496663	rs10812604	MOB3B	T	T > G	−0.416 (N.A.)	−0.343	0.70 (Cer)	0.2252	0.2861	0.3555
27497990	rs10967965	MOB3B	T	A > T	0.505 (N.A.)	0.360	−0.96 (C.H.)	0.1742	0.2185	0.1547
27499629	rs2492812	MOB3B	A	C > A	0.454 (N.A.)	0.334	−0.88 (C.H.)	0.2091	0.2419	0.2486
27508491	rs1537712	MOB3B	T	C > T	0.454 (N.A.)	0.332	−0.75 (Cer)	0.2323	0.2753	0.2915
27509213	rs12554036	MOB3B	T	G > T	0.484 (N.A.)	0.345	−1.0 (C.H.)	0.2054	0.2335	0.1796
27514964	rs4609281	MOB3B	G	G > T	−0.432 (N.A.)	−0.364	0.71 (Cer)	0.2417	0.2865	0.3509
27515969	rs774354	MOB3B	G	A > G	0.466 (N.A.)	0.345	−0.80 (C.H.)	0.2354	0.2754	0.2909
27516592	rs774352	MOB3B	C	T > C	0.466 (N.A.)	0.345	−0.80 (C.H.)	0.2354	0.2755	0.2910
27525753	rs10967973	IFNK	A	A > G	−0.602 (Cer)	−0.602	0.66 (Cer)	0.3570	0.4640	0.5599
27526049	rs700782	IFNK	T	C > T	0.498 (N.A.)	0.370	−0.98 (C.H.)	0.2109	0.2440	0.2461
27527514	rs2453552	MOB3B	G	T > G	0.501 (N.A.)	0.36	−1.0 (C.H.)	0.2110	0.2450	0.2650
27541043	rs7469146	C9orf72	C	C > T	−0.542 (Cer)	−0.542	0.65 (Cer)	0.3736	0.5119	0.5916
27543384	rs3849943	C9orf72	C	C > T	−0.485 (N.A.)	−0.336	0.96 (Cer, C.H.)	0.2005	0.2362	0.2178
27545962	rs700791	C9orf72	A	C > A	0.521 (N.A.)	0.388	−1.1 (C.H.)	0.1958	0.2293	0.1979
27579562	rs1982915	Intergenic	G	A > G	0.407 (Cer)	0.407	−0.49 (Cer)	0.4212	0.4982	0.4745
27580676	rs12350076	Intergenic	C	A > C	0.408 (Cer)	0.408	−0.51 (Cer)	0.4262	0.5104	0.4964
27581241	rs2783010	Intergenic	C	T > C	0.418 (Cer)	0.418	−0.56 (Cer)	0.4273	0.4645	0.4808
27582313	rs10967993	Intergenic	C	T > C	0.424 (Cer)	0.424	−0.54 (F.C.)	0.4025	0.5044	0.5032

[a] In bold is the allele that is associated with higher C9orf72 RNA expression levels, as reported by GTEx database. Ref = reference allele; Alt = Alternate allele; eQTL = expression quantitative trait loci; sQTL = spliced quantitative trait loci; NES = normalized effect size; AF = Allele frequency; AJ = Ashkenazi Jews; N.A. = Nucleus accumbens; Cer = Cerebellum; C.H. = Cerebellar hemisphere; F.C. = Frontal Cortex.

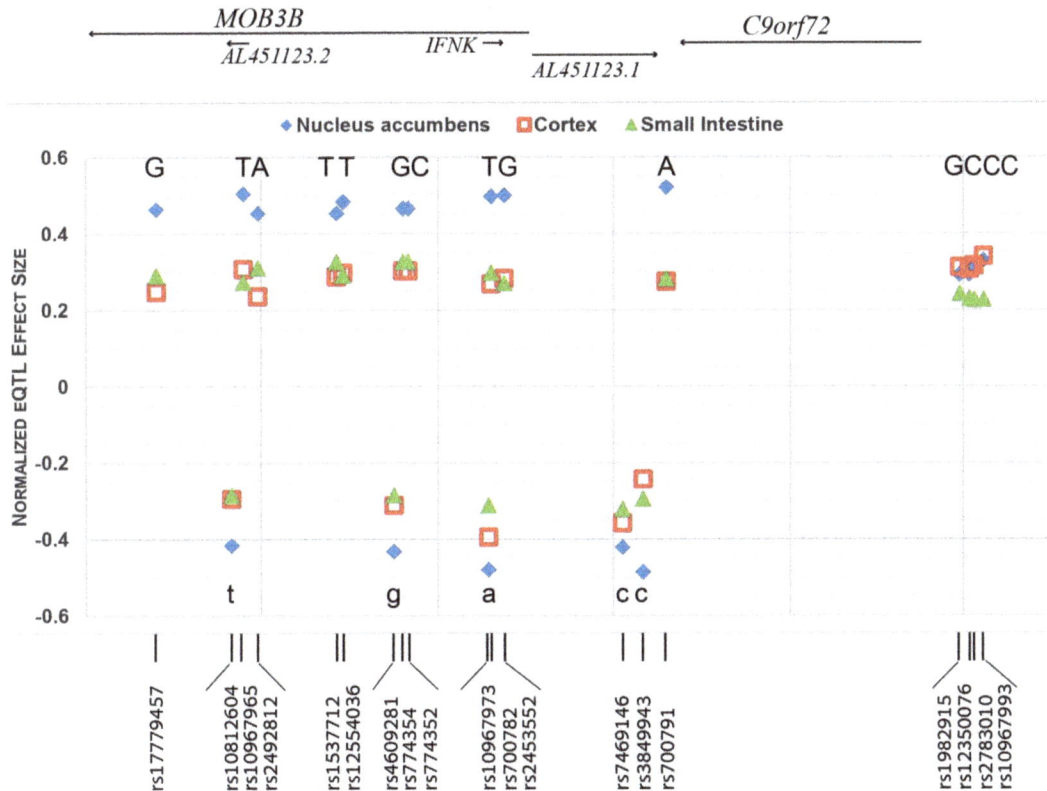

Figure 2. Higher levels of *C9orf72* RNA expression for all 19 SNVs within the risk-haplotype: Normalized eQTL effect size in Nucleus accumbens (dimond), Cortex (square), and small intestine (triangle). Upper panel and upper-case letters are the alternate allele observed in the risk haplotype, associated with higher levels of *C9orf72* expression; lower panel and lower-case letters are the reference allele observed in the risk haplotype, associated with higher levels of *C9orf72* expression.

We also attempted to define the correlation between the existence of risk-haplotype and number of repeats, by looking at all alleles (n = 1428 alleles): 100% negative correlation existed between the risk-haplotype and the 2-repeats' allele (with zero percent risk-haplotype), and 100% positive correlation was observed in carriers of 14–60 repeats (100% carried the risk-haplotype, Supplement Figure S1). Thus, we used 14 repeats as the best assessor for carrying the risk-haplotype and re-calculated the association of 14 repeats and higher with PD in our cohorts. Among all alleles in PD-NC, 2.3% had 14–60 repeats, compared to only 1.3% in controls, showing a trend toward significance (Figure 3a, OR = 1.75, CI = 0.96–3.19, p = 0.069, uncorrected), with no significance observed in PD-carriers (Figure 3b, OR = 1.46, CI = 0.72–2.97, p = 0.296). Further analysis showed a significant association in PD-NC when the cutoff of 13 repeats or 12 repeats was selected (OR = 1.70, CI = 1.07–2.72, p = 0.0257 and OR = 1.51, CI = 1.01–2.25, p = 0.044, respectively, Figure 3a, uncorrected), while there was no significant association in PD-carriers (OR = 1.46, CI = 0.72–2.97, p = 0.296; OR = 1.39, CI = 0.79–2.42, p = 0.249 and OR = 1.32, CI = 0.83–2.12, p = 0.245, respectively, Figure 3b). No significance was shown at 11 repeats cutoff, both in PD-NC and PD-carriers (Figure 3, OR = 1.22, CI = 0.89–1.69, p = 0.214 and OR = 1.20, CI = 0.82–1.74, p = 0.345, respectively).

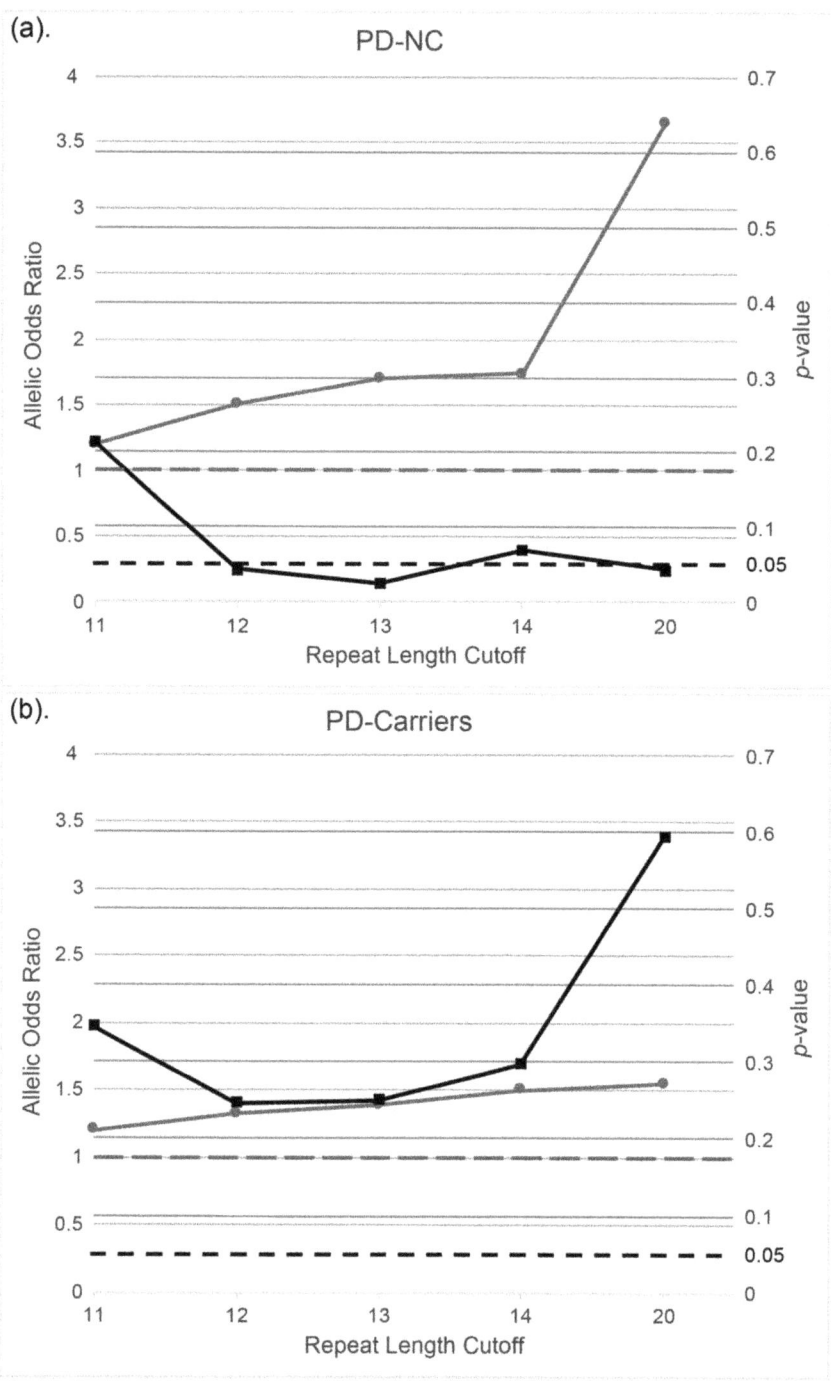

Figure 3. The effect of each cutoff of *C9orf72*-repeats-size on Parkinsons' disease risk in non-carriers of mutations ((**a**), PD-NC) and in carriers of mutation ((**b**), PD-Carriers; see Methods). Grey lines (circles) depict the Odds ratio and black lines (squares) depict the *p*-value. Dashed grey lines represent Odd Ratio of 1.0, and dashed black lines represent *p*-value of 0.05.

4. Discussion

More than 40 diseases, most of which affect the nervous system, were identified with a genetic basis of expansions of simple short DNA sequence (reviewed by Reference [31]). Among these diseases are myotonic dystrophy, Huntington's disease, spinocerebeller ataxia (SCA), spinal and bulbar muscular atrophy (SBMA), and Fragile-X syndrome. A common finding in these disorders is the correlation of the number of repeats with clinical phenotypes and penetrance. Interestingly, emerging studies suggest that expanded repeats and intermediate repeats can cause or act as risk factors for different neurological diseases, depending on the number of repeats. This was suggested for the *FMR1* gene, when over 200 CCG repeats cause mental retardation (Fragile-X syndrome), while the premutations of 45–200 repeats are a risk factor for Fragile-X- tremor/ataxia syndrome (FXTAS) in males and premature ovarian failure 1 (FXPOI) in females. Other examples are *ATXN2*, in which more than 34 repeats cause SCA2, while 29–33 are risk factor for ALS [32,33], and *ATXN1*, in which more than 38 repeat cause SCA1, while ≥33 are risk factor for ALS, mostly in *C9orf72* expansion carriers [34]. The latter is an example of the complexity of these mechanisms, as some of these risk alleles are more significant in specific subgroups of patients. To see if a similar phenomenon exists also for *C9orf72* intermediate repeats, we analyzed Parkinson's disease patients of Ashkenazi origin, in a stratified manner, in carriers of mutations in *LRRK2*, *GBA*, and/or *SMPD1* genes, and in patients that do not carry these mutations. We showed that the expanded alleles (>60 repeats) had no association with PD, as shown by other groups and in a meta-analysis [16–22]; however, intermediate-size repeats of 20–60 are significantly enriched in PD-NC, increasing the risk for PD, while, in PD-carriers, there was no effect. The high significant odds ratio of 3.68 in non-carriers may be due to the exclusion of those patients who carry known risk alleles in LRRK2, GBA, or SMPD1, as we believe that in these patients the risk for PD is likely influenced by these mutations and not by the C9orf72 intron 1 hexanucleotide repeat numbers. Of note is that Xi et al. also reported that intermediate repeats of 20–29 were found only in PD-NC, and not in PD-patients that carry the *LRRK2* p.G2019S mutation [19]. Based on these observations, we propose that *C9orf72* G_4C_2 hexanucleotide repeats in intron 1 act as a risk factor for PD when number of repeats are intermediates. Although the exact definition of intermediate- and pathogenic-length size is still in debate, and may differ in different populations, we believe that the presence of alleles lower than the 60 repeats, in the intermediate range, should not be dismissed as a potential risk for PD.

How may intermediate numbers of repeats affect the risk to develop PD?

In ALS, although the role of *C9orf72* expansions is not yet fully established, three main mechanisms are proposed to contribute to its pathogenicity: *C9orf72* loss-of-function, generation of toxic RNA aggregates, and short peptide accumulation [12,14]. As G_4C_2 repeats in the *C9orf72* gene are located within intron 1 and are near the promoter region, they may lead to changes in promoter regulation, depending on the number of repeats. Indeed, it was shown that ALS/FTD-expanded alleles are highly methylated and lead to lower levels of *C9orf72* mRNA and protein [35–37].

Do these same mechanisms contribute to Parkinson's disease risk? We demonstrated here that the risk-haplotype, which is shared by the intermediate *C9orf72* G_4C_2 repeats, includes many SNVs that have the same effect of increasing RNA expression levels of *C9orf72*, as reported in GTEx. We, therefore, suggest that the mechanism involved in the effect of intermediate-size repeats on PD-increased-risk, could be the higher expression of *C9orf72*. This mechanism was recently suggested for a different neurodegenerative disease, Corticobasal Degeneration (CBD), a rare neurodegenerative disease that shares some similar clinical features with PD [38]. Researchers showed a significant enrichment of intermediate repeats in autopsy-proven CBD, as well as increased *C9orf72*-RNA expression levels in human brain tissues and in CRISPR/cas9 knock-in iPSC cells, but no association with pathologic RNA foci or dipeptides aggregates.

One important question is whether intermediate-repeats-sizes or the increase of *C9orf72* expression, which is associated with the risk-haplotype, may drive the risk for PD.

As these two mostly go together, this question should be answered in an experimental set-up that separates the two events. Cali et al. tried to answer this question by knocking-in 28 repeats into iPSCs that normally have either 2 or 6 repeats (suggestive of cells that do not carry the risk haplotype) and demonstrated that these knocked-in cells show higher expression of *C9orf72* [38]. As GTEx data suggest higher expression of *C9orf72* for all 19 variants within the risk-haplotype, it is tempting to suggest that PD-risk may be determined by the level of *C9orf72* expression, mediated by both the risk-haplotype and the number of repeats, a hypothesis that needs further evaluation. This hypothesis raises other questions: does the overall level of *C9orf72* expression depend on the effect of the total number of repeats in both alleles, and whether the genomic region of the *C9orf72*-risk-haplotype might expand to a larger interval than the minimal linkage disequilibrium of 107 Kb, as suggested by gnomAD database? In addition, the GTEx data show that the risk-haplotype effect on *C9orf72* RNA expression levels is not uniform in all tissues. The effect size is large and significant mostly in brain tissues, as well as in the small intestine, but much smaller in whole blood and lymphocytes.

What could be the effect of intermediate-size repeats on cellular expression? Cali et al. performed a comparative genes expression analysis between cells with intermediate repeats and cells with low number of repeats [38], demonstrating upregulation of genes that are enriched for vesicle trafficking and protein degradation pathways, including golgi vesicle transport, response to ER-stress, and autophagy, pathways which are involved in PD.

In summary, our stratified analysis suggests that intermediate- size hexanucleotide repeats in *C9orf72* are a risk factor for PD in individuals who do not carry common AJ founder mutations in *LRRK2*, *GBA*, or *SMPD1*. These results should be interpreted with caution as no correction for multiple comparisons was performed, and similar analyses should be performed on a larger cohort of PD patients. However, we propose a model in which the risk for PD may be driven not only by the number of repeats, but also by the genotypes of SNVs within the risk-haplotype, affecting *C9orf72* RNA expression levels. Further studies are needed to elucidate the possible role of *C9orf72* in PD pathogenesis.

Supplementary Materials: The following are available online at https://www.mdpi.com/article/10.3390/genes12081210/s1, Table S1. Allele frequencies of G_4C_2 hexanucleotide repeat lengths in *C9orf72* in Parkinson's disease patients of Ashkenazi Jewish origin and controls. Figure S1. Number of alleles observed for each *C9orf72*-G_4C_2-repeat-size with risk or non-risk haplotypes in the Ashkenazi Jews (PDs and controls).

Author Contributions: Conceptualization, H.K., O.G., A.O.-U.; methodology, H.K., O.G., A.O.-U.; validation, H.K., O.G.; formal analysis, H.K., O.G.; investigation, H.K., O.G.; resources, M.G.-W., A.B.-S., T.G., A.T., A.M., N.G.; data curation, H.K., O.G., A.O.-U.; writing—original draft preparation, H.K., O.G., A.O.-U.; writing—review and editing, O.G., M.G.-W., A.B.-S., T.G., A.T., A.M., N.G., A.O.-U.; visualization, H.K., O.G.; supervision, O.G., A.O.-U.; project administration, M.G.-W., A.O.-U.; funding acquisition, A.O.-U. All authors have read and agreed to the published version of the manuscript.

Funding: This research was funded by Kahn Foundation and Michael J Fox Foundation.

Institutional Review Board Statement: The study was conducted according to the guidelines of the Declaration of Helsinki and approved by the Institutional Review Board (The Institutional Helsinki Committee of the Tel Aviv Sourasky Medical Center, Tel Aviv, Israel; protocol code 05-069, date of approval 8 September 2005), and by the Israeli National Supreme Helsinki Committee for Genetic Studies (protocol code 920050035, date of approval 8 September 2005).

Informed Consent Statement: Informed consent was obtained from all subjects involved in the study.

Data Availability Statement: The data presented in this study are available in the Supplementary Materials Table S1.

Acknowledgments: This work was performed in partial fulfillment of the requirements for a Ph.D. degree by Hila Kobo at the Sackler Faculty of Medicine, Tel Aviv University, Israel. We thank Haike Reznik-Wolf for assistance with Asuragene genotyping.

Conflicts of Interest: Authors H.K., O.G., M.G.-W. and A.B.-S. declare no conflicts of interest. Author T.G. has served as an advisor for Cytora, Synnerva, Teva, Medison, and Allergan, received honoraria from AbbVie and Neuroderm, research support from Parkinson's Foundation, University Tel-Aviv, Phonetica Ltd., and Israel innovation authority and travel support from AbbVie, Medison, Medtronic, and Allergan. Author AT has received honoraria from AbbVie Inc. and research grants from Michael J Fox Foundation. Author AM has received consulting fees from NeuroDerm and research grants from Michal J Fox Foundation, the Israeli Science Foundation and the US Department of Defense. Author NG serves as consultant to Intec Pharma, NeuroDerm, Denali, Abbvie, Sanofi-Genzyme, Biogen, Vibrant, BOL, LTI, Idorsia and Neuron23, Pharma2B, receives payment for lectures at Abbvie, Sanofi-Genzyme and Movement Disorder Society, received research support from the Michael J Fox Foundation, the National Parkinson Foundation, and the Israel Science Foundation, as well as from Teva NNE program, Biogen, The Aufzien Center, and the Sieratzki family Foundation at Tel-Aviv University, serves on the advisory board of LTI, NeuroDerm, Sionara, Sanofi-Genzyme, Biogen, Denali, Intec Pharma, Idorsia, and own stocks in Lysosomal Therapeutic Ltd., Vibrant, BOL, and serves as a member of the Editorial Board for the Journal of Parkinson's Disease. Author AOU has received research support from Michael J Fox Foundation and Chaya Charitable Fund and has received honoraria from Sanofi Genzyme. The funders had no role in the design of the study; in the collection, analyses, or interpretation of data; in the writing of the manuscript, or in the decision to publish the results.

Abbreviations

AAE: age at enrollment; AAO, age-at-onset, ALS, amyotrophic lateral sclerosis; C9orf72, chromosome 9 open reading frame 72; CI, confidence interval; FTD, frontotemporal dementia; GBA, glucosidase β acid; LRRK2, leucine-rich repeat kinase 2; SMPD1, sphingomyelin phosphodiesterase 1; N, number of individuals; OR, odds ratio; PD, Parkinson's disease; SD, standard deviation; y, years.

References

1. De Lau, L.M.; Breteler, M.M. Epidemiology of Parkinson's disease. *Lancet Neurol.* **2006**, *5*, 525–535. [CrossRef]
2. Blauwendraat, C.; Nalls, M.A.; Singleton, A.B. The genetic architecture of Parkinson's disease. *Lancet Neurol.* **2020**, *19*, 170–178. [CrossRef]
3. Corti, O.; Lesage, S.; Brice, A. What genetics tells us about the causes and mechanisms of Parkinson's disease. *Physiol. Rev.* **2011**, *91*, 1161–1218. [CrossRef]
4. Farrer, M.J. Genetics of Parkinson disease: Paradigm shifts and future prospects. *Nat. Rev. Genet.* **2006**, *7*, 306–318. [CrossRef] [PubMed]
5. Gasser, T.; Hardy, J.; Mizuno, Y. Milestones in PD genetics. *Mov. Disord.* **2011**, *26*, 1042–1048. [CrossRef]
6. Lesage, S.; Brice, A. Parkinson's disease: From monogenic forms to genetic susceptibility factors. *Hum. Mol. Genet.* **2009**, *18*, R48–R59. [CrossRef]
7. Lubbe, S.J.; Escott-Price, V.; Gibbs, J.R.; Nalls, M.A.; Bras, J.; Price, T.R.; Nicolas, A.; Jansen, I.E.; Mok, K.Y.; Pittman, A.M.; et al. Additional rare variant analysis in Parkinson's disease cases with and without known pathogenic mutations: Evidence for oligogenic inheritance. *Hum. Mol. Genet.* **2016**, *25*, 5483–5489. [CrossRef]
8. Van Heesbeen, H.J.; Smidt, M.P. Entanglement of Genetics and Epigenetics in Parkinson's Disease. *Front. Neurosci.* **2019**, *13*, 277. [CrossRef] [PubMed]
9. DeJesus-Hernandez, M.; Mackenzie, I.R.; Boeve, B.F.; Boxer, A.L.; Baker, M.; Rutherford, N.J.; Nicholson, A.M.; Finch, N.A.; Flynn, H.; Adamson, J.; et al. Expanded GGGGCC hexanucleotide repeat in noncoding region of C9ORF72 causes chromosome 9p-linked FTD and ALS. *Neuron* **2011**, *72*, 245–256. [CrossRef] [PubMed]
10. Renton, A.E.; Majounie, E.; Waite, A.; Simon-Sanchez, J.; Rollinson, S.; Gibbs, J.R.; Schymick, J.C.; Laaksovirta, H.; van Swieten, J.C.; Myllykangas, L.; et al. A hexanucleotide repeat expansion in C9ORF72 is the cause of chromosome 9p21-linked ALS-FTD. *Neuron* **2011**, *72*, 257–268. [CrossRef] [PubMed]
11. Liu, Y.; Yu, J.T.; Zong, Y.; Zhou, J.; Tan, L. C9ORF72 mutations in neurodegenerative diseases. *Mol. Neurobiol.* **2014**, *49*, 386–398. [CrossRef] [PubMed]
12. Shu, L.; Sun, Q.; Zhang, Y.; Xu, Q.; Guo, J.; Yan, X.; Tang, B. The Association between C9orf72 Repeats and Risk of Alzheimer's Disease and Amyotrophic Lateral Sclerosis: A Meta-Analysis. *Parkinsons Dis.* **2016**, *2016*, 5731734. [CrossRef] [PubMed]
13. Dols-Icardo, O.; Garcia-Redondo, A.; Rojas-Garcia, R.; Sanchez-Valle, R.; Noguera, A.; Gomez-Tortosa, E.; Pastor, P.; Hernandez, I.; Esteban-Perez, J.; Suarez-Calvet, M.; et al. Characterization of the repeat expansion size in c9orf72 in amyotrophic lateral sclerosis and frontotemporal dementia. *Hum. Mol. Genet.* **2014**, *23*, 749–754. [CrossRef]
14. Haeusler, A.R.; Donnelly, C.J.; Rothstein, J.D. The expanding biology of the C9orf72 nucleotide repeat expansion in neurodegenerative disease. *Nat. Rev. Neurosci.* **2016**, *17*, 383–395. [CrossRef] [PubMed]

15. Boeve, B.F.; Boylan, K.B.; Graff-Radford, N.R.; DeJesus-Hernandez, M.; Knopman, D.S.; Pedraza, O.; Vemuri, P.; Jones, D.; Lowe, V.; Murray, M.E.; et al. Characterization of frontotemporal dementia and/or amyotrophic lateral sclerosis associated with the GGGGCC repeat expansion in C9ORF72. *Brain J. Neurol.* **2012**, *135*, 765–783. [CrossRef]
16. Cooper-Knock, J.; Frolov, A.; Highley, J.R.; Charlesworth, G.; Kirby, J.; Milano, A.; Hartley, J.; Ince, P.G.; McDermott, C.J.; Lashley, T.; et al. C9ORF72 expansions, parkinsonism, and Parkinson disease: A clinicopathologic study. *Neurology* **2013**, *81*, 808–811. [CrossRef]
17. Nuytemans, K.; Bademci, G.; Kohli, M.M.; Beecham, G.W.; Wang, L.; Young, J.I.; Nahab, F.; Martin, E.R.; Gilbert, J.R.; Benatar, M.; et al. C9ORF72 intermediate repeat copies are a significant risk factor for Parkinson disease. *Ann. Hum. Genet.* **2013**, *77*, 351–363. [CrossRef]
18. Majounie, E.; Abramzon, Y.; Renton, A.E.; Keller, M.F.; Traynor, B.J.; Singleton, A.B. Large C9orf72 repeat expansions are not a common cause of Parkinson's disease. *Neurobiol. Aging* **2012**, *33*, 2527.e1–2527.e2. [CrossRef]
19. Xi, Z.; Zinman, L.; Grinberg, Y.; Moreno, D.; Sato, C.; Bilbao, J.M.; Ghani, M.; Hernandez, I.; Ruiz, A.; Boada, M.; et al. Investigation of c9orf72 in 4 neurodegenerative disorders. *Arch. Neurol.* **2012**, *69*, 1583–1590. [CrossRef]
20. Lesage, S.; Le Ber, I.; Condroyer, C.; Broussolle, E.; Gabelle, A.; Thobois, S.; Pasquier, F.; Mondon, K.; Dion, P.A.; Rochefort, D.; et al. C9orf72 repeat expansions are a rare genetic cause of parkinsonism. *Brain J. Neurol.* **2013**, *136*, 385–391. [CrossRef]
21. Jiao, B.; Guo, J.F.; Wang, Y.Q.; Yan, X.X.; Zhou, L.; Liu, X.Y.; Zhang, F.F.; Zhou, Y.F.; Xia, K.; Tang, B.S.; et al. C9orf72 mutation is rare in Alzheimer's disease, Parkinson's disease, and essential tremor in China. *Front. Cell. Neurosci.* **2013**, *7*, 164. [CrossRef]
22. Theuns, J.; Verstraeten, A.; Sleegers, K.; Wauters, E.; Gijselinck, I.; Smolders, S.; Crosiers, D.; Corsmit, E.; Elinck, E.; Sharma, M.; et al. Global investigation and meta-analysis of the C9orf72 (G4C2)n repeat in Parkinson disease. *Neurology* **2014**, *83*, 1906–1913. [CrossRef]
23. Bourinaris, T.; Houlden, H. C9orf72 and its Relevance in Parkinsonism and Movement Disorders: A Comprehensive Review of the Literature. *Mov. Disord. Clin. Pract.* **2018**, *5*, 575–585. [CrossRef]
24. Orr-Urtreger, A.; Shifrin, C.; Rozovski, U.; Rosner, S.; Bercovich, D.; Gurevich, T.; Yagev-More, H.; Bar-Shira, A.; Giladi, N. The LRRK2 G2019S mutation in Ashkenazi Jews with Parkinson disease: Is there a gender effect? *Neurology* **2007**, *69*, 1595–1602. [CrossRef]
25. Gan-Or, Z.; Giladi, N.; Rozovski, U.; Shifrin, C.; Rosner, S.; Gurevich, T.; Bar-Shira, A.; Orr-Urtreger, A. Genotype-phenotype correlations between GBA mutations and Parkinson disease risk and onset. *Neurology* **2008**, *70*, 2277–2283. [CrossRef] [PubMed]
26. Goldstein, O.; Gana-Weisz, M.; Nefussy, B.; Vainer, B.; Nayshool, O.; Bar-Shira, A.; Traynor, B.J.; Drory, V.E.; Orr-Urtreger, A. High frequency of C9orf72 hexanucleotide repeat expansion in amyotrophic lateral sclerosis patients from two founder populations sharing the same risk haplotype. *Neurobiol. Aging* **2017**. [CrossRef]
27. Goldstein, O.; Nayshool, O.; Nefussy, B.; Traynor, B.J.; Renton, A.E.; Gana-Weisz, M.; Drory, V.E.; Orr-Urtreger, A. OPTN 691_692insAG is a founder mutation causing recessive ALS and increased risk in heterozygotes. *Neurology* **2016**, *86*, 446–453. [CrossRef] [PubMed]
28. Goldstein, O.; Gana-Weisz, M.; Cohen-Avinoam, D.; Shiner, T.; Thaler, A.; Cedarbaum, J.M.; John, S.; Lalioti, M.; Gurevich, T.; Bar-Shira, A.; et al. Revisiting the non-Gaucher-GBA-E326K carrier state: Is it sufficient to increase Parkinson's disease risk? *Mol. Genet. Metab.* **2019**, *128*, 470–475. [CrossRef] [PubMed]
29. Vacic, V.; Ozelius, L.J.; Clark, L.N.; Bar-Shira, A.; Gana-Weisz, M.; Gurevich, T.; Gusev, A.; Kedmi, M.; Kenny, E.E.; Liu, X.; et al. Genome-wide mapping of IBD segments in an Ashkenazi PD cohort identifies associated haplotypes. *Hum. Mol. Genet.* **2014**, *23*, 4693–4702. [CrossRef] [PubMed]
30. Consortium, G.T. The GTEx Consortium atlas of genetic regulatory effects across human tissues. *Science* **2020**, *369*, 1318–1330. [CrossRef]
31. Paulson, H. Repeat expansion diseases. *Handb. Clin. Neurol.* **2018**, *147*, 105–123. [CrossRef] [PubMed]
32. Elden, A.C.; Kim, H.J.; Hart, M.P.; Chen-Plotkin, A.S.; Johnson, B.S.; Fang, X.; Armakola, M.; Geser, F.; Greene, R.; Lu, M.M.; et al. Ataxin-2 intermediate-length polyglutamine expansions are associated with increased risk for ALS. *Nature* **2010**, *466*, 1069–1075. [CrossRef]
33. Lee, T.; Li, Y.R.; Ingre, C.; Weber, M.; Grehl, T.; Gredal, O.; de Carvalho, M.; Meyer, T.; Tysnes, O.B.; Auburger, G.; et al. Ataxin-2 intermediate-length polyglutamine expansions in European ALS patients. *Hum. Mol. Genet.* **2011**, *20*, 1697–1700. [CrossRef] [PubMed]
34. Lattante, S.; Pomponi, M.G.; Conte, A.; Marangi, G.; Bisogni, G.; Patanella, A.K.; Meleo, E.; Lunetta, C.; Riva, N.; Mosca, L.; et al. ATXN1 intermediate-length polyglutamine expansions are associated with amyotrophic lateral sclerosis. *Neurobiol. Aging* **2017**. [CrossRef]
35. Xi, Z.; Zhang, M.; Bruni, A.C.; Maletta, R.G.; Colao, R.; Fratta, P.; Polke, J.M.; Sweeney, M.G.; Mudanohwo, E.; Nacmias, B.; et al. The C9orf72 repeat expansion itself is methylated in ALS and FTLD patients. *Acta Neuropathol.* **2015**, *129*, 715–727. [CrossRef] [PubMed]
36. Belzil, V.V.; Bauer, P.O.; Prudencio, M.; Gendron, T.F.; Stetler, C.T.; Yan, I.K.; Pregent, L.; Daughrity, L.; Baker, M.C.; Rademakers, R.; et al. Reduced C9orf72 gene expression in c9FTD/ALS is caused by histone trimethylation, an epigenetic event detectable in blood. *Acta Neuropathol.* **2013**, *126*, 895–905. [CrossRef]

37. Waite, A.J.; Baumer, D.; East, S.; Neal, J.; Morris, H.R.; Ansorge, O.; Blake, D.J. Reduced C9orf72 protein levels in frontal cortex of amyotrophic lateral sclerosis and frontotemporal degeneration brain with the C9ORF72 hexanucleotide repeat expansion. *Neurobiol. Aging* **2014**, *35*, 1779.e5–1779.e13. [CrossRef] [PubMed]
38. Cali, C.P.; Patino, M.; Tai, Y.K.; Ho, W.Y.; McLean, C.A.; Morris, C.M.; Seeley, W.W.; Miller, B.L.; Gaig, C.; Vonsattel, J.P.G.; et al. C9orf72 intermediate repeats are associated with corticobasal degeneration, increased C9orf72 expression and disruption of autophagy. *Acta Neuropathol.* **2019**, *138*, 795–811. [CrossRef] [PubMed]

MDPI
St. Alban-Anlage 66
4052 Basel
Switzerland
www.mdpi.com

Genes Editorial Office
E-mail: genes@mdpi.com
www.mdpi.com/journal/genes

Disclaimer/Publisher's Note: The statements, opinions and data contained in all publications are solely those of the individual author(s) and contributor(s) and not of MDPI and/or the editor(s). MDPI and/or the editor(s) disclaim responsibility for any injury to people or property resulting from any ideas, methods, instructions or products referred to in the content.

www.ingramcontent.com/pod-product-compliance
Lightning Source LLC
LaVergne TN
LVHW070428100526
838202LV00014B/1547